STUDENT'S SOLUTIONS MANUAL
to accompany

Laurence D. Hoffmann and Gerald L. Bradley

CALCULUS

For Business, Economics, and the Social and Life Sciences

FOURTH EDITION

Dr. Stanley M. Lukawecki
Clemson University

McGraw-Hill Publishing Company

New York St. Louis San Francisco Auckland Bogotá
Caracas Hamburg Lisbon London Madrid Mexico Milan
Montreal New Delhi Oklahoma City Paris San Juan
São Paulo Singapore Sydney Tokyo Toronto

Student's Solutions Manual to Accompany Hoffmann and Bradley:
CALCULUS FOR BUSINESS, ECONOMICS, AND THE SOCIAL AND LIFE SCIENCES.

3 4 5 6 7 8 9 0 EDW EDW 8 9 4 3 2 1 0 9

ISBN 0-07-029336-8

The editor was Robert Weinstein;
the production supervisor was Leroy A. Young.
Edwards Brothers, Inc., was printer and binder.

CONTENTS

CHAPTER 1	**FUNCTIONS, GRAPHS, AND LIMITS**	1
CHAPTER 2	**DIFFERENTIATION: BASIC CONCEPTS**	54
CHAPTER 3	**DIFFERENTIATION: FURTHER TOPICS**	102
CHAPTER 4	**EXPONENTIAL AND LOGARITHMIC FUNCTIONS**	156
CHAPTER 5	**ANTIDIFFERENTIATION**	197
CHAPTER 6	**FURTHER TOPICS IN INTEGRATION**	220
CHAPTER 7	**DIFFERENTIAL EQUATIONS**	258
CHAPTER 8	**LIMITS AT INFINITY AND IMPROPER INTEGRALS**	284
CHAPTER 9	**FUNCTIONS OF SEVERAL VARIABLES**	308
CHAPTER 10	**DOUBLE INTEGRALS**	348
CHAPTER 11	**INFINITE SERIES AND TAYLOR APPROXIMATION**	375
CHAPTER 12	**TRIGONOMETRIC FUNCTIONS**	401
APPENDIX		
	ALGEBRA REVIEW	423

CHAPTER 1

FUNCTIONS, GRAPHS, AND LIMITS

---

1 FUNCTIONS

2 GRAPHS

3 LINEAR FUNCTIONS

4 INTERSECTIONS OF GRAPHS:

BREAK-EVEN ANALYSIS AND MARKET EQUILIBRIUM

5 FUNCTIONAL MODELS

6 LIMITS AND CONTINUITY

REVIEW PROBLEMS

Chapter 1, Section 1

1.   If $f(x) = 3x^2 + 5x - 2$, then

$$f(1) = 3(1)^2 + 5(1) - 2 = 6$$
$$f(0) = 3(0)^2 + 5(0) - 2 = -2$$
$$f(-2) = 3(-2)^2 + 5(-2) - 2 = 0$$

3.   If $g(x) = x + \frac{1}{x}$, then

$$g(-1) = -1 + \frac{1}{-1} = -2$$
$$g(1) = 1 + \frac{1}{1} = 2$$
$$g(2) = 2 + \frac{1}{2} = \frac{5}{2}$$

5.   If $h(t) = \sqrt{t^2 + 2t + 4}$, then

$$h(2) = \sqrt{(2)^2 + 2(2) + 4} = \sqrt{12} = 2\sqrt{3} \simeq 3.46$$
$$h(0) = \sqrt{(0)^2 + 2(0) + 4} = \sqrt{4} = 2$$
$$h(-4) = \sqrt{(-4)^2 + 2(-4) + 4} = \sqrt{12} = 2\sqrt{3} \simeq 3.46$$

7.   If $f(t) = (2t - 1)^{-3/2} = \frac{1}{(2t-1)^{3/2}}$, then

$$f(1) = [2(1) - 1]^{-3/2} = (1)^{-3/2} = 1$$

$$f(5) = [2(5) - 1]^{-3/2} = (9)^{-3/2} = \frac{1}{[(9)^{1/2}]^3} = \frac{1}{(3)^3} = \frac{1}{27}$$

$$f(13) = [2(13) - 1]^{-3/2} = (25)^{-3/2} = \frac{1}{[(25)^{1/2}]^3} = \frac{1}{(5)^3} = \frac{1}{125}$$

9.   If $f(x) = x - |x - 2|$, then

$$f(1) = 1 - |1 - 2| = 1 - |-1| = 1 - 1 = 0$$
$$f(2) = 2 - |2 - 2| = 2 - |0| = 2 - 0 = 2$$
$$f(3) = 3 - |3 - 2| = 3 - |1| = 3 - 1 = 2$$

11.   If

$$f(x) = \begin{cases} 3 & \text{if } t < -5 \\ t + 1 & \text{if } -5 \leq t \leq 5 \\ \sqrt{t} & \text{if } t > 5 \end{cases}$$

then

$$f(-6) = 3$$
$$f(-5) = -5 + 1 = -4$$
$$f(16) = \sqrt{16} = 4$$

13.   Since the denominator of $g(x) = \frac{x^2+5}{x-2}$ cannot be 0, the domain of $g(x)$ consists of all real numbers x such that $x + 2 \neq 0$ or equivalently, $x \neq -2$.

15.   Since negative numbers do not have square roots, the domain of $y = \sqrt{x - 5}$ consist of all real numbers x such that $x - 5 \geq 0$, or equivalently, $x \geq 5$.

17.   Since $t^2 + 9 > 0$ for all real numbers t, the domain of $g(t) = \sqrt{t^2 + 9}$ is the set of all real numbers.

19.   Since negative numbers do not have square roots, the domain of the function $f(t) = (2t - 4)^{3/2} = (\sqrt{2t - 4})^3$ consists of all t such that $2t - 4 \geq 0$, or equivalently, such that $t \geq 2$.

21.   Since negative numbers do not have square roots and division by 0 is not possible, the domain of $f(x) = (x^2 - 9)^{-1/2} = \frac{1}{\sqrt{x^2-9}}$ consists of all real numbers x such that $x^2 - 9 > 0$, or equivalently, such that $(x + 3)(x - 3) > 0$. This product is zero at $x = -3$ and $x = 3$ and the signs of the factors for other values of x are tabulated below:

Interval	$x + 3$	$x - 3$	$(x + 3)(x - 3)$
$x < -3$	neg	neg	positive
$-3 < x < 3$	pos	neg	negative
$x > 3$	pos	pos	positive

Hence the domain of $f(x)$ consists of all real numbers x such that $x < -3$ or such that $x > 3$. Another way of expressing this domain is $|x| > 3$.

23.   Since division by 0 is not possible, the domain of $g(t) = \frac{1}{|t-1|}$ consists of all real numbers t such that $t - 1 \neq 0$ or equivalently, such that $t \neq 1$.

25. The cost of manufacturing q units is $C(q) = q^3 - 30q^2 + 400q + 500$ dollars. Hence:

(a) The cost of manufacturing 20 units is

$$C(20) = (20)^3 - 30(20)^2 + 400(20) + 500 = \$4,500$$

(b) The cost of manufacturing the 20th unit is

$$C(20) - C(19) = 4,500 - [(19)^3 - 30(19)^2 + 400(19) + 500]$$
$$= 4,500 - 4,129 = \$371$$

27. The temperature t hours past midnight is $C(t) = -\frac{1}{6}t^2 + 4t + 10$. Hence:

(a) The temperature at 2:00 P.M. (14 hours after midnight) is

$$C(14) = -\frac{1}{6}(14)^2 + 4(14) + 10 = 33\frac{1}{3} \text{ degrees Celsius}.$$

(b) The difference in temperature between 6:00 P.M. (18 hours past midnight) and 9:00 P.M. (21 hours past midnight) is

$$C(21) - C(18) = \left[-\frac{1}{6}(21)^2 + 4(21) + 10\right] - \left[-\frac{1}{6}(18)^2 + 4(18) + 10\right]$$
$$= 20\frac{1}{2} - 28 = -7\frac{1}{2} \text{ degrees}.$$

29. The time required to traverse the maze on the nth trial is $f(n) = 3 + \frac{12}{n}$ minutes.

(a) The domain of $f(n)$ consists of all real numbers n except $n = 0$ (since division by 0 is not possible).

(b) The values that n can actually take in the context of the practical problem are the positive integers; $1, 2, 3, \ldots$.

(c) The time required to traverse the maze on the 3rd trial is

$$f(3) = 3 + \frac{12}{3} = 7 \text{ minutes}.$$

(d) The trials on which the rat traverses the maze in 4 minutes or less are those for which $f(n) \leq 4$, that is, for which

$$3 + \frac{12}{n} \leq 4, \quad \frac{12}{n} \leq 1, \quad \text{or} \quad n \geq 12$$

Thus the 12th trial is the first on which the rat traverses the maze in 4 minutes or less.

(e) As the number n of trials increases, the time required to traverse the maze decreases and approaches 3 minutes (since $\frac{12}{n}$ approaches zero), which means the rat will never be able to traverse the maze in less than 3 minutes.

31. The number of worker-hours required to distribute telephone books to x percent of the homes is $f(x) = \frac{600x}{300-x}$.

    (a) The domain of $f(x)$ consists of all real numbers x except $x = 300$ (since division by zero is not possible).

    (b) Since x is a percentage, the values of x that are meaningful in the context of this problem are those for which $0 \leq x \leq 100$.

    (c) The number of worker-hours needed to distribute telephone books to 50 percent of the households is

$$f(50) = \frac{600(50)}{300 - 50} = 120$$

    (d) The number of worker-hours needed to reach the entire community (100 percent of the households) is

$$f(100) = \frac{600(100)}{300 - 100} = 300$$

    (e) The percentage of households receiving telephone books after 150 worker-hours is the value of x for which $f(x) = 150$, that is, for which

$$\frac{600x}{300 - x} = 150, \quad \frac{4x}{300 - x} = 1, \quad 4x = 300 - x, \quad \text{or} \quad x = 60$$

33. The height of the ball after t seconds is $H(t) = -16t^2 + 256$ feet. Hence:

    (a) After 2 seconds, the height of the ball is

$$H(2) = -16(2)^2 + 256 = 192 \text{ feet}.$$

    (b) During the 3rd second, the ball will travel

$$H(2) - H(3) = 192 - [-16(3)^2 + 256] = 80 \text{ feet}.$$

    (c) The height of the building is $H(0) = 256$ feet.

    (d) The ball hits the ground when $H(t) = 0$, that is, when

$$-16t^2 + 256 = 0, \quad 16t^2 = 256, \quad t^2 = 16, \quad \text{or} \quad t = 4 \text{ seconds}.$$

35. If $g(u) = 3u^2 + 2u - 6$ and $h(x) = x + 2$, then

$$g[h(x)] = g(x + 2) = 3(x + 2)^2 + 2(x + 2) - 6 = 3x^2 + 14x + 10$$

37. If $g(u) = (u - 1)^3 + 2u^2$ and $h(x) = x + 1$, then

$$g[h(x)] = g(x + 1) = [(x + 1) - 1]^3 + 2(x + 1)^2 = x^3 + 2x^2 + 4x + 2$$

39. If $g(u) = \frac{1}{u^2}$ and $h(x) = x - 1$, then

$$g[h(x)] = g(x - 1) = \frac{1}{(x - 1)^2}$$

41. If $g(u) = \sqrt{u+1}$ and $h(x) = x^2 - 1$, then

$$g[h(x)] = g(x^2 - 1) = \sqrt{(x^2 - 1) + 1} = \sqrt{x^2} = |x|$$

43. If $f(x) = 2x^2 - 3x + 1$, then

$$f(x - 2) = 2(x - 2)^2 - 3(x - 2) + 1 = 2x^2 - 11x + 15$$

45. If $f(x) = (x + 1)^5 - 3x^2$, then

$$f(x - 1) = [(x - 1) + 1]^5 - 3(x - 1)^2 = x^5 - 3x^2 + 6x - 3$$

47. If $f(x) = \sqrt{x}$, then $f(x^2 + 3x - 1) = \sqrt{x^2 + 3x - 1}$.

49. If $f(x) = \frac{x-1}{x}$, then $f(x + 1) = \frac{(x+1)-1}{x+1} = \frac{x}{x+1}$.

51. $f(x) = \sqrt{3x - 5}$ can be written as $g[h(x)]$, where

$$g(u) = \sqrt{u} \quad \text{and} \quad h(x) = 3x - 5$$

53. $f(x) = \frac{1}{x^2+1}$ can be written as $g[h(x)]$, where

$$g(u) = \frac{1}{u} \quad \text{and} \quad h(x) = x^2 + 1$$

55. $f(x) = \sqrt{x + 3} - \frac{1}{(x+4)^3} = \sqrt{x + 3} - \frac{1}{((x+3)+1)^3}$ can be written as $g[h(x)]$, where

$$g(u) = \sqrt{u} - \frac{1}{(u + 1)^3} \quad \text{and} \quad h(x) = x + 3$$

57. If $C(q) = q^2 + q + 900$ dollars and $q(t) = 25t$ hours, then:
     (a) $C[q(t)] = C(25t) = (25t)^2 + 25t + 900 = 625t^2 + 25t + 900$.
     (b) The cost at the end of the 3rd hour is

$$C[q(3)] = 625(3)^2 + 25(3) + 900 = \$6,600$$

     (c) Set $C[q(t)]$ equal to 11,000 and solve for t to get

$$11,000 = 625t^2 + 25t + 900$$
$$625t^2 + 25t - 10,100 = 0$$
$$25(25t^2 + t - 404) = 0$$
$$25(25t + 101)(t - 4) = 0 \quad \text{or} \quad t = 4 \text{ hours}$$

Chapter 1, Section 2

1.   Some points on the graph of $f(x) = x$ are shown below. The graph is a straight line.

x	0	1	−1	2	−2
$f(x)$	0	1	−1	2	−2

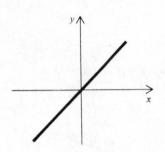

Graph for Problem 1.

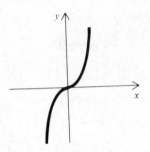

Graph for Problem 3.

3.   Some points on the graph of $f(x) = x^3$ are shown below.

x	0	1/2	−1/2	1	−1	2	−2
$f(x)$	0	1/8	−1/8	1	−1	8	−8

5.   Since $x = 0$ is not in the domain of $f(x) = \frac{1}{x}$, the graph breaks into two pieces at $x = 0$. Some points on the graph are shown below.

x	1/8	−1/8	1/2	−1/2	2	−2	8	−8
$f(x)$	8	−8	2	−2	1/2	−1/2	1/8	−1/8

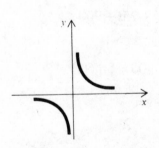

Graph for Problem 5.

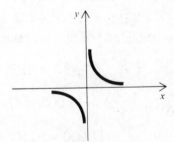

Graph for Problem 7.

7.   Since $x = 0$ is not in the domain of $f(x) = \frac{1}{x^3}$, the graph breaks into two pieces

at $x = 0$. Some points on the graph are shown below.

x	1/3	−1/3	1	−1	3	−3
$f(x)$	27	−27	1	−1	1/27	−1/27

9.   Some points on the graph of $f(x) = 2x - 1$ are shown below. The graph is a straight line.

x	0	1	−1	2	−2
$f(x)$	−1	1	−3	3	−5

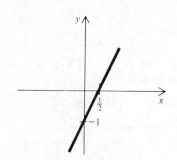   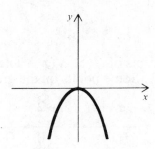

Graph for Problem 9.   Graph for Problem 11.

11.   Some points on the graph of $f(x) = -x^2$ are shown below. (The graph is a parabola and is the reflection across the x axis of the familiar graph $y = x^2$.)

x	0	1	−1	2	−2	3	−3
$f(x)$	0	−1	−1	−4	−4	−9	−9

13.   Some points on the graph of

$$f(x) = \begin{cases} x - 1 & \text{if } x \leq 0 \\ x + 1 & \text{if } x > 0 \end{cases}$$

are shown below. In compiling the table, the formula $f(x) = x - 1$ was used when $x \leq 0$ and the formula $f(x) = x + 1$ was used when $x > 0$.

x	−3	−2	−1	0	1	2	3
$f(x)$	−4	−3	−2	−1	2	3	4

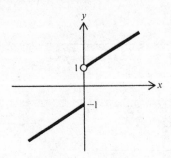

Graph for Problem 13.

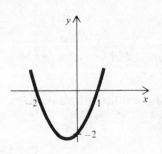

Graph for Problem 15.

15. The x intercepts of $f(x) = (x-1)(x+2)$ are $x = 1$ and $x = -2$. The y intercept is $f(0) = -2$. Some points on the graph are shown below.

x	-2	1	0	-3	-1	2	3
$f(x)$	0	0	-2	4	-2	4	10

17. The y intercept of $f(x) = x^2 - x - 6$ is $f(0) = -6$. To find the x intercepts (and to simplify the calculations), begin by factoring the function as $f(x) = (x-3)(x+2)$. The x intercepts are $(3,0)$ and $(-2,0)$. Some other points on the graph are shown below.

x	-4	-3	-2	-1	0	1	2	3	4	5
$f(x)$	14	6	0	-4	-6	-6	-4	0	6	14

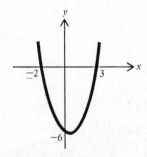

Graph for Problem 17.

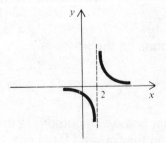

Graph for Problem 19.

19. Since $x = 2$ is not in the domain of $f(x) = \frac{1}{x-2}$, the graph breaks into two pieces at $x = 2$. Some points on the graph are shown below.

x	0	1	3/2	3/4	3	5/2	5/4
$f(x)$	$-1/2$	-1	-2	-4	1	2	4

21. The function $f(x) = \frac{x^2 - 2x}{x+1}$ has a discontinuity at $x = -1$, where its denominator is zero. From the factored form $f(x) = \frac{x(x-2)}{x+1}$ you can see that the x intercepts are $(0, 0)$ and $(2, 0)$. Some other points on the graph of f are shown below.

x	-10	-3	-2	$-3/2$	-1	$-1/2$	0	1	2	3	10
$f(x)$	-13.3	-7.5	-8	-10.5		1.5	0	-0.5	0	0.75	7.3

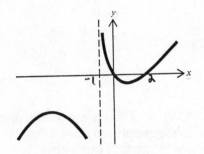   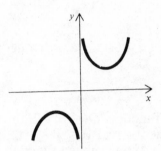

Graph for Problem 21.   Graph for Problem 23.

23. Since $x = 0$ is not in the domain of $f(x) = x + \frac{1}{x}$, the graph breaks into two pieces at $x = 0$. Some points on the graph are shown below. (Notice that the absolute value of $f(x)$ is large when x is close to zero.)

x	$\pm 1/10$	$\pm 1/2$	± 1	± 2	± 10
$f(x)$	± 10.1	± 2.5	± 2	± 2.5	± 0.1

25. The monthly profit is

$$P(x) = (\text{number of recorders sold})(\text{price} - \text{cost}) = (120 - x)(x - 20).$$

The x intercepts of the graph are $x = 120$ and $x = 20$, and the y intercept is 2,400. Some points on the graph are shown below.

x	120	20	0	30	60	90
$f(x)$	0	0	2400	900	2400	2100

The graph suggests that the profit is greatest when $x = 70$, that is, when 70 recorders are sold.

Note: $P(x)$ is quadratic function. Completing the square gives

$$P(x) = -(x - 70)^2 + 2,500$$

from which it is clear that the maximum value is $2,500 and occurs when $x = 70$.

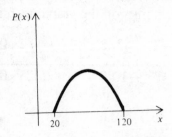

Graph for Problem 25.

27. (a) The function $D(p) = -200p + 12{,}000 = -200(p - 60)$ represents the demand on the interval $0 \le p \le 60$. Some points on the graph are shown below. The graph of the demand function is a line segment in the first quadrant.

P	0	20	40	60
$D(p)$	12,000	8000	4000	0

(b) The total monthly expenditure is

$$E(p) = (\text{price per unit})(\text{demand})$$
$$= p(-200p + 12{,}000) = -200p(p - 60)$$

(c) Some points on the graph of the expenditure function are shown below. (The graph is a parabola.)

P	0	20	40	60
$E(p)$	0	16,000	16,000	0

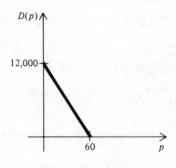

The graph of D(p).

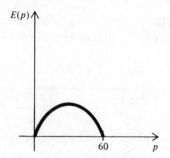

The graph of E(p).

(d) The p intercepts of $E(p) = -200p(p - 60)$ are $p = 0$ and $p = 60$. $E(0) = 0$ since the price is zero. $E(60) = 0$ since demand drops to zero when the price is $60.

(e)   The graph suggests that the expenditure will be greatest when the price is $p = 30$ dollars.

Note: Completing the square gives

$$E(p) = -200(p - 30)^2 + 18,000$$

from which it is clear that the maximum expenditure is \$18,000 and occurs when $p = 30$.

29.   The function $f(x) = \frac{600x}{300-x}$ gives the number of worker-hours required to distribute phone books to x percent of the homes. Since $x = 300$ is not in the domain of $f(x)$, the graph breaks into two pieces at $x = 300$. Some points on the graph are shown below. Since x stands for a percentage, only the portion of the graph for $0 \le x \le 100$ is relevant.

x	0	100	290	310	1100	-900
$f(x)$	0	300	17,400	$-18,600$	-825	-450

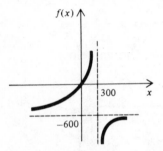

Graph for Problem 29.

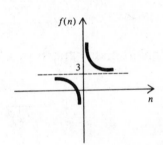

Graph for Problem 31.

31.   (a)   The time (in minutes) required to traverse the maze on the nth trial is $f(n) = 3 + \frac{12}{n}$. Since $n = 0$ is not in the domain, the graph breaks into two pieces at $n = 0$. Some points on the graph are shown below.

n	1	2	12	100	-1	-2	-12	-100
$f(n)$	15	9	4	3.12	-9	-3	2	2.88

      (b)   Since n stands for the number of the trial, the graph is relevant only for $n = 1, 2, 3, \ldots$.

      (c)   If n is large, $\frac{12}{n}$ is small and so $f(n) = 3 + \frac{12}{n}$ is close to 3. This means that as the rat traverses the maze repeatedly, the time required approaches 3 minutes. The rat will never be able to traverse the maze in less than 3 minutes.

33.   The cost function $C(x) = 20x + \frac{2000}{x}$ is defined for all x except $x = 0$ and has a practical interpretation if x is a positive integer. Some points on the graph are shown below. (To get a continuous curve, the graph was sketched for $x > 0$.)

x	1	2	5	8	10	20
$C(x)$	2020	1040	500	410	400	500

The graph suggests that the cost will be minimal if $x = 10$, that is, if 10 machines are used.

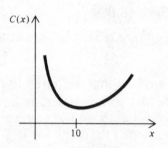

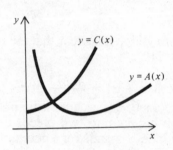

Graph for Problem 33.   Graph for Problem 35.

35. If the total cost of producing x units is $C(x) = x^2 + 4x + 16$, then the average cost per unit is

$$A(x) = \frac{C(x)}{x} = \frac{x^2 + 4x + 16}{x} = x + 4 + \frac{16}{x}$$

Some points on the graphs of $C(x)$ and $A(x)$ are shown below.

x	0	1	2	3
$C(x)$	16	21	28	37

x	1	2	4	8	16
$A(x)$	21	14	12	14	21

37. (a) Each y value in the table for $y = -x^2$ is the negative of the corresponding y value in the table for $y = x^2$. Hence the points on the graph of $y = -x^2$ are the reflections across the x axis of the points of the graph of $y = x^2$.

x	0	1	2	3	-3	-2	-1
$y = x^2$	0	1	4	9	9	4	1
$y = -x^2$	0	-1	-4	-9	-9	-4	-1

(b) In general, if $g(x) = -f(x)$, then, as in part (a), the graph of $g(x)$ is the reflection across the x axis of the graph of $f(x)$.

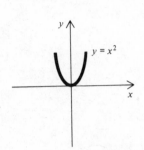

Graph of $y = x^2$.

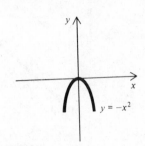

Graph of $y = -x^2$.

39. (a) Some points on the graphs of $y = x^2$ and $y = (x - 2)^2$ are shown below.

x	0	−1	1	−2	2	−3	3
$y = x^2$	0	1	1	4	4	9	9
x		2	1	3	0	4	−1	5
$y = (x - 2)^2$		0	1	1	4	4	9	9

Notice that the y values in the second table are identical to the y values in the first table, but they occur when x is 2 units greater than in the first table. This means that the points in the first table are shifted 2 units to the right to get the points in the second table. Hence the graph of the function $y = (x - 2)^2$ can be obtained by shifting the graph of $y = x^2$ to the right by 2 units.

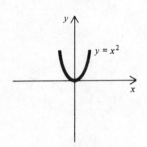

Graph of $y = x^2$.

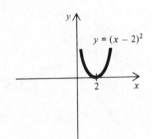

Graph of $y = (x - 2)^2$.

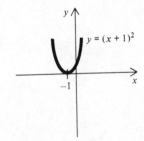

Graph of $y = (x + 1)^2$.

(b) Similarly, the graph of $y = (x + 1)^2$ is the graph of $y = x^2$ shifted -1 unit to the right, that is, 1 unit to the left.

(c) In general, the graph of $g(x) = f(x - c)$ will be the graph of $f(x)$ shifted c units to the right if c is positive and to the left if c is negative.

41. Let the point (x_1, y_1) be labeled P and the point (x_2, y_2) be labeled Q as shown in the figure.

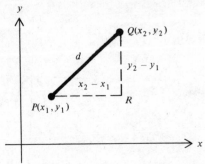

Right triangle for the distance formula.

In the right triangle PQR, the length of PR is $x_2 - x_1$ and that of QR is $y_2 - y_1$. By the pythagorean theorem,

$$d^2 = (x_2 - x_1)^2 + (y_2 - y_1)^2.$$

Since the distance is taken to be positive, it follows that

$$d = \sqrt{(x_2 - x_1)^2 + (y_2 - y_1)^2}$$

and the distance formula is proved.

Chapter 1, Section 3

In Problems 1 through 5, the slope m of the line joining the points $P_1 = (x_1, y_1)$ and $P_2 = (x_2, y_2)$ is $m = \frac{y_2 - y_1}{x_2 - x_1}$ if $x_1 \neq x_2$.

1.   If $P_1 = (2, -3)$ and $P_2 = (0, 4)$, then $m = \frac{4 - (-3)}{0 - 2} = -\frac{7}{2}$.

3.   If $P_1 = (2, 0)$ and $P_2 = (0, 2)$, then $m = \frac{2 - 0}{0 - 2} = -1$.

5.   If $P_1 = (2, 6)$ and $P_2 = (2, -4)$, then the slope is undefined since the denominator is zero in the formula for the slope. (The corresponding line is vertical.)

In Problems 7 through 15, use the fact that if the equation of a line is in the form $y = mx + b$ then m is the slope and b is the y intercept.

7.   If $y = 5x + 2$, then $m = 5$ and $b = 2$.

9.   If $x + y = 2$, then $y = -x + 2$ and so $m = -1$ and $b = 2$.

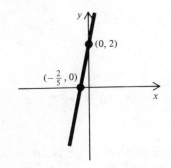   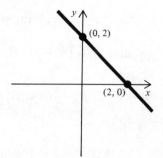

Graph for Problem 7.   Graph for Problem 9.

11.   If $2x - 4y = 12$, then $y = \frac{-2}{-4}x + \frac{12}{-4} = \frac{1}{2}x - 3$ and so $m = \frac{1}{2}$ and $b = -3$.

13.   If $4x = 2y + 6$, then $2y = 4x - 6$ or $y = \frac{4}{2}x - \frac{6}{2} = 2x - 3$ and so $m = 2$ and $b = -3$.

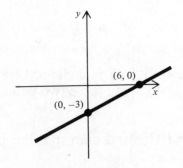   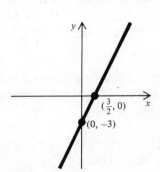

Graph for Problem 11.   Graph for Problem 13.

15.   If $y = 2$, then $y = 0x + 2$ and so $m = 0$ and $b = 2$. (The corresponding line is horizontal.)

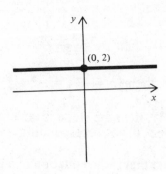

Graph for Problem 15.

In Problems 17 through 27, use the fact that the equation of the line with slope m that passes through the point $P = (x_0, y_0)$ is $y - y_0 = m(x - x_0)$.

17.   If $m = 1$ and $P = (2, 0)$, then $y - 0 = 1(x - 2)$ or $y = x - 2$.

19.   If $m = -\frac{1}{2}$ and $P = (5, -2)$, then

$$y - (-2) = -\frac{1}{2}(x - 5), \quad y + 2 = -\frac{1}{2}x + \frac{5}{2}, \quad \text{or} \quad y = -\frac{1}{2}x + \frac{1}{2}$$

21.   Since the line is parallel to the x axis, it is horizontal and its slope is $m = 0$. The given point is $P = (2, 5)$. Hence the equation of the line is $y - 5 = 0(x - 2)$ or $y = 5$.

23.   From the points $(1, 0)$ and $(0, 1)$, the slope is $m = \frac{1-0}{0-1} = -1$ and, with $P = (1, 0)$, the equation of the line is $y - 0 = -1(x - 1)$ or $y = -x + 1$. Note: The same equation would result if $P = (0, 1)$ were used.

25.   From the points $(-2, 3)$ and $(0, 5)$, the slope is $m = \frac{5-3}{0-(-2)} = 1$ and, with $P = (-2, 3)$, the equation of the line is

$$y - 3 = 1[x - (-2)], \quad y - 3 = x + 2, \quad \text{or} \quad y = x + 5$$

27.   The slope of the line through the points $(1, 5)$ and $(1, -4)$ is undefined (since the denominator of the slope formula is zero). The line joining these two points is the vertical line $x = 1$.

29.   (a)   Let x denote the number of miles driven and $C(x)$ the corresponding cost (in dollars). Then,

$$C(x) = (\text{cost per mile})(\text{number of miles}) + \text{fixed cost}$$
$$= 0.14x + 20$$

$C(x)$ is a linear function with slope 0.14 and vertical intercept 20 and represents the cost if $x \geq 0$.

(b) The rental cost of a 50-mile trip is $C(50) = 0.14(50) + 20 = \$27$.

(c) Set $C(x)$ equal to 45.20 and solve for x to get

$$45.20 = 0.14x + 20$$
$$25.20 = 0.14x$$
$$x = \frac{25.20}{0.14} = 180 \text{ miles}$$

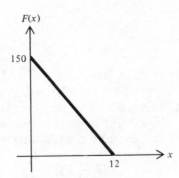

Graph for Problem 31.

31. (a) Let x denote the number of weeks elapsed and $y = F(x)$ the corresponding membership fee. The function relating x and y is linear. Since the fee is \$150 at the start of the summer (when $x = 0$) and the fee is \$0 at the end of the summer (when $x = 12$), the graph passes through the points $(0, 150)$ and $(12, 0)$. The corresponding slope is

$$m = \frac{150 - 0}{0 - 12} = -12.5$$

and the y intercept is 150. Hence the equation for the membership fee is

$$F(x) = -12.5x + 150 \quad \text{for } 0 \leq x \leq 12.$$

(b) The fee after 5 weeks is $F(5) = -12.5(5) + 150 = \$87.50$.

33. (a) Let x denote the age in years of the machinery and $V(x)$ its corresponding value. It is given that V is a linear function of x. At the time of purchase, $x = 0$ and $V(0) = 20,000$. Ten years later, $x = 10$ and $V(10) = 1,000$. The slope of the line through $(0, 20,000)$ and $(10, 1000)$ is

$$m = \frac{1,000 - 20,000}{10 - 0} = -1,900$$

and the vertical intercept is 20,000. Hence the equation of the line is

$$V(x) = -1,900x + 20,000.$$

The value of the machine will be reduced to zero when $V(x) = 0$, that is, when $x = \frac{200}{19}$. Thus $V(x)$ gives the value of the machine for $0 \leq x \leq \frac{200}{19}$.

(b)   The value after 4 years is $V(4) = -1,900(4) + 20,000 = \$12,400$.

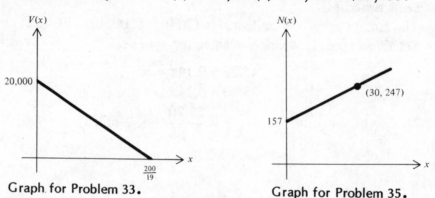

Graph for Problem 33.   Graph for Problem 35.

35.   (a)   Let x denote the number of days since the reduced rate went into effect and $N(x)$ the corresponding number of vehicles qualifying for the reduced rate. Since the number of qualifying vehicles is increasing at a constant rate, N is a linear function of x. Since $N(0) = 157$ (when the program began) and $N(30) = 247$ (30 days later), the line passes through $(0, 157)$ and $(30, 247)$ and its slope is

$$m = \frac{247 - 157}{30 - 0} = 3$$

The vertical intercept is 157 and so the equation is

$$N(x) = 3x + 157 \quad \text{for } x \geq 0$$

(b)   Fourteen days from now, 44 days will have elapsed since the program began and the number of cars qualifying will be $N(44) = 3(44) + 157 = 289$.

37.   (a)   The original value of the book is \$3 and the value doubles every 10 years. A table of values is shown below.

Age	0	10	20	30	40
Value	3	6	12	24	48

At the end of 30 years the book will be worth \$24 and at the end of 40 years it will be worth \$48.

(b)   If the relationship were linear, the slopes of line segments between any pairs of points would be equal. For the pair $(0, 3)$ and $(10, 6)$, the slope is

$$m = \frac{6 - 3}{10 - 0} = \frac{3}{10}$$

and for the pair $(0, 3)$ and $(20, 12)$ the slope is

$$m = \frac{12 - 3}{20 - 0} = \frac{9}{20}$$

Since the slopes are different, the points do not lie on a straight line and the relationship is not linear.

39. In the figure below, the slope of L_1 is $m_1 = \frac{b}{a}$, the slope of L_2 is $m_2 = \frac{c}{a}$, and L_1 is perpendicular to L_2.

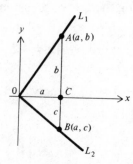

Perpendicular lines for Problem 39.

From the distance formula from Section 2, Problem 41,

$$\text{Length of } OA = \sqrt{a^2 + b^2} \quad \text{and length of } OB = \sqrt{a^2 + c^2}$$

The length of AB is $b - c$. By the pythagorean theorem applied to the right triangle AOB,

$$(a^2 + b^2) + (a^2 + c^2) = (b - c)^2 \quad \text{or} \quad 2a^2 + b^2 + c^2 = b^2 - 2bc + c^2$$

Hence,

$$2a^2 = -2bc, \quad \frac{bc}{a^2} = -1, \quad \text{or} \quad \left(\frac{b}{c}\right)\left(\frac{c}{a}\right) = -1$$

But $\frac{b}{a} = m_1$ and $\frac{c}{a} = m_2$ and so

$$m_1 m_2 = -1 \quad \text{or} \quad m_1 = -\frac{1}{m_2}$$

Chapter 1, Section 4

1.   The graphs of $y = 3x + 5$ and $y = -x + 3$ suggest that there is a point of intersection in the second quadrant. Setting the two expressions for y equal to each other yields

$$3x + 5 = -x + 3, \quad 4x = -2, \quad \text{or} \quad x = -\frac{1}{2}$$

When $x = -\frac{1}{2}$, $y = 3\left(-\frac{1}{2}\right) + 5 = \frac{7}{2}$. Hence $\left(-\frac{1}{2}, \frac{7}{2}\right)$ is the point of intersection.

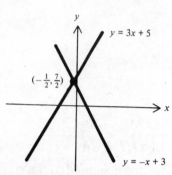

Graph for Problem 1.

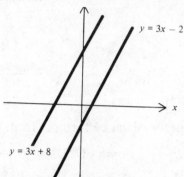

Graph for Problem 3.

3.   The graphs of $y = 3x + 8$ and $y = 3x - 2$ are parallel lines with slope 3. Hence there are no points of intersection.

5.   The graphs of $y = x^2 - x$ and $y = x - 1$ suggest that there is a point of intersection at $(1, 0)$. Setting the two expressions for y equal to each other yields

$$x^2 - x = x - 1, \quad x^2 - 2x + 1 = 0, \quad (x-1)^2 = 0, \quad \text{or} \quad x = 1$$

When $x = 1$, $y = (1)^2 - 1 = 0$. Hence the point of intersection is $(1, 0)$.

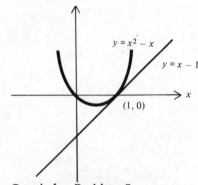

Graph for Problem 5.

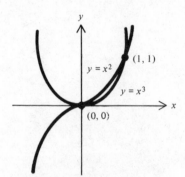

Graph for Problem 7.

7.   The graphs of $y = x^3$ and $y = x^2$ suggest that there is a point of intersection at $(0, 0)$ and at another point in the first quadrant. Setting the two expressions for y equal to each other yields

$$x^3 = x^2, \quad x^3 - x^2 = 0, \quad x^2(x - 1) = 0, \quad \text{or} \quad x = 0 \text{ and } x = 1$$

When $x = 0$, $y = (0)^2 = 0$ and when $x = 1$, $y = (1)^2 = 1$. Hence the points of intersection are $(0,0)$ and $(1,1)$.

9.   The graphs of $y = x^2 + 2$ and $y = x$ suggest that there are no points of intersection. Setting the two expressions for y equal to each other yields

$$x^2 + 2 = x \quad \text{or} \quad x^2 - x + 2 = 0$$

which is a quadratic equation with no real roots since the discriminant is

$$b^2 - 4ac = (-1)^2 - 4(1)(2) = -7 < 0$$

Hence there are no points of intersection.

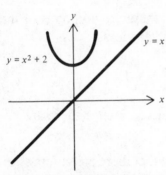

Graph for Problem 9.

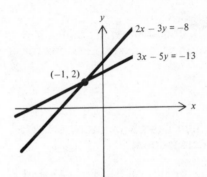

Graph for Problem 11.

11. The graphs of $2x - 3y = -8$ and $3x - 5y = -13$ suggest that there is a point of intersection in the second quadrant. Solving the two equations for y yields

$$y = \frac{2}{3}x + \frac{8}{3} \quad \text{and} \quad y = \frac{3}{5}x + \frac{13}{5}$$

Setting the two expressions for y equal to each other yields

$$\frac{2}{3}x + \frac{8}{3} = \frac{3}{5}x + \frac{13}{5}, \quad 10x + 40 = 9x + 39, \quad \text{or} \quad x = -1$$

When $x = -1$, $y = \frac{2}{3}(-1) + \frac{8}{3} = 2$. Hence the point of intersection is $(-1, 2)$.

13. The graphs of $y = \frac{1}{x^2}$ and $y = 4$ suggest that there are points of intersection in the first and second quadrants. Setting the two expressions for y equal to each other yields

$$\frac{1}{x^2} = 4, \quad x^2 = \frac{1}{4}, \quad \text{or} \quad x = \pm\frac{1}{2}.$$

When $x = \frac{1}{2}$, $y = \frac{1}{(1/2)^2} = 4$ and when $x = -\frac{1}{2}$, $y = \frac{1}{(-1/2)^2} = 4$. Hence the points of intersection are $\left(\frac{1}{2}, 4\right)$ and $\left(-\frac{1}{2}, 4\right)$.

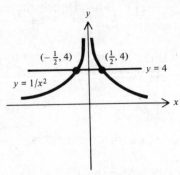

Graph for Problem 13.

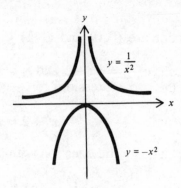

Graph for Problem 15.

15. The graphs of $y = \frac{1}{x^2}$ and $y = -x^2$ suggest that there are no points of intersection. Setting the two expressions for y equal to each other yields

$$\frac{1}{x^2} = -x^2, \quad \text{or} \quad x^4 = -1$$

which has no solutions since x^4 cannot be negative. Hence there are no points of intersection.

17. The graphs of $y = x^2 - 2x$ and $y = x - 1$ suggest that there are points of intersection in the first and fourth quadrants. Setting the two expressions for y equal to each other yields

$$x^2 - 2x = x - 1 \quad \text{or} \quad x^2 - 3x + 1 = 0$$

which is a quadratic equation that can be solved with the quadratic formula

$$x = \frac{-b \pm \sqrt{b^2 - 4ac}}{2a}$$

In this case, $a = 1$, $b = -3$, and $c = 1$. Hence,

$$x = \frac{-(-3) \pm \sqrt{(-3)^2 - 4(1)(1)}}{2(1)} = \frac{3 \pm \sqrt{9 - 4}}{2} = \frac{3 \pm \sqrt{5}}{2}$$

When $x = \frac{3+\sqrt{5}}{2}$.

$$y = \frac{3 + \sqrt{5}}{2} - 1 = \frac{3 + \sqrt{5}}{2} - \frac{2}{2} = \frac{1 + \sqrt{5}}{2}$$

and when $x = \frac{3-\sqrt{5}}{2}$,

$$y = \frac{3 - \sqrt{5}}{2} - 1 = \frac{3 - \sqrt{5}}{2} - \frac{2}{2} = \frac{1 - \sqrt{5}}{2}.$$

Hence the points of intersection are $\left(\frac{3+\sqrt{5}}{2}, \frac{1+\sqrt{5}}{2}\right)$ and $\left(\frac{3-\sqrt{5}}{2}, \frac{1-\sqrt{5}}{2}\right)$.

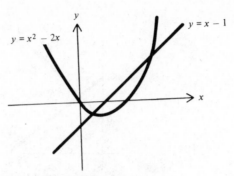

Graph for Problem 17.

19. (a) Let x denote the number of kayaks made and sold, $C(x)$ the corresponding cost, and $R(x)$ the corresponding revenue. Then,

$$R(x) = \text{(unit price)(number sold)} = 175x$$

and

$$C(x) = \text{(unit cost)(number made)} + \text{overhead} = 25x + 600.$$

The students break even when $R(x) = C(x)$, that is, when

$$175x = 25x + 600 \quad \text{or} \quad x = 4.$$

Hence 4 kayaks must be sold to break even.

(b) If $P(x)$ is the profit generated by the sale of x kayaks, then

$$P(x) = R(x) - C(x) = 175x - (25x + 600) = 150x - 600$$

Profit will be $450 when $P(x) = 450$, that is, when

$$450 = 150x - 600, \quad 150x = 1050, \quad \text{or} \quad x = 7$$

that is, 7 kayaks must be sold for a profit of $450.

21. Let x denote the number of hours the courts are used and $C_1(x)$ and $C_2(x)$ the corresponding costs at the first and second club, respectively. Then

$$C_1(x) = 500 + 1x \quad \text{and} \quad C_2(x) = 440 + 1.75x.$$

The cost at the two clubs will be the same if

$$500 + x = 440 + 1.75x, \quad 0.75x = 60, \quad \text{or} \quad x = 80.$$

Thus if a player uses the courts for exactly 80 hours, the cost at the two clubs is the same. If a player uses the courts for less than 80 hours, the second club is the less expensive (since the graph of $C_2(x)$ is below that of $C_1(x)$ if $x < 80$), and if a player uses the courts for more than 80 hours, the first club is the less expensive.

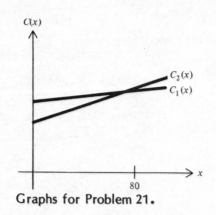

Graphs for Problem 21.

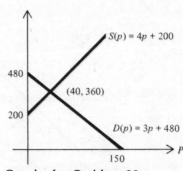

Graphs for Problem 23.

23. The equilibrium price is the value of p for which the supply function $S(p) = 4p + 200$ is equal to the demand function $D(p) = -3p + 480$, that is, for which,

$$4p + 200 = -3p + 480 \quad \text{or} \quad p = 40.$$

At the equilibrium price of \$40, the number of units supplied (and demanded) is $S(40) = 4(40) + 200 = 360$.

25. (a) The equilibrium price is the value of p for which the supply function $S(p) = p - 10$ is equal to the demand function $D(p) = \frac{5,600}{p}$, that is, for which

$$p - 10 = \frac{5,600}{p}, \quad p^2 - 10p = 5,600, \quad p^2 - 10p - 5,600 = 0,$$
$$(p - 80)(p + 70) = 0, \quad \text{or} \quad p = 80 \text{ and } p = -70$$

Only the positive value is meaningful in this context and so the equilibrium price is \$80. At this price, the number of units supplied (and demanded) is $S(80) = 80 - 10 = 70$.

(b) The graphs of the supply and demand functions are shown below.

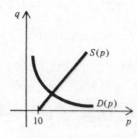

Supply and demand curves for Problem 25.

(c) The supply graph crosses the p axis when $S(p) = 0$, that is, when $p = 10$, indicating the supplier is unwilling to provide the commodity unless the selling price exceeds \$10.

27. Let x denote the time in hours the spy has been travelling. Then $x - \frac{2}{3}$ is the time the smugglers have been travelling (since 40 minutes is $\frac{2}{3}$ hours). The distance the spy travels is $72x$ kilometers and the corresponding distance the smugglers travel is $168 \left(x - \frac{2}{3}\right)$ kilometers. They reach the same point when

$$72x = 168 \left(x - \frac{2}{3}\right), \quad 72x = 168x - 112, \quad \text{or} \quad x = \frac{112}{96} = 1\frac{1}{6}.$$

At the end of $1\frac{1}{6}$ hours (1 hours and 10 minutes), the spy (and smugglers) have traveled $72 \left(\frac{7}{6}\right) = 84$ kilometers, and hence the spy escapes since freedom is reached at the border after only 83.8 kilometers.

Chapter 1, Section 5

1.   Let x denote the width of a rectangular garden and $A(x)$ the corresponding area. Since the length is given as twice the width, the length is $2x$ and the area is

$$A(x) = (\text{length})(\text{width}) = (2x)(x) = 2x^2 .$$

3.   Let x denote the selling price of the book and $P(x)$ the corresponding profit function. Then

$$P(x) = (\text{number of books sold})(\text{profit per book})$$

The profit per book is $x - 3$, since the cost of each book is \$3. To determine the number of books sold, first notice that $15 - x$ is the number of \$1 reductions, since the initial price of the book is \$15. For each of these \$1 reductions, 20 more books are sold, so the total number of books sold is $200 + 20(15 - x)$. Putting it all together,

$$P(x) = [200 + 20(15 - x)](x - 3) = (500 - 20x)(x - 3)$$
$$= 20(25 - x)(x - 3)$$

The graph of $P(x)$ suggests that the profit is maximal when $x = 14$, that is, when the books are sold for \$14 apiece. <u>Remark</u>: In Chapter 3 you will learn how to use calculus to show that $x = 14$ is indeed the optimal selling price. Without calculus, you could complete the square to get

$$P(x) = -20(x - 14)^2 + 2{,}420$$

from which it is clear that the maximum profit is \$2,420 and is generated when the price is $x = \$14$.

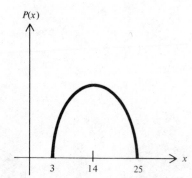

Profit function for Problem 3.

5.   Let x denote the number of people in the bus and $R(x)$ the corresponding revenue. Then,

$$R(x) = (\text{number of people})(\text{cost per person})$$

For each person in excess of 35, the original cost of \$60 is reduced by $\frac{1}{2}$ of a dollar. Since $x - 35$ is the excess, the cost per person is $60 - (x - 35)\left(\frac{1}{2}\right)$. Putting it all together,

$$R(x) = x\left[60 - (x - 35)\left(\frac{1}{2}\right)\right] = -\frac{1}{2}x(x - 155)$$

The graph suggests that the revenue will be maximal for some value of x near 77 or 78. Remark: In Chapter 3 you will learn how to use calculus to determine that the maximal revenue occurs when $x = 77.5$. Without calculus, you could complete the square to get

$$R(x) = -\frac{1}{2}\left(x - \frac{155}{2}\right)^2 + \frac{24{,}025}{8}$$

from which it is clear that the maximum value of $R(x)$ is $\frac{24{,}025}{8} = \$3{,}003.125$ and is generated when $x = 77.5$.

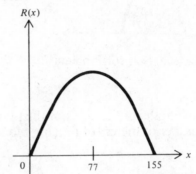

Revenue function for Problem 5.

7.   Let x denote the number of days after July 1 and $R(x)$ the corresponding revenue (in dollars). Then,

$$R(x) = (\text{number of bushels sold})(\text{price per bushel})$$

Since the crop increases at the rate of 1 bushel per day and 80 bushels were available on July 1, the number of bushels sold after x days is $80 + x$. Since the price per bushel decreases by 0.02 dollars per day and was \$2 on July 1, the price per bushel after x days is $2 - 0.02x$ dollars. Putting it all together,

$$R(x) = (80 + x)(2 - 0.02x) = 0.02(100 - x)(80 + x).$$

The graph of $R(x)$ suggests that the revenue will be greatest for some value of x near 10. Remark: In Chapter 3 you will learn how to use calculus to determine that $R(x)$ is greatest when $x = 10$. Without calculus, you could complete the square to get

$$R(x) = -0.02(x - 10)^2 + 162$$

from which it is clear that the maximum revenue is \$162 and is generated $x = 10$ days after July 1, that is, on July 11.

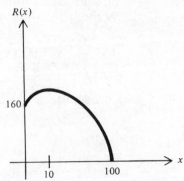

Revenue function for Problem 7.

9. Let x and y denote the lengths of the sides of the playground as shown in the figure and let P denote the number of meters of fencing required to enclose the playground.

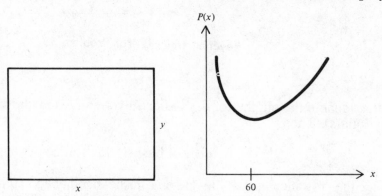

Rectangular playground and perimeter function for Problem 9.

Then
$$P = 2x + 2y$$
which is a function of two variables. Since the area is 3,600 square meters,
$$xy = 3,600 \quad \text{or} \quad y = \frac{3,600}{x}.$$
Substituting for y in the formula for P yields
$$P(x) = 2x + 2\left[\frac{3,600}{x}\right] = 2x + \frac{7,200}{x}$$
The graph suggests that $P(x)$ is minimal for some value of x near 60. In Chapter 3 you will use calculus to show that $P(x)$ is indeed minimized when $x = 60$. If $x = 60$, $y = \frac{3,600}{60} = 60$. Hence the playground of minimal perimeter is a square, 60 meters by 60 meters.

11. Let x denote the length of a side of the square base and y the height of the box as shown in the figure. Let C denote the cost of constructing the box. Then,

    $C = $ (cost per square meter of base and top)(area of base and top)
    $+$ (cost per square meter of sides)(area of sides)

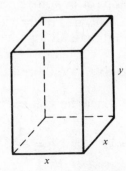

Rectangular box for Problem 11.

The combined area of the base and top is $2x^2$ and the combined area of the sides if $4xy$. Hence,

$$C = 2(2x^2) + 1(4xy) = 4x^2 + 4xy$$

which is a function of two variables. Since the volume is 250 cubic meters,

$$x^2 y = 250 \quad \text{or} \quad y = \frac{250}{x^2}.$$

Substituting for y in the formula for C yields

$$C(x) = 4x + 4x \left[\frac{250}{x^2} \right] = 4x^2 + \frac{1,000}{x}$$

13. Let x denote the length of the side of one of the removed squares and $V(x)$ the volume of the resulting box as shown in the figure. Then,

$$V(x) = (\text{area of base})(\text{height})$$

$$= (18 - 2x)(18 - 2x)(x) = 4x(9 - x)^2$$

The graph suggests that the volume will be greatest when x is close to 3 inches. In Chapter 3 you will learn how to use calculus to find this optimal value exactly.

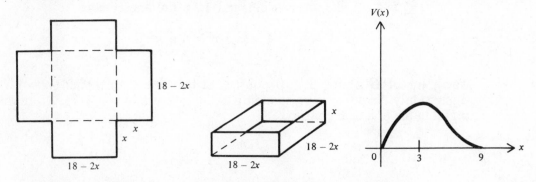

Cardboard sheet, folded box, and volume function for Problem 13.

15. Let r denote the radius of the circular top and h the height of the can. Let C denote the total cost of constructing the can. Then,

$$C = (\text{cost of top and bottom}) + (\text{cost of side})$$

Imagine the top and bottom of the can removed and the side cut and spread out as shown in the figure. The area of the top (or bottom) is πr^2 and so the cost of the top and bottom is $0.04(2\pi r^2) = 0.08\pi r^2$, The area of the side is $2\pi rh$ and so the cost of the side is $0.02(2\pi rh) = 0.04\pi rh$. Putting it all together,

$$C = 0.08\pi r^2 + 0.04\pi rh$$

which is a function of two variables. Since the volume is to be 4π,

$$\pi r^2 h = 4\pi \quad \text{or} \quad h = \frac{4}{r^2}.$$

Substituting for h in the formula for C yields

$$C(r) = 0.08\pi r^2 + 0.04\pi r \left[\frac{4}{r^2}\right] = 0.08\pi r^2 + \frac{0.16\pi}{r} = 0.08\pi \left(r^2 + \frac{2}{r}\right)$$

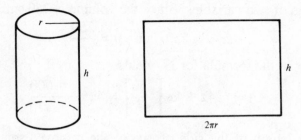

Cylindrical can for Problem 15.

17. (a) Let x denote the number of people in the group and $C(x)$ the corresponding total admission charge for the group. If $x \geq 50$, the group is charged $1x$ dollars, and if $0 \leq x < 50$, the group is charged $1.5x$ dollars. Hence,

$$C(x) = \begin{cases} 1.5x & \text{if } 0 \leq x < 50 \\ x & \text{if } x \geq 50 \end{cases}$$

The graph of $C(x)$ for $x \geq 0$ consists of two line segments as shown in the figure. Only the points corresponding to the integers $x = 0, 1, 2, \ldots$ are meaningful in the practical context.

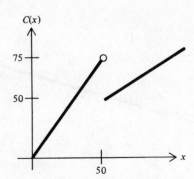

Cost function for Problem 17.

(b) $C(49) = 1.5(49) = \$73.50$ and $C(50) = \$50$. Hence, by recruiting one additional member, a group of 49 saves $73.50 - 50 = \$23.50$.

19.   Let x denote the weight (in ounces) of the letter and $C(x)$ the corresponding cost (in cents). Then

$$C(x) = \begin{cases} 20 & \text{if } 0 < x \le 1 \\ 37 & \text{if } 1 < x \le 2 \\ 54 & \text{if } 2 < x \le 3 \\ 71 & \text{if } 3 < x \le 4 \end{cases}$$

The graph of $C(x)$ consists of four horizontal line segments.

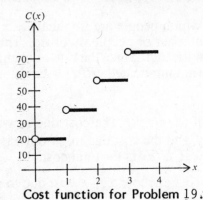

Cost function for Problem 19.

21.   (a)   Let x denote the taxable income and $f(x)$ the corresponding income tax. Then

$$f(x) = \begin{cases} 2{,}160 + .23(x - 16{,}190) = .23x - 1563.70 & \text{if } 16{,}190 < x \le 19{,}640 \\ 2{,}954 + .26(x - 19{,}640) = .26x - 2152.40 & \text{if } 19{,}640 < x \le 25{,}360 \\ 4{,}441 + .30(x - 25{,}360) = .30x - 3167 & \text{if } 25{,}360 < x \le 31{,}080 \end{cases}$$

The graph of $f(x)$ consists of three disconnected line segments.

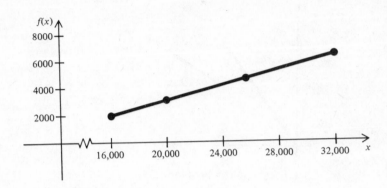

(b)   The slopes of the line segments are .23 , .26 , and .30 , respectively. The slopes increase so the taxable income increases. In practical terms, the higher the taxable income the higher the rates at which the tax is computed.

23.   Let R denote the rate of population growth and p the population size. Since R is directly proportional to p, $R(p) = kp$, where k is the constant of proportionality.

25.   Let R denote the rate at which temperature changes, M the temperature of the medium, and t the temperature of the object. Then $M - t$ is the difference in temperature between the medium and object. Since the rate of change is directly proportional to the difference, $R(t) = k(M - t)$, where k is the constant of proportionality.

27.   Let R denote the rate at which people are implicated, x the number of people implicated, and n the total number of people involved. Then $n - x$ is the number of people involved but not implicated. Since the rate of change is jointly proportional to those implicated and those not implicated, $R(x) = kx(n - x)$, where k is the constant of proportionality.

29.   Let x denote the speed of the truck. The cost due to wages is $\frac{k_1}{s}$, where k_1 is a constant of proportionality, and the cost due to gasoline is $k_2 s$, where k_2 is another constant of proportionality, If $C(s)$ is the total cost,

$$C(s) = \text{cost due to wages} + \text{cost due to gasoline}$$

$$= \frac{k_1}{s} + k_2 s$$

31.   The speed of the car C is 60 miles per hour and the speed of the truck T is 30 miles per hour. At some time $t > 0$, car C has traveled $60t$ miles and truck T has traveled $30t$ miles. The situations for $t = 0$ and for $t > 0$ are shown in the figure.
       Let $D(t)$ denote the distance between T and C at time t. At time t, C is $60t$ miles from the intersection and T is $300 - 30t$ miles from the intersection. By the pythagorean theorem,

$$[D(t)]^2 = (60t)^2 + (300 - 30t)^2 \quad \text{or}$$

$$D(t) = \sqrt{(60t)^2 + (300 - 30t)^2} = 30\sqrt{5t^2 - 20t + 100}$$

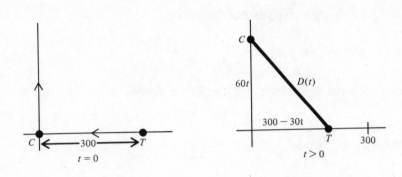

33. Let x denote the width of the printed portion and y the length of the printed portion as shown in the figure.

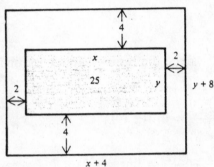

Rectangular poster for Problem 33.

Then $x+4$ is the width of the poster and $y+8$ is its length. The area A of the poster is

$$A = (x+4)(y+8)$$

which is a function of two variables. Since the area of the printed portion is to be 25 square centimeters,

$$xy = 25 \quad \text{or} \quad y = \frac{25}{x}.$$

Substituting for y in the area formula yields

$$A(x) = (x+4)\left(\frac{25}{x}+8\right) = 25 + 8x + \frac{100}{x} + 32 = 8x + \frac{100}{x} + 57.$$

Chapter 1, Section 6

1. $\lim\limits_{x \to a} f(x) = b$ (even though $f(a)$ is not defined).

3. $\lim\limits_{x \to a} f(x) = b$ (even though $f(a) = c$).

5. $\lim\limits_{x \to a} f(x)$ does not exist since as x approaches a from the left, the function becomes unbounded.

7. $\lim\limits_{x \to 2}(3x^2 - 5x + 2) = 3 \lim\limits_{x \to 2} x^2 - 5 \lim\limits_{x \to 2} x + \lim\limits_{x \to 2} 2 = 3(2)^2 - 5(2) + 2 = 4$

9. $\lim\limits_{x \to 0}(x^5 - 6x^4 + 7) = \lim\limits_{x \to 0} x^5 - 6 \lim\limits_{x \to 0} x^4 + \lim\limits_{x \to 0} 7 = 0 - 6(0) + 7 = 7$

11. $\lim\limits_{x \to 3}(x - 1)^2(x + 1) = \lim\limits_{x \to 3}(x - 1)^2 \lim\limits_{x \to 3}(x + 1) = (3 - 1)^2(3 + 1) = 16$

13. $\lim\limits_{x \to 2} \dfrac{x+1}{x+2} = \dfrac{\lim\limits_{x \to 2}(x+1)}{\lim\limits_{x \to 2}(x+2)} = \dfrac{3}{4}$

15. $\lim\limits_{x \to 5} \dfrac{x+3}{x-5}$ does not exist since the limit of the denominator is zero while the limit of the numerator is not zero.

17. $\lim\limits_{x \to 1} \dfrac{x^2-1}{x-1} = \lim\limits_{x \to 1} \dfrac{(x+1)(x-1)}{x-1} = \lim\limits_{x \to 1}(x + 1) = 2$

19. $\lim\limits_{x \to 5} \dfrac{x^2-3x-10}{x-5} = \lim\limits_{x \to 5} \dfrac{(x-5)(x+2)}{x-5} = \lim\limits_{x \to 5}(x + 2) = 7$

21. $\lim\limits_{x \to 4} \dfrac{(x+1)(x-4)}{(x-1)(x-4)} = \lim\limits_{x \to 4} \dfrac{x+1}{x-1} = \dfrac{\lim\limits_{x \to 4}(x+1)}{\lim\limits_{x \to 4}(x-1)} = \dfrac{5}{3}$

23. $\lim\limits_{x \to -2} \dfrac{x^2-x-6}{x^2+3x+2} = \lim\limits_{x \to -2} \dfrac{(x-3)(x+2)}{(x+1)(x+2)} = \lim\limits_{x \to -2} \dfrac{x-3}{x+1} = \dfrac{\lim\limits_{x \to -2}(x-3)}{\lim\limits_{x \to -2}(x+1)} = \dfrac{-5}{-1} = 5$

25. $\lim\limits_{x \to 4} \dfrac{\sqrt{x}-2}{x-4} = \lim\limits_{x \to 4} \dfrac{\sqrt{x}-2}{x-4} \dfrac{\sqrt{x}+2}{\sqrt{x}+2} = \lim\limits_{x \to 4} \dfrac{x-4}{(x-4)(\sqrt{x}+2)} = \lim\limits_{x \to 4} \dfrac{1}{\sqrt{x}+2} = \dfrac{1}{2+2} = \dfrac{1}{4}$

27. $\lim\limits_{x \to 1} \dfrac{x-1}{\sqrt{x}-1} = \lim\limits_{x \to 1} \dfrac{x-1}{\sqrt{x}-1} \dfrac{\sqrt{x}+1}{\sqrt{x}+1} = \lim\limits_{x \to 1} \dfrac{(x-1)(\sqrt{x}+1)}{x-1} = \lim\limits_{x \to 1}(\sqrt{x} + 1) = 1 + 1 = 2$

29.   If $f(x) = 5x^2 - 6x + 1$, then $f(2) = 9$ and $\lim\limits_{x \to 2} f(x) = 9$, and so f is continuous at $x = 2$.

31.   If $f(x) = \dfrac{x+2}{x+1}$, then $f(1) = \dfrac{3}{2}$ and

$$\lim\limits_{x \to 1} f(x) = \lim\limits_{x \to 1} \dfrac{x+2}{x+1} = \dfrac{\lim\limits_{x \to 1}(x + 2)}{\lim\limits_{x \to 1}(x + 1)} = \dfrac{3}{2}.$$

Hence f is continuous at $x = 1$.

33. If $f(x) = \frac{x+1}{x-1}$, $f(1)$ is undefined since the denominator is zero, and hence f is not continuous at $x = 1$.

35. If $f(x) = \frac{\sqrt{x}-2}{x-4}$, $f(4)$ is undefined since the denominator is zero, and hence f is not continuous at $x = 4$.

37. If

$$f(x) = \begin{cases} x+1 & \text{if } x \leq 2 \\ 2 & \text{if } x > 2 \end{cases}$$

then $f(2) = 2 + 1 = 3$ and $\lim\limits_{x \to 2} f(x)$ must be determined. As x approaches 2 from the right,

$$\lim_{x \to 2} f(x) = \lim_{x \to 2} 2 = 2$$

and as x approaches 2 from the left,

$$\lim_{x \to 2} f(x) = \lim_{x \to 2} (x + 1) = 3.$$

Hence, the limit does not exist (since different limits are obtained from the left and the right), and so f is not continuous at $x = 2$.

39. If

$$f(x) = \begin{cases} x+1 & \text{if } x < 0 \\ x-1 & \text{if } x \geq 0 \end{cases}$$

then $f(0) = -1$ and $\lim\limits_{x \to 0} f(x)$ must be determined. As x approaches 0 from the right,

$$\lim_{x \to 0} f(x) = \lim_{x \to 0} (x - 1) = -1$$

and as x approaches 0 from the left,

$$\lim_{x \to 0} f(x) = \lim_{x \to 0} (x + 1) = 1.$$

Hence, the limit does not exist (since different limits are obtained from the left and the right), and so f is not continuous at $x = 0$.

41. If

$$f(x) = \begin{cases} \frac{x^2-1}{x+1} & \text{if } x < -1 \\ x^2 - 3 & \text{if } x \geq -1 \end{cases}$$

then $f(-1) = -2$ and $\lim\limits_{x \to -1} f(x)$ must be determined. As x approaches -1 from the right,

$$\lim_{x \to -1} f(x) = \lim_{x \to -1} (x^2 - 3) = -2$$

and as x approaches -1 from the left,

$$\lim_{x \to -1} f(x) = \lim_{x \to -1} \frac{x^2 - 1}{x + 1} = \lim_{x \to -1} \frac{(x - 1)(x + 1)}{x + 1} = \lim_{x \to -1} (x - 1) = -2.$$

Hence, the limit exists and is equal to $f(-1)$, and so f is continuous at $x = -1$.

43. The polynomial $f(x) = x^5 - x^3$ is continuous for all values of x.

45. $f(x) = \frac{3x-1}{2x-6}$ is not continuous at $x = 3$, where the denominator is zero.

47. $f(x) = \frac{x^2-1}{x+1}$ is not continuous at $x = -1$, where the denominator is zero.

49. $f(x) = \frac{x}{(x+5)(x-1)}$ is not continuous at $x = -5$ or at $x = 1$, where the denominator is zero.

51. $f(x) = \frac{x^2-2x+1}{x^2-x-2} = \frac{x^2-2x+1}{(x-2)(x+1)}$ is not continuous at $x = -1$ or at $x = 2$, where the denominator is zero.

53. The function

$$f(x) = \begin{cases} x^2 & \text{if } x \leq 2 \\ 9 & \text{if } x > 2 \end{cases}$$

is possibly not continuous only at $x = 2$. As x approaches 2 from the right,

$$\lim_{x \to 2} f(x) = \lim_{x \to 2} 9 = 9$$

and as x approaches 2 from the left,

$$\lim_{x \to 2} f(x) = \lim_{x \to 2} x^2 = 4.$$

Hence, the limit does not exist and so f is not continuous at $x = 2$.

55. If

$$f(x) = \begin{cases} Ax - 3 & \text{if } x < 2 \\ 3 - x + 2x^2 & \text{if } x \geq 2 \end{cases}$$

then $f(x)$ is continuous everywhere except possibly at $x = 2$ since $Ax - 3$ and $3 - x + 2x^2$ are polynomials. Since $f(2) = 3 - 2 + 2(2)^2 = 9$, in order that $f(x)$ be continuous at $x = 2$, A must be chosen so that $\lim_{x \to 2} f(x) = 9$.

As x approaches 2 from the right,

$$\lim_{x \to 2} f(x) = \lim_{x \to 2}(3 - x + 2x^2)$$
$$= \lim_{x \to 2} 3 - \lim_{x \to 2} x + 2(\lim_{x \to 2} x)^2$$
$$= 3 - 2 + 8 = 9$$

and as x approaches 2 from the left,

$$\lim_{x \to 2} f(x) = \lim_{x \to 2}(Ax - 3)$$
$$= A \lim_{x \to 2} x - \lim_{x \to 2} 3$$
$$= 2A - 3.$$

The $\lim\limits_{x\to 2} f(x) = 9$ exists when $2A - 3 = 9$ or $A = 6$. Thus $f(x)$ is continuous at $x = 2$ only when $A = 6$.

57. If

$$f(x) = \begin{cases} Ax^2 + 5x - 9 & \text{if } x < 1 \\ B & \text{if } x = 1 \\ (3 - x)(A - 2x) & \text{if } x > 1 \end{cases}$$

then $f(x)$ is continuous everywhere except possibly at $x = 1$ since $Ax^2 + 5x - 9$ and $(3 - x)(A - 2x)$ are polynomials.

Since $f(1) = B$, A and B need to be determined so that $\lim\limits_{x\to 1} f(x) = B$.

As x approaches 1 from the right,

$$\begin{aligned} \lim_{x\to 1} f(x) &= \lim_{x\to 1}(3 - x)(A - 2x) \\ &= \lim_{x\to 1}(3 - x)\lim_{x\to 1}(A - 2x) \\ &= (3 - 1)(A - 2(1)) \\ &= 2(A - 2) \end{aligned}$$

and as x approaches 1 from the left

$$\begin{aligned} \lim_{x\to 1} f(x) &= \lim_{x\to 1}(Ax^2 + 5x - 9) \\ &= A(\lim_{x\to 1} x)^2 + 5\lim_{x\to 1} x - \lim_{x\to 1} 9 \\ &= A(1)^2 + 5(1) - 9 \\ &= A - 4. \end{aligned}$$

The $\lim\limits_{x\to 1} f(x)$ exists whenever

$$\begin{aligned} 2(A - 2) &= A - 4 \\ 2A - 4 &= A - 4 \\ A &= 0. \end{aligned}$$

When $A = 0$, $\lim\limits_{x\to 1} f(x) = -4$ and thus, to have continuity at $x = 1$, $f(1) = B = -4$.
The function is continuous for all x when $A = 0$ and $B = -4$.

59. On the open interval $0 < x < 1$

$$f(x) = x\left(1 + \frac{1}{x}\right) = x + 1$$

since $x \neq 0$. Thus, $f(x)$, a polynomial on $0 < x < 1$, is continuous.
The function $f(x) = x\left(1 + \frac{1}{x}\right)$ is not continuous at $x = 0$ since $f(0)$ is not defined. The function $f(x)$ is defined at $x = 1$ since $f(1) = 1\left(1 + \frac{1}{1}\right) = 2$ and as x approaches 1 from the left

$$\begin{aligned} \lim_{x\to 1}\left(1 + \frac{1}{x}\right) &= \left(\lim_{x\to 1} x\right)\left(\lim_{x\to 1}\left(1 + \frac{1}{x}\right)\right) \\ &= 1\left(1 + \frac{1}{1}\right) \\ &= 2. \end{aligned}$$

The function $f(x)$ is thus continuous on $0 < x \leq 1$.

61.   On the open interval $0 < x < 3$

$$f(x) = \begin{cases} x(x-1) & \text{if } x < 3 \\ \frac{x^2-9}{x-3} & \text{if } x \geq 3 \end{cases}$$

can be rewritten as

$$f(x) = x(x-1) = x^2 - x,$$

a polynomial function and thus, continuous on $0 < x < 3$.

At the left hand end point $f(0) = 0(0-1) = 0$ and as x approaches 0 from the right

$$\begin{aligned} \lim_{x \to 0} f(x) &= \lim_{x \to 0} x(x-1) \\ &= (\lim_{x \to 0} x)(\lim_{x \to 0}(x-1)) \\ &= 0(0-1) \\ &= 0. \end{aligned}$$

Thus, $f(x)$ is continuous at the left hand end point $x = 0$.

At the right hand end point $f(3) = \frac{3^2-9}{3-3} = \frac{0}{0}$ and $f(x)$ is not continuous at the right hand end point $x = 3$ since $f(x)$ is not defined there.

63.   An alternate version of the intermediate value property is: A function $f(x)$ that is continuous on a closed interval $a \leq x \leq b$ takes on every value between $f(a)$ and $f(b)$ as x assumes all numbers between a and b.

In most cases, manufacturers' supply $S(p)$ increases and consumers' demand $D(p)$ decreases as the market price p increases. This means over the closed interval $p_1 \leq p \leq p_2$, $S(p_1) < S(p_2)$ and $D(p_1) > D(p_2)$.

It is given that $D(p_1) > S(p_1)$ and $S(p_2) > D(p_2)$.

The four inequalities $S(p_1) < S(p_2)$, $D(p_1) > D(p_2)$, $D(p_1) > S(p_1)$ and $S(p_2) > D(p_2)$ result in the four illustrated cases:

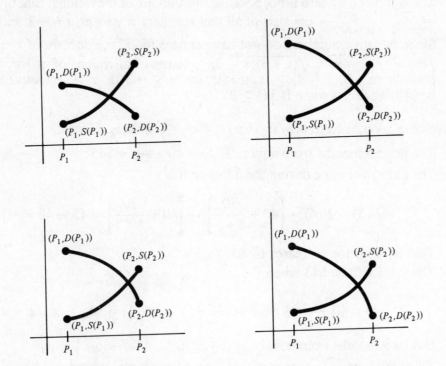

The intersections of the graphs is the equilibrium point p_e since $D(p_e) = S(p_e)$.

Chapter 1, Review Problems

1. (a) The domain of the quadratic function $f(x) = x^2 - 2x + 6$ consists of all real numbers x.

   (b) Since division by zero is not possible, the domain of the rational function $f(x) = \frac{x-3}{x^2+x-2} = \frac{x-3}{(x+2)(x-1)}$ consists of all real numbers x except $x = -2$ and $x = 1$.

   (c) Since negative numbers do not have square roots, the domain of the function $f(x) = \sqrt{x^2 - 9} = \sqrt{(x+3)(x-3)}$ consists of all values of x for which the product $(x+3)(x-3) \geq 0$, that is, for $x \leq -3$ or $x \geq 3$. Another way of describing this domain is $|x| \geq 3$.

2. The price x months from now is $P(x) = 40 + \frac{30}{x+1}$ dollars. Hence:

   (a) The price 5 months from now is $P(5) = 40 + \frac{30}{5+1} = \$45$.

   (b) The change in price during the 5th month is

   $$P(5) - P(4) = \left[40 + \frac{30}{5+1}\right] - \left[40 + \frac{30}{4+1}\right] = 45 - 46 = -1$$

   That is, the price decreases by $1.

   (c) The price will be $43 when

   $$43 = 40 + \frac{30}{x+1}, \quad 3 = \frac{30}{x+1}, \quad 3(x+1) = 30, \quad \text{or} \quad x = 9$$

   that is, 9 months from now.

   (d) Since the term $\frac{30}{x+1}$ gets very small as x gets large, the price $P(x) = 40 + \frac{30}{x+1}$ will approach $40 in the long run.

3. (a) If $g(u) = u^2 + 2u + 1$ and $h(x) = 1 - x$, then

   $$g[h(x)] = g(1-x) = (1-x)^2 + 2(1-x) + 1 = x^2 - 4x + 4$$

   (b) If $g(u) = \frac{1}{2u+1}$ and $h(x) = x + 2$, then

   $$g[h(x)] = g(x+2) = \frac{1}{2(x+2)+1} = \frac{1}{2x+5}$$

   (c) If $g(u) = \sqrt{1-u}$ and $h(x) = 2x + 4$, then

   $$g[h(x)] = g(2x+4) = \sqrt{1-(2x+4)} = \sqrt{-2x-3}$$

4. (a) If $f(x) = x^2 - x + 4$, then

   $$f(x-2) = (x-2)^2 - (x-2) + 4 = x^2 - 5x + 10$$

   (b) If $f(x) = \sqrt{x} + \frac{2}{x-1}$, then

   $$f(x^2+1) = \sqrt{x^2+1} + \frac{2}{(x^2+1)-1} = \sqrt{x^2+1} + \frac{2}{x^2}$$

(b)   If $f(x) = \sqrt{x} + \frac{2}{x-1}$, then

$$f(x^2 + 1) = \sqrt{x^2 + 1} + \frac{2}{(x^2 + 1) - 1} = \sqrt{x^2 + 1} + \frac{2}{x^2}$$

(c)   If $f(x) = x^2$, then

$$f(x + 1) - f(x) = (x + 1)^2 - x^2 = x^2 + 2x + 1 - x^2 = 2x + 1$$

5.   (a) $f(x) = (x^2 + 3x + 4)^5$ can be written as $g[h(x)]$, where

$$g(u) = u^5 \quad \text{and} \quad h(x) = x^2 + 3x + 4$$

(b) $f(x) = (3x + 1)^2 + \frac{5}{2(3x+2)^3}$ can be written as $g[h(x)]$, where

$$g(u) = u^2 + \frac{5}{2(u + 1)^3} \quad \text{and} \quad h(x) = 3x + 1$$

6.   (a)   Since the smog level Q is related to the variable p by the equation

$$Q(p) = \sqrt{0.5p + 19.4}$$

and the variable p is related to the variable t by the equation

$$p(t) = 8 + 0.2t^2$$

it follows that the composite function

$$Q[p(t)] = \sqrt{0.5(8 + 0.2t^2) + 19.4}$$
$$= \sqrt{4 + 0.1t^2 + 19.4}$$
$$= \sqrt{23.4 + 0.1t^2}$$

expresses the smog level as a function of the variable t .

(b)   The smog level 3 years from now will be

$$Q[p(3)] = \sqrt{23.4 + 0.1(3^2)} = \sqrt{24.3} \simeq 4.93 \text{ units}$$

(c)   Set $Q[p(t)]$ equal to 5 and solve for t to get

$$5 = \sqrt{23.4 + 0.1t^2}$$
$$25 = 23.4 + 0.1t^2$$
$$1.6 = 0.1t^2$$
$$t^2 = \frac{1.6}{0.1} = 16 \quad \text{or} \quad t = 4 \text{ years from now.}$$

7.   If the graph of $y = 3x^2 - 2x + c$ passes through the point $(2, 4)$, then $y = 4$ and $x = 2$ and so

$$4 = 3(2)^2 - 2(2) + c, \quad 4 = 12 - 4 + c, \quad \text{or} \quad c = -4$$

8.   (a)   Some points on the graph of $f(x) = x^2 + 2x - 8 = (x + 4)(x - 2)$ are shown below.

x	-6	-5	-4	-3	-2	-1	0	1	2	3	4
$f(x)$	16	7	0	-5	-8	-9	-8	-5	0	7	16

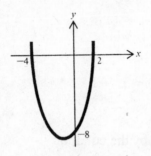

Graph for Problem 8a.   Graph for Problem 8b.

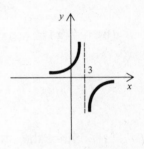

Graph for Problem 8c.

(b)   Some points on the graph of

$$f(x) = \begin{cases} 1/x^2 & \text{if } x < 0 \\ x & \text{if } 0 \le x < 3 \\ 4 & \text{if } x \ge 3 \end{cases}$$

are shown below.

x	-3	-2	-1	$-1/2$	0	1	2	3	4	5	6
$f(x)$	$1/9$	$1/4$	1		4	0	1	2	4	4	4	4

(c)   Since $x = 3$ is not in the domain of $f(x) = -\frac{2}{x-3}$, the graph breaks into two pieces at $x = 3$. Some points on the graph are shown below.

x	-97	1	2	2.99	3.01	4	5	103
$f(x)$	$.02$	1	2	200	-200	-2	-1	$-.02$

(d)   Since $x = 0$ is not in the domain of $f(x) = x + \frac{2}{x}$, the graph breaks into two pieces at $x = 0$. Some points on the graph are shown below.

x	± 0.01	± 1	± 2	± 4	± 100
$f(x)$	± 200.01	± 3	± 3	± 4.5	± 100.02

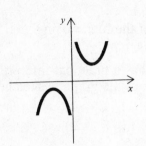

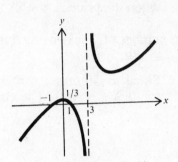

Graph for Problem 8d.   Graph for Problem 8e.

(e)   The function $f(x) = \frac{x^2-1}{x-3}$ has a discontinuity at $x = 3$ where its denominator is zero. From the factored form $f(x) = \frac{(x+1)(x-1)}{x-3}$ you can see that the x intercepts at $(-1,0)$ and $(1,0)$. Some other points on the graph of f are shown below.

x	-50	-3	-2	-1	0	1	2	3	4	5	50
$f(x)$	-47.2	-1.3	-0.6	0	$1/3$	0	-3		15	12	53.2

9.   (a)   The function $D(p) = -50p+800$ is linear with slope -50 and vertical intercept 800 and represents demand for $0 \le p \le 16$.

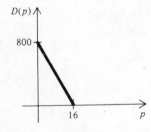

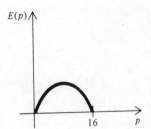

(b)   The total monthly expenditure is

$$E(p) = (\text{price per unit})(\text{demand})$$
$$= p(-50p + 800) = -50p(p - 16)$$

Since the expenditure is assumed to be nonnegative, the relevant interval is $0 \le p \le 16$.

(c)   The graph suggests that the expenditure will be greatest if $p = 8$.
Remark: In Chapter 3 you will learn how to use calculus to find the optimal price. Without calculus, you could complete the square to get

$$E(p) = -50(p - 8)^2 + 3{,}200$$

from which it is clear that the greatest expenditure is \$3,200 and is generated when the price is $p = \$8$.

10. The number of weeks needed to reach x percent of the fund raising goal is given by $f(x) = \frac{10x}{150-x}$.

(a) Since x denotes a percentage, the function has a practical interpretation only for $0 \leq x \leq 100$. The corresponding portion of the graph is sketched below.

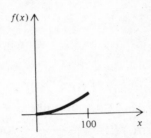

Graph for Problem 10a.

(b) The number of weeks needed to reach 50 percent of the goal is

$$f(50) = \frac{10(50)}{150 - 50} = 5$$

(c) The number of weeks needed to reach 100 percent of the goal is

$$f(100) = \frac{10(100)}{150 - 100} = 20$$

11. (a) If $y = 3x + 2$, then $m = 3$ and $b = 2$.

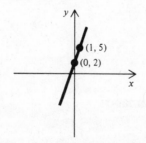

Graph for Problem 11a.

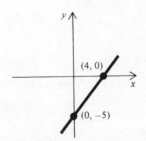

Graph for Problem 11b.

(b) If $5x - 4y = 20$, then $y = \frac{5}{4}x - 5$, and so $m = \frac{5}{4}$ and $b = -5$.

(c)   If $2y + 3x = 0$, then $y = -\frac{3}{2}x + 0$, and so $m = -\frac{3}{2}$ and $b = 0$.

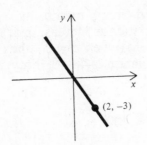

Graph for Problem 11c.

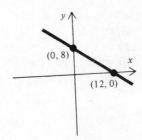

Graph for Problem 11d.

(d)   If $\frac{x}{3} + \frac{y}{2} = 4$, then $\frac{y}{2} = -\frac{x}{3} + 4$ or $y = -\frac{2}{3}x + 8$ so $m = -\frac{2}{3}$ and $b = 8$.

12.   Since $m = 5$ and $b = -4$, the equation of the line is $y = 5x - 4$.

13.   Since $m = -2$ and the point $(1, 3)$ is on the line, the equation of the line is
$$y - 3 = -2(x - 1) \quad \text{or} \quad y = -2x + 5.$$

14.   From the points $(2, 4)$ and $(1, -3)$, the slope is $m = \frac{-3-4}{1-2} = 7$ and with $P = (2, 4)$, the equation of the line is
$$y - 4 = 7(x - 2) \quad \text{or} \quad y = 7x - 10.$$
The same equation would result if $P = (1, -3)$ were used.

15.   (a)   Let x denote the time in months since the beginning of the year and $P(x)$ the corresponding price (in cents) of gasoline. Since the price increases at a constant rate of 2 cents per gallon per month, p is a linear function of x with slope $m = 2$. Since the price on June first (when $x = 5$) is 103 cents, the graph passes through $(5, 103)$. The equation is therefore
$$P - 103 = 2(x - 5) \quad \text{or} \quad P(x) = 2x + 93$$

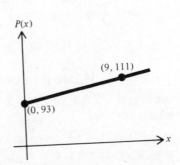

Price function for Problem 15.

(b)   The price at the beginning of the year was $P(0) = 93$ cents per gallon.

(c)   The price on October first (when $x = 9$) is $P(9) = 2(9) + 93 = 111$, that is, $1.11 per gallon.

16.   (a)   Let x denote the number of months (measured from 3 months ago) and $C(x)$ the corresponding circulation. Since the circulation is increasing at a constant rate, $C(x)$ is a linear function. Three months ago (when $x = 0$), the circulation was 3,200, and today (when $x = 3$), the circulation is 4,400. From the points $(0, 3200)$ and $(3, 4400)$, the slope is

$$m = \frac{4,400 - 3,200}{3 - 0} = 400$$

and the vertical intercept is 3,200. Hence the equation of the line is

$$C(x) = 400x + 3,200$$

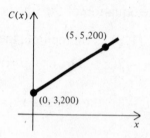

Circulation function for Problem 16.

(b)   Two months from now (that is, when $x = 5$) the circulation will be

$$C(5) = 400(5) + 3,200 = 5,200 .$$

17.   (a)   The graphs of $y = -3x + 5$ and $y = 2x - 10$ seem to intersect in the fourth quadrant. Setting the two expressions for y equal to each other yields

$$-3x + 5 = 2x - 10 \quad \text{or} \quad x = 3 .$$

When $x = 3$, $y = -3(3) + 5 = -4$. Hence the point of intersection is $(3, -4)$.

(b)   The graphs of $y = x + 7$ and $y = -2 + x$ are parallel lines with slope 1. Hence there are no points of intersection.

(c)   The graphs of $y = x^2 - 1$ and $y = 1 - x^2$ seem to intersect at the points $(-1, 0)$ and $(1, 0)$. Setting the two expressions for y equal to each other yields

$$x^2 - 1 = 1 - x^2 , \quad x^2 = 1 , \quad \text{or} \quad x = \pm 1 .$$

When $x = \pm 1$, $y = (\pm 1)^2 - 1 = 0$. Hence the points of intersection are $(-1, 0)$ and $(1, 0)$.

(d)   The graphs of $y = x^2$ and $y = 15 - 2x$ seem to intersect in the first and second quadrants. Setting the two expressions for y equal to each other yields

$$x^2 = 15 - 2x, \quad x^2 + 2x - 15 = 0,$$
$$(x + 5)(x - 3) = 0, \quad x = -5 \text{ and } x = 3.$$

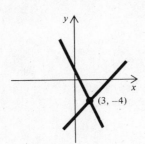

Graph for Problem 17a.

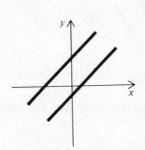

Graph for Problem 17b.

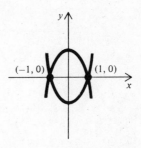

Graph for Problem 17c.

When $x = -5$ $y = 25$, and when $x = 3$, $y = 9$. Hence the points of intersection are $(-5, 25)$ and $(3, 9)$.

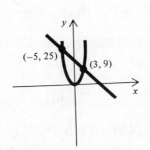

Graph for Problem 17d.

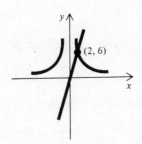

Graph for Problem 17e.

(e)   The graphs of $y = \frac{24}{x^2}$ and $y = 3x$ seem to intersect in the first quadrant. Setting the two expressions for y equal to each other yields

$$\frac{24}{x^2} = 3x, \quad 3x^3 = 24, \quad x^3 = 8, \quad \text{or} \quad x = 2$$

When $x = 2$, $y = 3(2) = 6$. Hence the point of intersection is $(2, 6)$.

18.   Let x denote time measured in half-hour units. The corresponding costs for the first and second plumbers, respectively, are

$$C_1(x) = 25 + 16x \quad \text{and} \quad C_2(x) = 31 + 14x$$

Setting the two cost functions equal to each other yields

$$31 + 14x = 25 + 16x \quad \text{or} \quad x = 3$$

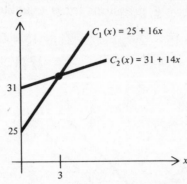

Cost functions for Problem 18.

Thus, if a plumber is to be used for exactly 3 half-hours ($1\frac{1}{2}$ hours), the two plumbers charge the same. If a plumber is needed for fewer than $1\frac{1}{2}$ hours, the first plumber is the less expensive since the graph of C_1 is initially below that of C_2, and if a plumber is needed for more than $1\frac{1}{2}$ hours, the second plumber is the less expensive.

19. (a) Let x denote the number of units manufactured and sold and $C(x)$ and $R(x)$ the corresponding cost and revenue functions, respectively. Then

$$C(x) = 4{,}500 + 50x \quad \text{and} \quad R(x) = 80x$$

For the manufacturer to break even, the cost must be equal to the revenue. That is,

$$4{,}500 + 50x = 80x \quad \text{or} \quad x = 150 \text{ units}$$

(b) Let $P(x)$ denote the profit from the manufacture and sale of x units. Then,

$$P(x) = R(x) - C(x) = 80x - (4{,}500 + 50x) = 30x - 4{,}500$$

When 200 units are sold, the profit is

$$P(200) = 30(200) - 4{,}500 = \$1{,}500$$

(c) The profit will be \$900 when

$$900 = 30x - 4{,}500 \quad \text{or} \quad x = 180$$

that is, when 180 units are manufactured and sold.

20. Let x denote the selling price of a bookcase (in dollars) and $P(x)$ the corresponding profit. Then

$$P(x) = (\text{number of units sold})(\text{profit per unit}) = (50 - x)(x - 10)$$

The graph suggests that the profit will be greatest when x is approximately \$30. Remark: In Chapter 3 you will learn how to use calculus to find the optimal price exactly. Without calculus, you could complete the square to get

$$P(x) = -(x - 30)^2 + 400$$

from which it is clear that the maximum profit is \$400 and occurs when $x = \$30$.

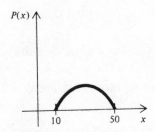

Profit function for Problem 20.

21.   Let x denote the selling price (in dollars) and $P(x)$ the corresponding profit function. Then

$$P(x) = (\text{number of units sold})(\text{profit per unit})$$

Since the cost is \$50, the profit per unit is $x - 50$. Since the number of \$1 reductions is $80 - x$, the number of \$5 reductions is $\frac{80-x}{5}$, and so

$$\text{Number of units sold } = 40 + 10\left[\frac{80 - x}{5}\right] = 200 - 2x$$

Putting it all together,

$$P(x) = (200 - 2x)(x - 50) = 2(100 - x)(x - 50)$$

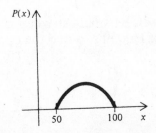

Profit function for Problem 21.

The graph suggests that the profit will be greatest when x is approximately \$75. <u>Remark</u>: In Chapter 3 you will learn how to use calculus to find the optimal price exactly. Without calculus, you could complete the square to get

$$P(x) = -2(x - 75)^2 + 1{,}250$$

from which it is clear that the maximum profit is \$1,250 and occurs when $x = \$75$.

22. Let r denote the radius, h the height, and V the volume of the can. Then,

$$V = \pi r^2 h$$

To write h in terms of r, use the fact that the cost of constructing the can is to be 80 cents. That is,

$$80 = \text{cost of bottom} + \text{cost of side}$$

where

$$\text{cost of bottom} = (\text{cost per square inch})(\text{area}) = 3\pi r^2$$

and

$$\text{cost of side} = (\text{cost per square inch})(\text{area}) = 2(2\pi rh) = 4\pi rh$$

Hence,

$$80 = 3\pi r^2 + 4\pi rh, \quad 4\pi rh = 80 - 3\pi r^2 \quad \text{or} \quad h = \frac{20}{\pi r} - \frac{3r}{4}$$

Now substitute this expression for h into the formula for V to get

$$V(r) = \pi r^2 \left[\frac{20}{\pi r} - \frac{3r}{4} \right] = 20r - \frac{3}{4}\pi r^3$$

23. Let x denote the number of machines used and $C(x)$ the corresponding cost function. Then,

$$C(x) = (\text{set up cost}) + (\text{operating cost})$$
$$= 80(\text{number of machines}) + 5.76(\text{number of hours})$$

Since 400,000 medals are to be produced and each of the x machines can produce 200 medals per hour,

$$\text{Number of hours} = \frac{400,000}{200x} = \frac{2,000}{x}$$

Putting it all together,

$$C(x) = 80x + 5.76 \left[\frac{2,000}{x} \right] = 80x + \frac{11,520}{x}$$

The graph suggests that the cost will be smallest when x is approximately 12. In Chapter 3 you will learn how to use calculus to find the optimal number of machines exactly.

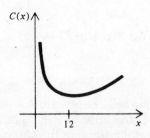

Cost function for Problem 23.

24. Let x denote the taxable income and $f(x)$ the corresponding income tax. Then

$$f(x) = \begin{cases} 2{,}045 + .20(x - 16{,}190) = .20x - 1{,}193 & \text{if } 16{,}190 < x \le 19{,}640 \\ 2{,}735 + .24(x - 19{,}640) = .24x - 1{,}978.6 & \text{if } 19{,}640 < x \le 25{,}360 \\ 4{,}108 + .28(x - 25{,}360) = .28x - 2{,}992.8 & \text{if } 25{,}360 < x \le 31{,}080 \end{cases}$$

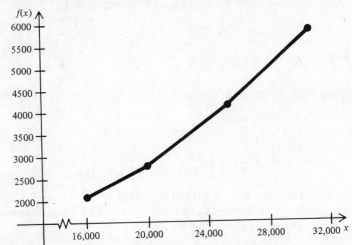

25. Let x denote the number of relevant facts recalled, n the total number of relevant facts in the person's memory, and $R(x)$ the rate of recall. Then $n - x$ is the number of relevant facts not recalled. Hence $R(x) = k(n - x)$, where k is a constant of proportionality.

26. (a) $\displaystyle\lim_{x\to 2} \frac{x^2-3x}{x+1} = \frac{\lim_{x\to 2}(x^2-3x)}{\lim_{x\to 2}(x+1)} = \frac{4-6}{3} = -\frac{2}{3}$

   (b) $\displaystyle\lim_{x\to 1} \frac{x^2+x-2}{x^2-1} = \frac{\lim_{x\to 1}(x^2+x-2)}{\lim_{x\to 1}(x^2-1)} = \frac{0}{0}$

which means simplify before taking the limits:

$$\lim_{x\to 1} \frac{x^2+x-2}{x^2-1} = \lim_{x\to 1} \frac{(x+2)(x-1)}{(x+1)(x-1)} = \lim_{x\to 1} \frac{x+2}{x+1} = \frac{3}{2}$$

   (c) $\displaystyle\lim_{x\to 1} \left(\frac{1}{x^2} - \frac{1}{x}\right) = 1 - 1 = 0$

   (d) $\displaystyle\lim_{x\to 2} \frac{x^3-8}{2-x} = \frac{\lim_{x\to 2}(x^3-8)}{\lim_{x\to 2}(2-x)} = \frac{0}{0}$

which means simplify before taking the limit:

$$\lim_{x\to 2} \frac{x^2 - 8}{2 - x} = \lim_{x\to 2} \frac{(x - 2)(x^2 + 2x + 4)}{-(x - 2)}$$

$$= \lim_{x\to 2} \frac{x^2 + 2x + 4}{-1}$$

$$= -\lim_{x\to 2}(x^2 + 2x + 4) = -12$$

27. (a) $f(x) = 5x^3 - 3x + \sqrt{x}$ is not continuous for $x < 0$ since square roots of negative numbers do not exist.

(b) $f(x) = \frac{x^2-1}{x+3}$ is not continuous at $x = -3$ since $f(-3) = \frac{-10}{0}$ and division by 0 is undefined.

(c) $g(x) = \frac{x^3+5x}{(x-2)(2x+3)}$ is not continuous at $x = 2$ and $x = -3/2$ since $g(2) = \frac{18}{0}$ and $g(-3/2) = \frac{-87/8}{0}$ and division by 0 is undefined.

(d) $h(x) = \begin{cases} x^3 + 2x - 33 & \text{if } x < 3 \\[2mm] \frac{x^2-6x+9}{x-3} & \text{if } x \geq 3 \end{cases}$

is not continuous at $x = 3$ since $h(3) = \frac{9-18+9}{3-3} = \frac{0}{0}$ and division by 0 is undefined.

28.   If

$$f(x) = \begin{cases} 2x + 3 & \text{if } x < 1 \\ Ax - 1 & \text{if } x \geq 1 \end{cases}$$

then $f(x)$ is continuous everywhere except possibly at $x = 1$ since $2x+3$ and $Ax-1$ are polynomials. Since $f(1) = A - 1$, in order that $f(x)$ be continuous at $x = 1$, A must be chosen so that $\lim\limits_{x \to 1} f(x) = A - 1$.

As x approaches 1 from the right

$$\begin{aligned} \lim_{x \to 1} f(x) &= \lim_{x \to 1}(Ax - 1) \\ &= A - 1 \end{aligned}$$

and as x approaches 1 from the left

$$\begin{aligned} \lim_{x \to 1} f(x) &= \lim_{x \to 1}(2x + 3) \\ &= 5 \end{aligned}$$

The $\lim\limits_{x \to 1} f(x)$ exists whenever $A - 1 = 5$ or $A = 6$ and furthermore, for $A = 6$

$$\begin{aligned} \lim_{x \to 1} f(x) &= 5 \\ f(1) &= 6 - 1 = 5. \end{aligned}$$

Thus, $f(x)$ is continuous at $x = 1$ only when $A = 6$.

29.   If

$$f(x) = \begin{cases} \frac{x^2-1}{x+1} & \text{if } x < -1 \\[2mm] Ax^2 + x - 3 & \text{if } x \geq -1 \end{cases}$$

then $f(x)$ is continuous everywhere except possibly at $x = -1$ since $\frac{x^2-1}{x+1}$ is a rational function and $Ax^2 + x - 3$ is a polynomial. Since $f(-1) = A - 4$, in order that $f(x)$ be continuous at $x = -1$, A must be chosen so that $\lim\limits_{x \to -1} f(x) = A - 4$.

As x approaches -1 from the right,

$$\begin{aligned} \lim_{x \to -1} f(x) &= \lim_{x \to -1}(Ax^2 + x - 3) \\ &= A - 4 \end{aligned}$$

and as x approaches -1 from the left

$$\lim_{x \to -1} f(x) = \lim_{x \to -1} \frac{x^2 - 1}{x + 1}$$
$$= \lim_{x \to -1} \frac{(x + 1)(x - 1)}{x + 1}$$
$$= \lim_{x \to -1} (x - 1)$$
$$= -2 .$$

The $\lim_{x \to -1} f(x)$ exists whenever $A - 4 = -2$ or $A = 2$ and furthermore, for $A = 2$

$$\lim_{x \to -1} f(x) = -2$$

and

$$f(-1) = 2 - 4 = -2 .$$

Thus, $f(x)$ is continuous at $x = -1$ only when $A = 2$.

CHAPTER 2

DIFFERENTIATION: BASIC CONCEPTS

---

1 THE DERIVATIVE

2 TECHNIQUES OF DIFFERENTIATION

3 RATE OF CHANGE AND MARGINAL ANALYSIS

4 APPROXIMATION BY DIFFERENTIALS

5 THE CHAIN RULE

6 IMPLICIT DIFFERENTIATION

7 HIGHER-ORDER DERIVATIVES

REVIEW PROBLEMS

Chapter 2, Section 1

1. If $f(x) = 5x - 3$, then

$$
\begin{aligned}
\frac{f(x + \Delta x) - f(x)}{\Delta x} &= \frac{[5(x + \Delta x) - 3] - [5x - 3]}{\Delta x} \\
&= \frac{5x + 5\Delta x - 3 - 5x + 3}{\Delta x} \\
&= \frac{5\Delta x}{\Delta x} \\
&= 5
\end{aligned}
$$

As $\Delta x \to 0$, this difference quotient approaches 5. Thus, $f'(x) = 5$, and hence the slope of the line tangent of $f(x)$ at $(2, f(2)) = (2, 7)$ is $f'(2) = 5$.

3.   If $f(x) = 2x^2 - 3x + 5$, then

$$\frac{f(x + \Delta x) - f(x)}{\Delta x} = \frac{[2(x + \Delta x)^2 - 3(x + \Delta x) + 5] - [2x^2 - 3x + 5]}{\Delta x}$$

$$= \frac{2x^2 + 4x\Delta x + 2(\Delta x)^2 - 3x - 3\Delta x + 5 - 2x^2 + 3x - 5}{\Delta x}$$

$$= \frac{4x\Delta x + 2(\Delta x)^2 - 3\Delta x}{\Delta x}$$

$$= 4x + 2\Delta x - 3$$

As $\Delta x \to 0$, this difference quotient approaches $4x - 3$. Thus, $f'(x) = 4x - 3$, and the slope of the line tangent to $y = f(x)$ at $(0, f(0)) = (0, 5)$ is $f'(0) = -3$.

5.   If $f(x) = \frac{2}{x}$, then

$$\frac{f(x + \Delta x) - f(x)}{\Delta x} = \frac{\left[\frac{2}{x + \Delta x}\right] - \left[\frac{2}{x}\right]}{\Delta x}$$

$$= \left[\frac{\frac{2}{x + \Delta x} - \frac{2}{x}}{\Delta x}\right]\left[\frac{x(x + \Delta x)}{x(x + \Delta x)}\right]$$

$$= \frac{2x - 2(x + \Delta x)}{\Delta x(x)(x + \Delta x)}$$

$$= \frac{-2}{x(x + \Delta x)}$$

As $\Delta x \to 0$, this difference quotient approaches $-\frac{2}{x^2}$. Thus, $f'(x) = -\frac{2}{x^2}$, and the slope of the line tangent to $f(x)$ at $(\frac{1}{2}, f(\frac{1}{2})) = (\frac{1}{2}, 4)$ is $f'(\frac{1}{2}) = -8$.

7.   If $f(x) = \sqrt{x}$, then

$$\frac{f(x + x) - f(x)}{\Delta x} = \frac{\left[\sqrt{x + \Delta x}\right] - \left[\sqrt{x}\right]}{\Delta x}$$

$$= \left[\frac{\sqrt{x + \Delta x} - \sqrt{x}}{\Delta x}\right]\left[\frac{\sqrt{x + \Delta x} + \sqrt{x}}{\sqrt{x + \Delta x} + \sqrt{x}}\right]$$

$$= \frac{(x + \Delta x) - x}{\Delta x\left[\sqrt{x + \Delta x} + \sqrt{x}\right]}$$

$$= \frac{1}{\sqrt{x + \Delta x} + \sqrt{x}}$$

As $\Delta x \to 0$, this difference quotient approaches $\frac{1}{2\sqrt{x}}$. Thus, $f'(x) = \frac{1}{2\sqrt{x}}$, and the slope of the line tangent to $y = f(x)$ at $(9, f(9)) = (9, 3)$ is $f'(9) = \frac{1}{6}$.

9. If $f(x) = x^3 - x$, then

$$\frac{f(x + \Delta x) - f(x)}{\Delta x} = \frac{[(x + \Delta x)^3 - (x + \Delta x)] - [x^3 - x]}{\Delta x}$$

$$= \frac{x^3 + 3x^2\Delta x + 3x(\Delta x)^2 + (\Delta x)^3 - x - \Delta x - x^3 + x}{\Delta x}$$

$$= 3x^2 + 3x(\Delta x) + (\Delta x)^2 - 1$$

As $\Delta x \to 0$, this difference quotient approaches $3x^2 - 1$. Thus, $f'(x) = 3x^2 - 1$. The slope m of the line tangent to $f(x)$ at $(-2, f(-2)) = (-2, -6)$ is $f'(-2) = 11$, and the equation of this tangent line is

$$y - (-6) = 11(x - (-2)) \quad \text{or} \quad y = 11x + 16$$

11. If $f(x) = 2\sqrt{x}$, then

$$\frac{f(x + \Delta x) - f(x)}{\Delta x} = \frac{\left[2\sqrt{x + \Delta x}\right] - \left[2\sqrt{x}\right]}{\Delta x}$$

$$= 2\left[\frac{\sqrt{x + \Delta x} - \sqrt{x}}{\Delta x}\right]\left[\frac{\sqrt{x + \Delta x} + \sqrt{x}}{\sqrt{x + \Delta x} + \sqrt{x}}\right]$$

$$= \frac{2\Delta x}{\Delta x\left[\sqrt{x + \Delta x} + \sqrt{x}\right]}$$

$$= \frac{2}{\sqrt{x + \Delta x} + \sqrt{x}}$$

As $\Delta x \to 0$, this difference quotient approaches $\frac{1}{\sqrt{x}}$. Thus, $f'(x) = \frac{1}{\sqrt{x}}$. The slope m of the line tangent to $f(x)$ at $(4, f(4)) = (4, 4)$ is $f'(4) = \frac{1}{2}$, and the equation of this tangent line is

$$y - 4 = \frac{1}{2}(x - 4) \quad \text{or} \quad y = \frac{1}{2}x + 2$$

13. (a) If $f(x) = x^3$, then $f(1) = 1$ and $f(1.1) = (1.1)^3 = 1.331$. The slope of the secant line joining the points $(1, 1)$ and $(1.1, 1.331)$ on the graph of f is

$$m = \frac{\Delta y}{\Delta x} = \frac{1.331 - 1}{1.1 - 1} = \frac{0.331}{0.1} = 3.31$$

(b) If $f(x) = x^3$,

$$\frac{f(x + \Delta x) - f(x)}{\Delta x} = \frac{(x + \Delta x)^3 - x^3}{\Delta x}$$

$$= \frac{x^3 + 3x^2\Delta x + 3x(\Delta x)^2 + (\Delta x)^3 - x^3}{\Delta x}$$

$$= 3x^2 + 3x\Delta x + (\Delta x)^2$$

As $\Delta x \to 0$, this difference quotient approaches $3x^2$. Thus $f'(x) = 3x^2$. The slope m of the line tangent to $f(x)$ at $(1, f(1)) = (1, 1)$ is $f'(1) = 3$. Notice that this slope was approximated by the slope of the secant in part (a).

15. The graph of $y = f(x) = x^2 - 3x = x(x - 3)$ is shown below.

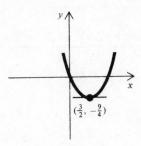

Graph for Problem 15.

The lowest point occurs when the derivative (the slope of the tangent) is zero. To find the derivative:

$$\frac{f(x + \Delta x) - f(x)}{\Delta x} = \frac{[(x + \Delta x)^2 - 3(x + \Delta x)] - [x^2 - 3x]}{\Delta x}$$

$$= \frac{x^2 + 2x(\Delta x) + (\Delta x)^2 - 3x - 3(\Delta x) - x^2 + 3x}{\Delta x}$$

$$= 2x + \Delta x - 3$$

As $\Delta x \to 0$, this difference quotient approaches $2x - 3$, and so $f'(x) = 2x - 3$. The derivative is zero when $2x - 3 = 0$ or $x = \frac{3}{2}$. It follows that the lowest point on the graph is $(\frac{3}{2}, f(\frac{3}{2})) = (\frac{3}{2}, -\frac{9}{4})$.

17. The graph of $y = f(x) = x^3 - x^2 = x^2(x - 1)$ is shown below.

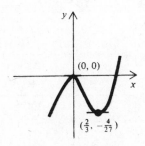

Graph for Problem 17.

When the derivative is zero, the graph has a horizontal tangent. To find these points, first compute the derivative.

$$\frac{f(x + \Delta x) - f(x)}{\Delta x} = \frac{[(x + \Delta x)^3 - (x + \Delta x)^2] - [x^3 - x^2]}{\Delta x}$$

$$= \frac{x^3 + 3x^2(\Delta x) + 3x(\Delta x)^2 + (\Delta x)^3 - x^2 - 2x(\Delta x) - (\Delta x)^2 - x^3 + x^2}{\Delta x}$$

$$= 3x^2 + 3x(\Delta x) + (\Delta x)^2 - 2x - \Delta x$$

As $\Delta x \to 0$, this difference quotient approaches $3x^2 - 2x$, and so $f'(x) = 3x^2 - 2x$. The derivative is zero when

$$3x^2 - 2x = 0, \quad x(3 - 2x) = 0, \quad \text{or} \quad x = 0 \quad \text{and} \quad x = \frac{2}{3}$$

Hence there is a horizontal tangent at $(\frac{2}{3}, f(\frac{2}{3})) = (\frac{2}{3}, -\frac{4}{27})$ where the graph has a valley and at $(0, f(0)) = (0, 0)$ where the graph has a peak.

19. The derivative is the slope of the tangent. Hence if the derivative is positive on an interval, the slope of the tangent is positive and the graph is rising as x increases as shown in the figures below.

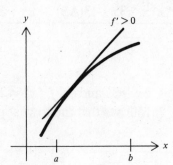 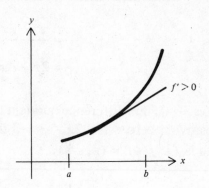

Graphs with f'(x) > 0 for Problem 19.

21. (a) From Example 1.2 in the text, $\frac{d}{dx}(x^2) = 2x$. If $f(x) = x^2 - 3$, then

$$\frac{f(x + \Delta x) - f(x)}{\Delta x} = \frac{[(x + \Delta x)^2 - 3] - [x^2 - 3]}{\Delta x}$$

$$= \frac{x^2 + 2x(\Delta x) + (\Delta x)^2 - x^2}{\Delta x}$$

$$= 2x + \Delta x$$

As $\Delta x \to 0$, this difference quotient approaches $2x$. Hence $\frac{d}{dx}(x^2 - 3) = 2x$.

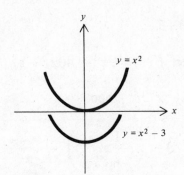

Parallel curves for Problem 21.

The graph of $y = x^2 - 3$ is the graph of $y = x^2$ lowered by 3 units. Thus the graphs are parallel and their tangents have the same slopes for any value of x. This accounts geometrically for the fact that their derivatives are identical.

(b)   The graph of $y = x^2 + 5$ is also parallel to the graph of $y = x^2$. Hence,

$$\frac{d}{dx}(x^2 + 5) = \frac{d}{dx}(x^2) = 2x$$

23.   (a)   By Problem 13, $\frac{d}{dx}(x^3) = 3x^2$, and by Example 1.2, $\frac{d}{dx}(x^2) = 2x$.

     (b)   There seems to be a pattern: the derivative of x raised to a power is the power times x raised to the power decreased by 1. According to this pattern,

$$\frac{d}{dx}(x^4) = 4x^{4-1} = 4x^3 \quad \text{and} \quad \frac{d}{dx}(x^{27}) = 27x^{27-1} = 27x^{26}$$

Chapter 2, Section 2

1.   If $y = x^2 + 2x + 3$, then $\frac{dy}{dx} = 2x + 2$.

3.   If $f(x) = x^9 - 5x^8 + x + 12$, then $f'(x) = 9x^8 - 40x^7 + 1$.

5.   If $y = \frac{1}{x} + \frac{1}{x^2} - \frac{1}{\sqrt{x}} = x^{-1} + x^{-2} - x^{-1/2}$, then

$$\frac{dy}{dx} = -1x^{-2} + (-2x^{-3}) - \left(-\frac{1}{2}x^{-3/2}\right) = -\frac{1}{x^2} - \frac{2}{x^3} + \frac{1}{2\sqrt{x^3}}$$

7.   If $f(x) = \sqrt{x^3} + \frac{1}{\sqrt{x^3}} = x^{3/2} + x^{-3/2}$ then

$$f'(x) = \frac{3}{2}x^{1/2} - \frac{3}{2}x^{-5/2} = \frac{3}{2}\sqrt{x} - \frac{3}{2\sqrt{x^5}}$$

9.   If $y = -\frac{x^2}{16} + \frac{2}{x} - x^{3/2} + \frac{1}{3x^2} + \frac{x}{3} = -\frac{1}{16}x^2 + 2x^{-1} - x^{3/2} + \frac{1}{3}x^{-2} + \frac{1}{3}x$, then

$$\frac{dy}{dx} = -\frac{1}{16}(2x) + 2(-1x^{-2}) - \frac{3}{2}x^{1/2} + \frac{1}{3}(-2x^{-3}) + \frac{1}{3}(1)$$

$$= -\frac{1}{8}x - \frac{2}{x^2} - \frac{3}{2}x^{1/2} - \frac{2}{3x^2} + \frac{1}{3}$$

11.   If $f(x) = (2x + 1)(3x - 2)$, then

$$f'(x) = (2x + 1)\frac{d}{dx}(3x - 2) + (3x - 2)\frac{d}{dx}(2x + 1)$$

$$= (2x + 1)(3) + (3x - 2)(2) = 12x - 1$$

13.   If $y = 10(3x + 1)(1 - 5x)$, then

$$\frac{dy}{dx} = 10\left[\frac{d}{dx}(3x + 1)(1 - 5x)\right]$$

$$= 10\left[(3x + 1)\frac{d}{dx}(1 - 5x) + (1 - 5x)\frac{d}{dx}(3x + 1)\right]$$

$$= 10[(3x + 1)(-5) + (1 - 5x)(3)] = -300x - 20$$

15.   If $f(x) = \frac{1}{3}(x^5 - 2x^3 + 1)$, then,

$$f'(x) = \frac{1}{3}\left[\frac{d}{dx}(x^5 - 2x^3 + 1)\right] = \frac{1}{3}(5x^4 - 6x^2)$$

17. If $y = \frac{x+1}{x-2}$, then

$$\frac{dy}{dx} = \frac{(x-2)\frac{d}{dx}(x+1) - (x+1)\frac{d}{dx}(x-2)}{(x-2)^2}$$

$$= \frac{(x-2)(1) - (x+1)(1)}{(x-2)^2}$$

$$= \frac{-3}{(x-2)^2}$$

19. If $f(x) = \frac{x}{x^2-2}$, then

$$f'(x) = \frac{(x^2-2)\frac{d}{dx}(x) - x\frac{d}{dx}(x^2-2)}{(x^2-2)^2}$$

$$= \frac{(x^2-2)(1) - x(2x)}{(x^2-2)^2} = \frac{-x^2-2}{(x^2-2)^2}$$

21. If $y = \frac{3}{x+5}$, then

$$\frac{dy}{dx} = \frac{(x+5)\frac{d}{dx}(3) - 3\frac{d}{dx}(x+5)}{(x+5)^2} = \frac{(x+5)(0) - 3(1)}{(x+5)^2} = \frac{-3}{(x+5)^2}$$

23. If $f(x) = \frac{x^2-3x+2}{2x^2+5x-1}$, then

$$f'(x) = \frac{(2x^2+5x-1)\frac{d}{dx}(x^2-3x+2) - (x^2-3x+2)\frac{d}{dx}(2x^2+5x-1)}{(2x^2+5x-1)^2}$$

$$= \frac{(2x^2+5x-1)(2x-3) - (x^2-3x+2)(4x+5)}{(2x^2+5x-1)^2}$$

$$= \frac{11x^2 - 10x - 7}{(2x^2+5x-1)^2}$$

25. If $y = (2x+1)(x-3)(1-4x) = (2x+1)[(x-3)(1-4x)]$, apply the product rule twice to get

$$\frac{dy}{dx} = (2x+1)\frac{d}{dx}[(x-3)(1-4x)] + [(x-3)(1-4x)]\frac{d}{dx}(2x+1)$$

$$= (2x+1)\left[(x-3)\frac{d}{dx}(1-4x) + (1-4x)\frac{d}{dx}(x-3)\right] + (x-3)(1-4x)(2)$$

$$= (2x+1)[(x-3)(-4) + (1-4x)(1)] + (x-3)(1-4x)(2)$$

$$= (2x+1)(-4x+12+1-4x) + 2(13x-3-4x^2)$$

$$= (2x+1)(-8x+13) + 26x - 6 - 8x^2$$

$$= -16x^2 + 18x + 13 + 26x - 6 - 8x^2$$

$$= -24x^2 + 44x + 7$$

27. If $y = (x^2 + 1)(1 - x^3)$, then

$$\frac{dy}{dx} = (x^2 + 1)\frac{d}{dx}(1 - x^3) + (1 - x^3)\frac{d}{dx}(x^2 + 1)$$
$$= (x^2 + 1)(-3x^2) + (1 - x^3)(2x)$$
$$= -5x^4 - 3x^2 + 2x$$

At the given point $(1, 0)$, $x = 1$ and so the slope of the tangent is $\frac{dy}{dx} = -5(1)^4 - 3(1)^2 + 2(1) = -6$. Hence the equation of the tangent line is

$$y - 0 = -6(x - 1) \quad \text{or} \quad y = -6x + 6$$

29. If $f(x) = 1 - \frac{1}{x} + \frac{2}{\sqrt{x}} = 1 - x^{-1} + 2x^{-1/2}$, then

$$f'(x) = -(-1x^{-2}) + 2\left(-\frac{1}{2}x^{-3/2}\right) = \frac{1}{x^2} - \frac{1}{x^{3/2}}$$

At the given point $(4, \frac{7}{4})$, $x = 4$ and the slope of the tangent is

$$f'(4) = \frac{1}{(4)^2} - \frac{1}{(4)^{3/2}} = \frac{1}{16} - \frac{1}{8} = -\frac{1}{16}$$

Hence the equation of the tangent line is

$$y - \frac{7}{4} = -\frac{1}{16}(x - 4) \quad \text{or} \quad y = -\frac{1}{16}x + 2$$

31. If $f(x) = x - \frac{1}{x^2} = x - x^{-2}$, then

$$f'(x) = 1 - (-2x^{-3}) = 1 + \frac{2}{x^3}$$

and $f'(1) = 3$ is the slope of the tangent line at the point $(1, f(1)) = (1, 0)$. The equation of the tangent line is

$$y - 0 = 3(x - 1) \quad \text{or} \quad y = 3x - 3$$

33. If $f(x) = (x^3 - 2x^2 + 3x - 1)(x^5 - 4x^2 + 2)$, then

$$f'(x) = (x^3 - 2x^2 + 3x - 1)\frac{d}{dx}(x^5 - 4x^2 + 2) + (x^5 - 4x^2 + 2)\frac{d}{dx}(x^3 - 2x^2 + 3x - 1)$$
$$= (x^3 - 2x^2 + 3x - 1)(5x^4 - 8x) + (x^5 - 4x^2 + 2)(3x^2 - 4x + 3)$$

and so $f'(0) = (-1)(0) + (2)(3) = 6$ is the slope of the tangent line at the point $(0, f(0)) = (0, -2)$. The equation of the tangent line is

$$y - (-2) = 6(x - 0) \quad \text{or} \quad y = 6x - 2$$

35.　(a)　If $y = \frac{2x-3}{x^3}$, the quotient rule gives

$$\frac{dy}{dx} = \frac{x^3 \frac{d}{dx}(2x-3) - (2x-3)\frac{d}{dx}(x^3)}{x^6}$$

$$= \frac{x^3(2) - (2x-3)(3x^2)}{x^6}$$

$$= \frac{-4x^3 + 9x^2}{x^6} = \frac{-4x+9}{x^4}$$

(b)　If y is rewritten as $y = (2x - 3)(x^{-3})$, the product rule gives

$$\frac{dy}{dx} = (2x-3)\frac{d}{dx}(x^{-3}) + (x^{-3})\frac{d}{dx}(2x-3)$$

$$= (2x-3)(-3x^{-4}) + x^{-3}(2)$$

$$= \frac{-3(2x-3)}{x^4} + \frac{2}{x^3} = \frac{-3(2x-3)+2x}{x^4} = \frac{-4x+9}{x^4}$$

(c)　If y is rewritten as $y = 2x^{-2} - 3x^{-3}$, the power rule gives

$$\frac{dy}{dx} = -4x^{-3} + 9x^{-4} = -\frac{4}{x^3} + \frac{9}{x^4} = \frac{-4x+9}{x^4}$$

37.　The graph of the function $f(x) = x^2 - 4x - 5 = (x-5)(x+1)$ is sketched below.

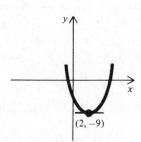

Graph for Problem 37.

At the lowest point of the graph, the tangent line is horizontal and the slope of the tangent is zero. To find this point, set $f'(x)$ equal to zero and solve for x. Since $f'(x) = 2x - 4$, $f'(x)$ is zero when $x = 2$. Hence the lowest point is $(2, f(2)) = (2, -9)$.

39.　If the lowest point on the graph of $f(x) = ax^2 + bx$ is $(3, -8)$, then $f(3) = -8$ and $f'(3) = 0$. If $f(3) = -8$, then

$$f(3) = a(3)^2 + b(3) = 9a + 3b = -8$$

and, since $f'(x) = 2ax + b$, $f'(3) = 0$ implies

$$2a(3) + b = 6a + b = 0$$

To find a and b, solve this pair of equations. From the second equation, $b = -6a$, which, substituted into the first equation, gives

$$9a + 3(-6a) = -8 \quad \text{or} \quad a = \frac{8}{9}$$

If $a = \frac{8}{9}$, then $b = -6(\frac{8}{9}) = -\frac{16}{3}$.

41. First find the x coordinates of the points (x, y) on the graph of the function $f(x) = x^2 - 4x + 25$ with the property that the tangent at (x, y) also passes through the origin $(0, 0)$. One way to do this is to use the fact that there are two ways to calculate the slope of such a line:

$$\text{Slope} = f'(x) = 2x - 4$$

and

$$\text{Slope} = \frac{\Delta y}{\Delta x} = \frac{y - 0}{x - 0} = \frac{y}{x}$$

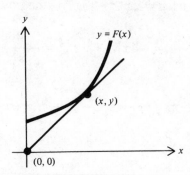

Illustration for Problem 41.

Setting the two expressions for the slope equal to each other yields

$$2x - 4 = \frac{y}{x} \quad \text{or} \quad 2x^2 - 4x = y$$

But $y = f(x) = x^2 - 4x + 25$, so

$$2x^2 - 4x = x^2 - 4x + 25, \quad x^2 = 25, \quad \text{or} \quad x = \pm 5$$

If $x = 5$, the slope of the tangent is $f'(5) = 6$, and so the equation of the corresponding tangent is

$$y - 0 = 6(x - 0) \quad \text{or} \quad y = 6x$$

If $x = -5$, the slope of the tangent is $f'(-5) = -14$, and so the equation of the corresponding tangent is

$$y - 0 = -14(x - 0) \quad \text{or} \quad y = -14x$$

43. (a) Let $E(p)$ denote the total monthly expenditure. Then,

$$E(p) = (\text{monthly demand})(\text{price per unit})$$

The monthly demand is $D(p) = -200p + 12{,}000$ units when the price is p dollars per unit. Hence,

$$E(p) = (-200p + 12{,}000)(p) = -200p(p - 60)$$

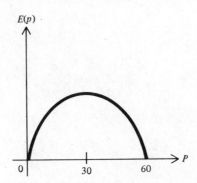

Expenditure function for Problem 43.

(b) Consumer expenditure will be greatest at the peak of the graph, that is, where the tangent line is horizontal and the derivative is zero. Since $E'(p) = -400p + 12{,}000$ is zero when $p = 30$ it follows that the expenditure is greatest when the market price is \$30. The corresponding (maximal) expenditure is $E(30) = \$180{,}000$.

45. Let $h(x) = \dfrac{f(x)}{g(x)}$. Then

$$
\begin{aligned}
\frac{h(x+\Delta x) - h(x)}{\Delta x} &= \frac{1}{\Delta x}\left[\frac{f(x+\Delta x)}{g(x+\Delta x)} - \frac{f(x)}{g(x)}\right] \\[2mm]
&= \frac{1}{\Delta x}\left[\frac{f(x+\Delta x)g(x) - f(x)g(x+\Delta x)}{g(x+\Delta x)g(x)}\right] \\[2mm]
&= \frac{1}{\Delta x}\left[\frac{f(x+\Delta x)g(x) - f(x)g(x) + f(x)g(x) - f(x)g(x+\Delta x)}{g(x+\Delta x)g(x)}\right] \\[2mm]
&= \frac{1}{\Delta x}\left[\frac{g(x)[f(x+\Delta x) - f(x)]}{g(x+\Delta x)g(x)} - \frac{f(x)[g(x+\Delta x) - g(x)]}{g(x+\Delta x)g(x)}\right] \\[2mm]
&= \frac{1}{g(x+\Delta x)}\left[\frac{f(x+\Delta x) - f(x)}{\Delta x}\right] \\[2mm]
&\quad - \frac{f(x)}{g(x+\Delta x)g(x)}\left[\frac{g(x+\Delta x) - g(x)}{\Delta x}\right]
\end{aligned}
$$

As $\Delta x \to 0$, $g(x+\Delta x) \to g(x)$ and

$$\frac{f(x+\Delta x) - f(x)}{\Delta x} \to f'(x) \quad \text{and} \quad \frac{g(x+\Delta x) - g(x)}{\Delta x} \to g'(x)$$

Hence, as $\Delta x \to 0$, the difference quotient approaches

$$\frac{1}{g(x)}[f'(x)] - \frac{f(x)}{[g(x)]^2}[g'(x)] = \frac{f'(x)g(x) - f(x)g'(x)}{[g(x)]^2}$$

and the quotient rule is verified.

Chapter 2, Section 3

1. The rate of change of $f(t) = t^3 - 3t + 5$ at any value of t is $f'(t) = 3t^2 - 3$ and so at $t = 2$ and 1, $f'(2) = 3(2)^2 - 3 = 9$; $f'(1) = 3(1)^2 - 3 = 0$, respectively.

3. (a) Since $C(t) = 100t^2 + 400t + 5,000$ is the circulation t years from now, the rate of change of the circulation t years from now is

    $$C'(t) = 200t + 400 \text{ newspapers per year}$$

    (b) The rate of change of the circulation 5 years from now is

    $$C'(5) = 200(5) + 400 = 1,400 \text{ newspapers per year}$$

    (c) The actual change in the circulation during the 6th year is

    $$C(6) - C(5) = [100(6)^2 + 400(6) + 5000] - [100(5)^2 + 400(5) + 5000]$$
    $$= 1,500 \text{ newspapers}$$

5. (a) Since $f(x) = -x^3 + 6x^2 + 15x$ is the number of radios assembled x hours after 8:00 A.M., the rate at which the radios are being assembled x hours after 8:00 A.M. is
    $$f'(x) = -3x^2 + 12x + 15 \text{ radios per hour}$$

    (b) The rate of assembly at 9:00 A.M. (when $x = 1$) is

    $$f'(1) = -3(1)^2 + 12(1) + 15 = 24 \text{ radios per hour}$$

    (c) The actual number of radios assembled between 9:00 A.M. (when $x = 1$) and 10:00 A.M. (when $x = 2$) is

    $$f(2) - f(1) = [-(2)^3 + 6(2)^2 + 15(2)] - [-(1)^3 + (1)^2 + 15(1)]$$
    $$= 26 \text{ radios}$$

7. (a) Since $P(t) = 20 - \frac{6}{t+1}$ is the population (in thousands) t years from now, the rate at which the population is changing t years from now is

    $$P'(t) = 0 - \frac{(t+1)(0) - 6(1)}{(t+1)^2} = \frac{6}{(t+1)^2}$$

    thousand per year.

    (b) One year from now, the rate of change will be

    $$P'(1) = \frac{6}{(1+1)^2} = 1.5 \text{ thousand per year}$$

    that is, 1,500 per year.

(c)   The actual population increase during the second year is

$$P(2) - P(1) = \left[20 - \frac{6}{2+1}\right] - \left[20 - \frac{6}{1+1}\right] = 1 \text{ thousand}$$

(d)   Nine years from now the rate of change will be

$$P'(9) = \frac{6}{100} \text{ thousand} = 60 \text{ people per year}$$

(e)   As t increases, $\frac{6}{(t+1)^2}$ approaches zero. Thus, the rate of population growth will approach zero in the long run.

9.   The speed of the first car C_1 is 60 kilometers per hour and the speed of the second car C_2 is 80 kilometers per hour. After t hours, C_1 has traveled $60t$ kilometers and C_2 has traveled $80t$ kilometers, as shown in the figure. Let $D(t)$ denote the distance between C_1 and C_2 at time t.

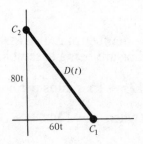

Relative position of cars for Problem 9.

By the pythagorean theorem,

$$[D(t)]^2 = (80t)^2 + (60t)^2 = 6,400t^2 + 3,600t^2 = 10,000t^2$$

and so

$$D(t) = \sqrt{10,000t^2} = 100t \text{ kilometers}$$

The rate of change of this distance is the derivative

$$D'(t) = 100 \text{ kilometers per hour}$$

and is independent of the time t.

11.   (a)   Since $C(q) = 3q^2 + q + 50$ is the total cost in dollars of manufacturing q units, the marginal cost is $C'(q) = 6q + 1$ dollars per unit. If output is increased from 40 to 41 units, the corresponding increase in cost is approximately $C'(40) = \$241$.

      (b)   The actual cost of producing the 41st unit is

$$C(41) - C(40) = [3(41)^2 + 41 + 50] - [3(40)^2 + 40 + 50] = \$244$$

13. (a) Since $R(q) = 240q + 0.05q^2$ is the total revenue in dollars when q units are sold, the marginal revenue is $R'(q) = 240 + 0.10q$ dollars per unit. The revenue generated by the sale of the 81st unit is approximately $R'(80) = 240 + 0.10(80) = \$248$.

(b) The actual additional revenue generated by the sale of the 81st unit is

$$R(81) - R(80) = [240(81) + 0.05(81)^2] - [240(80) + 0.05(80)^2]$$
$$= \$248.05$$

15. Since $Q(K) = 600K^{1/2}$ is the daily output if K thousand dollars is invested, the marginal output is

$$Q'(K) = 600\left(\frac{1}{2}K^{-1/2}\right) = \frac{300}{K^{1/2}}$$

units per thousand dollars. The approximate effect on output from increasing capital investment from \$900,000 ($K = 900$) to \$901,000 is

$$Q'(900) = \frac{300}{(900)^{1/2}} = \frac{300}{30} = 10 \text{ units}$$

17. If $f(t) = 3t^2 - 7t + 5$ then the percentage rate of change with respect to t is

$$100\frac{f'(t)}{f(t)} = 100\left(\frac{6t - 7}{3t^2 - 7t + 15}\right).$$

19. (a) Since $P(x) = 2x + 4x^{3/2} + 500$ is the population x months from now, the rate of population growth is

$$P'(x) = 2 + 4\left(\frac{3}{2}x^{1/2}\right) = 2 + 6x^{1/2} \text{ people per month}$$

Nine months from now, the population will be changing at the rate of $P'(9) = 2 + 6(9)^{1/2} = 20$ people per month.

(b) The percentage rate at which the population will be changing 9 months from now is

$$100\left[\frac{P'(9)}{P(9)}\right] = \frac{100(20)}{2(9) + 4(9)^{3/2} + 500}$$
$$= \frac{2,000}{5,126} \simeq 0.39 \text{ percent per month}$$

21. (a) Since $T(x) = 20x^2 + 40x + 600$ dollars is the average property tax x years after 1982, the rate at which the average property tax changes is $T'(x) = 40x + 40$ dollars per year. In 1988 (when $x = 6$) the rate of change is $T'(6) = 40(6) + 40 = \$280$ dollars per year.

(b)   The percentage rate at which the average property tax was increasing in 1988 was

$$100 \left[\frac{T'(6)}{T(6)} \right] = \frac{100(280)}{20(6)^2 + 40(6) + 600} \simeq 17.95 \text{ percent per year}$$

23.   (a)   Since your starting salary is \$24,000 and you get a raise of \$2,000 per year, your salary x years from now will be

$$S(x) = 24,000 + 2,000x \text{ dollars}$$

The percentage rate of change of this salary x years from now is

$$100 \left[\frac{S'(x)}{S(x)} \right] = 100 \left[\frac{2,000}{24,000 + 2,000x} \right] = \frac{100}{12 + x} \text{ percent per year}$$

(b)   The percentage rate of change after 1 year is $\frac{100}{13} \simeq 7.69$ percent per year.

(c)   In the long run, $\frac{100}{12+x}$ approaches zero. That is, the percentage rate of change of your salary will approach 0 (even though your salary will continue to increase at a constant rate).

25.   (a)   Since the initial speed is $S_0 = 0$ feet per second and the height of the building is $H_0 = 144$ feet,

$$H(t) = -16t^2 + S_0 t + H_0 = -16t^2 + 144$$

The stone hits the ground when $H(t) = 0$, that is, when

$$-16t^2 + 144 = 0, \quad t^2 = 9, \quad \text{or} \quad t = 3 \text{ seconds}$$

(b)   The speed with which the stone hits the ground is $H'(3)$. Since $H'(t) = -32t$, $H'(3) = -96$ feet per second, the minus sign indicating that the direction is down.

27.   (a)   If after 2 seconds the ball passes you on the way down, then $H(2) = H_0$, where $H(t) = -16t^2 + S_0 t + H_0$. Hence

$$-16(2)^2 + S_0(2) + H_0 = H_0, \quad -64 + 2S_0 = 0, \quad \text{or} \quad S_0 = 32$$

feet per second.

(b)   The height of the building is H_0 feet. From part (a) you know that $H(t) = -16t^2 + 32t + H_0$. Moreover, $H(4) = 0$ since the ball hits the ground after 4 seconds. Thus,

$$-16(4)^2 + 32(4) + H_0 = 0 \quad \text{or} \quad H_0 = 128 \text{ feet}$$

(c)   From parts (a) and (b) you know that $H(t) = -16t^2 + 32t + 128$ and so the speed of the ball is

$$H'(t) = -32t + 32 \text{ feet per second}$$

After 2 seconds, the speed will be $H'(2) = -32$ feet per second, where the minus sign indicates that the direction of motion is down.

(d)   The speed at which the ball hits the ground is $H'(4) = -96$ feet per second.

29.   Since y is a linear function of x, $y = mx + b$, where m and b are constants. The rate of change is $\frac{dy}{dx} = m$, which is a constant. Notice that this constant rate of change is the slope of the line.

31.   The total manufacturing cost C is a function of q (where q is the number of units produced) and q is a function of t (where t is the number of hours during which the factory operates). Hence:

(a) $\frac{dC}{dq}$ = rate of change of cost with respect to number of units produced

$$= \frac{\text{dollars}}{\text{unit}}$$

(b) $\frac{dq}{dt}$ = rate of change of units produced with respect to time

$$= \frac{\text{units}}{\text{hour}}$$

(c) $\frac{dC}{dq}\frac{dQ}{dt} = \frac{\text{dollars}}{\text{unit}}\frac{\text{units}}{\text{hour}} = \frac{\text{dollars}}{\text{hour}}$

= rate of change of cost with respect to time.

Chapter 2, Section 4

1.   The change in the function is approximately the derivative of the function times the change in its variable, that is, $f'(x)\Delta x$. Since $f(x) = x^2 - 3x + 5$ then $f'(x) = 2x - 3$. As x increases from 5 to 5.3, $\Delta x = .3$ and so

$$f'(3)\Delta x = [2(3) - 3](.3)$$
$$= .9 \text{ is an estimated change.}$$

3.   The percentage change in a function is $100\frac{\Delta f}{f(x)}$. Given $f(x) = x^2 + 2x - 9$ and x increases from 4 to 4.3 then $\Delta x = .3$, $\Delta f = f(4.3) - f(4) = [(4.3)^2 + 2(4.3) - 9] - [(4)^2 + 2(4) - 9] = 3.09$ and $f(4) = 15$. Thus,

$$\text{percentage change in } f = 100\left(\frac{\Delta f}{f(x)}\right)$$
$$= 100\left(\frac{3.09}{4}\right) = 77.25\%.$$

5.   Since the total cost is $C(q) = 0.1q^3 - 0.5q^2 + 500q + 200$, the change in cost resulting from an increase in production from 4 units to 4.1 units ($\Delta q = 0.1$) is

$$\Delta C = C(4.1) - C(4) \simeq C'(4)(0.1)$$

Since $C'(q) = 0.3q^2 - q + 500$ and $C'(4) = 500.80$, it follows that $\Delta C \simeq (500.80)(0.1) = 50.08$. That is, cost will increase by approximately \$50.08.

7.   Since the circulation will be $C(t) = 100t^2 + 400t + 5,000$ papers t years from now, the increase in circulation during the next 6 months ($\Delta t = 0.5$) will be

$$\Delta C = C(0.5) - C(0) \simeq C'(0)(0.5)$$

Since $C'(t) = 200t + 400$ and $C'(0) = 400$, it follows that $\Delta C \simeq 400(0.5) = 200$ newspapers.

9.   Since the average level of carbon monoxide in the air t years from now will be $Q(t) = 0.05t^2 + 0.1t + 3.4$ parts per million, the change in the carbon monoxide level during the next 6 months ($\Delta t = 0.5$) will be

$$\Delta Q = Q(0.5) - Q(0) \simeq Q'(0)(0.5)$$

Since $Q'(t) = 0.1t + 0.1$ and $Q'(0) = 0.1$, it follows that $\Delta Q \simeq 0.1(0.5) = 0.05$ parts per million.

11.   Since daily output is $Q(K) = 600K^{1/2}$ units when K thousand dollars is invested, the change in output due to an increase in capital investment from \$900,000 to \$900,800 ($\Delta K = 0.8$) is

$$\Delta Q = Q(900.8) - Q(900) \simeq Q'(900)(0.8)$$

Since $Q'(K) = 300K^{-1/2}$ and $Q'(900) = 10$, it follows that $\Delta Q \simeq 10(0.5) = 8$ units.

13. Since the speed of the blood is $S(R) = 1.8 \times 10^5 R^2$ centimeters per second, the difference between the calculated value and the true value of the speed for $R = 1.2 \times 10^{-2}$ and $\Delta R = \pm 5 \times 10^{-4}$ is

$$\Delta S = S(1.2 \times 10^{-2} \pm 5 \times 10^{-4}) - S(1.2 \times 10^{-2}) \simeq S'(1.2 \times 10^{-2})(\pm 5 \times 10^{-4})$$

Since $S'(R) = 3.6 \times 10^5 R$ and $S'(1.2 \times 10^{-2}) = (3.6 \times 10^5)(1.2 \times 10^{-2}) = 4.32 \times 10^3$, it follows that

$$\Delta S \simeq (4.32 \times 10^3)(\pm 5 \times 10^{-4}) = \pm 21.60 \times 10^{-1} = \pm 2.16 \text{ centimeters per second}.$$

15. The volume of a sphere is $V = \frac{4}{3}\pi r^3$. If the radius r is measured to be 6 inches with a possible error of up to 1 percent, then the maximum possible error in r is $\Delta r = \pm 0.01(6) = \pm 0.06$ inch, and the corresponding error in the measurement of the volume is

$$\Delta V = V(6 \pm 0.06) - V(6) \simeq V'(6)(\pm 0.06)$$

Since $V'(r) = 4\pi r^2$ and $V'(6) = 144\pi$, it follows that

$$\Delta V \simeq 144\pi(\pm 0.06) = \pm 8.64\pi \text{ cubic inches}$$

To express this error as a percentage, divide by the volume $V(6) = 288\pi$ and multiply by 100. That is,

$$\text{Percentage change } = 100\frac{\Delta V}{V(6)} \simeq \pm 100\frac{8.64\pi}{288\pi} = \pm 3 \text{ percent}$$

17. The total cost is $C(q) = \frac{1}{6}q^3 + 642q + 400$ dollars. The derivative is $C'(q) = \frac{1}{2}q^2 + 642$. To find the decrease in production that will reduce total cost by \$130, solve for Δq in the approximation formula $\Delta C \simeq C'(q)\Delta q$ with $\Delta C = -130$ and $q = 4$ to get

$$-130 \simeq C'(4)\Delta q = \left[\frac{1}{2}(4)^2 + 642\right]\Delta q$$

$$-130 \simeq 650\Delta q \quad \text{or} \quad \Delta q \simeq -\frac{130}{650} = -0.2 \text{ units}$$

That is, production should be decreased by approximately 0.2 units.

19. The annual earnings were $A(t) = 0.1t^2 + 10t + 20$ thousand dollars t years after 1984. The derivative is $A'(t) = 0.2t + 10$. At the beginning of the third quarter of 1988, $t = 4.5$. The change in t during this quarter is $\Delta t = 0.25$. Hence,

$$\text{Percentage change in } A \simeq 100\frac{A'(4.5)\Delta t}{A(4.5)}$$

$$= 100\frac{[0.2(4.5) + 10](0.25)}{0.1(4.5)^2 + 10(4.5) + 20}$$

$$= 4.066 \text{ percent}$$

21. The area of a circle is $A = \pi r^2$. If the radius is increased by 1 percent, the change in the radius is $\Delta r = 0.01r$ and the corresponding change in area is

$$\Delta A = A(r + 0.01r) - A(r) \simeq A'(r)(0.01r)$$

Since $A'(r) = 2\pi r$,

$$\Delta A \simeq 2\pi r(0.01r) = 0.02\pi r^2 = 0.2A$$

That is, the increase of 1 percent in the radius causes an increase in area of approximately 2 percent.

23. The speed of the blood is $S(R) = cR^2$. If there is a 1 percent error in the measurement of R, then $\Delta R = \pm 0.01R$, and

$$\text{Percentage change in } S \simeq 100\frac{S'(R)\Delta R}{S(R)} = 100\frac{2cR(\pm 0.01R)}{cR^2}$$

$$= \pm\frac{2cR^2}{cR^2} = \pm 2 \text{ percent}$$

25. Output is $Q(L) = 9,000L^{1/3}$. The derivative is $Q'(L) = 3,000L^{-2/3}$. The fact that L increases by 1.5 percent means that $\Delta L = 0.015L$. Hence,

$$\text{Percentage change in } Q \simeq 100\frac{Q'(L)\Delta L}{Q(L)}$$

$$= 100\frac{3,000L^{-2/3}(0.015L)}{9,000L^{1/3}}$$

$$= 100\frac{45L^{1/3}}{9,000L^{1/3}} = \frac{4,500}{9,000} = 0.5 \text{ percent}$$

27. The output is $Q(K) = 1,200K^{1/2}$. The derivative is $Q'(K) = 600K^{-1/2}$. We are given that the percentage change in Q is 1.2 percent, and the goal is to find the percentage change in K, which can be represented as $100\frac{\Delta K}{K}$. Apply the formula

$$\text{Percentage change in } Q \simeq 100\frac{Q'(K)\Delta K}{Q(K)}$$

with 1.2 on the left-hand side and solve for $100\frac{\Delta K}{K}$ as follows:

$$1.2 \simeq 100\frac{600K^{-1/2}\Delta K}{1,200K^{1/2}}$$

$$1.2 \simeq 100\frac{\Delta K}{2K} \quad \text{or} \quad 100\frac{\Delta K}{K} \simeq 2(1.2) = 2.4 \text{ percent}$$

That is, an increase in capital investment of approximately 2.4 percent is required to increase output by 1.2 percent.

29. The volume of a soccer ball is $V = \frac{4}{3}\pi r^3$. The inner diameter is $8\frac{1}{2}$ inches, and so the inner radius is $\frac{17}{4}$ inches. The leather is $\frac{1}{8}$ inch thick, that is, $\Delta r = \frac{1}{8}$. The volume of the leather shell is the change in volume due to an increase in radius from $r = \frac{17}{4}$ to $r = \frac{17}{4} + \frac{1}{8} = \frac{35}{8}$. That is,

$$\Delta V = V\left(\frac{35}{8}\right) - V\left(\frac{17}{4}\right) \simeq V'\left(\frac{17}{4}\right)\left(\frac{1}{8}\right)$$

Since $V'(r) = 4\pi r^2$ and $V'(\frac{17}{4}) = 72.25\pi$, it follows that

$$\Delta V \simeq 72.25\pi\left(\frac{1}{8}\right) \simeq 9.03\pi \simeq 28.37 \text{ cubic inches}$$

Chapter 2, Section 5

1.   If $y = u^2 + 1$ and $u = 3x - 2$, then $\frac{dy}{du} = 2u$, $\frac{du}{dx} = 3$, and

$$\frac{dy}{dx} = \frac{dy}{du}\frac{du}{dx} = (2u)(3) = 6u = 6(3x - 2)$$

3.   If $y = \sqrt{u} = u^{1/2}$ and $u = x^2 + 2x - 3$, then $\frac{dy}{du} = \frac{1}{2}u^{-1/2} = \frac{1}{2u^{1/2}}$, $\frac{du}{dx} = 2x + 2$, and

$$\frac{dy}{dx} = \frac{dy}{du}\frac{du}{dx} = \left[\frac{1}{2u^{1/2}}\right](2x + 2) = \frac{x + 1}{(x^2 + 2x - 3)^{1/2}}$$

5.   If $y = \frac{1}{u^2} = u^{-2}$ and $u = x^2 + 1$, then $\frac{dy}{du} = -2u^{-3}$, $\frac{dy}{dx} = 2x$, and

$$\frac{dy}{dx} = \frac{dy}{du}\frac{du}{dx} = \left[\frac{-2}{u^3}\right](2x) = \frac{-4x}{(x^2 + 1)^3}$$

7.   If $y = \frac{1}{\sqrt{u}} = u^{-1/2}$ and $u = x^2 - 9$, then $\frac{dy}{du} = -\frac{1}{2}u^{-3/2} = \frac{-1}{2u^{3/2}}$, $\frac{du}{dx} = 2x$, and

$$\frac{dy}{dx} = \frac{dy}{du}\frac{du}{dx} = \left[\frac{-1}{2u^{3/2}}\right](2x) = \frac{-x}{(x^2 - 9)^{3/2}}$$

9.   If $y = \frac{1}{u-1}$ and $u = x^2$, then

$$\frac{dy}{du} = \frac{(u - 1)(0) - (1)(1)}{(u - 1)^2} = \frac{-1}{(u - 1)^2} \quad \text{and} \quad \frac{du}{dx} = 2x$$

and

$$\frac{dy}{dx} = \frac{dy}{du}\frac{du}{dx} = \left[\frac{-1}{(u - 1)^2}\right](2x) = \frac{-2x}{(x^2 - 1)^2}$$

11. If $y = 3u^4 - 4u + 5$ and $u = x^3 - 2x - 5$, then $\frac{dy}{du} = 12u^3 - 4$, $\frac{du}{dx} = 3x^2 - 2$, and

$$\frac{dy}{dx} = \frac{dy}{du}\frac{du}{dx} = (12u^3 - 4)(3x^2 - 2)$$

When $x = 2$, $u = (2)^3 - 2(2) - 5 = -1$ and so

$$\frac{dy}{dx} = [12(-1)^3 - 4][3(2)^2 - 2] = -160$$

13. If $y = \sqrt{u} = u^{1/2}$ and $u = x^2 - 2x + 6$, then $\frac{dy}{du} = \frac{1}{2}u^{-1/2} = \frac{1}{2u^{1/2}}$, $\frac{du}{dx} = 2x - 2$, and

$$\frac{dy}{dx} = \frac{dy}{du}\frac{du}{dx} = \left[\frac{1}{2u^{1/2}}\right](2x - 2) = \frac{x - 1}{u^{1/2}}$$

When $x = 3$, $u = (3)^2 - 2(3) + 6 = 9$, and so

$$\frac{dy}{dx} = \frac{3-1}{(9)^{1/2}} = \frac{2}{3}$$

15. If $y = \frac{1}{u} = u^{-1}$ and $u = 3 - \frac{1}{x^2} = 3 - x^{-2}$, then $\frac{dy}{du} = -u^{-2} = \frac{-1}{u^2}$, $\frac{du}{dx} = -(-2x^{-3}) = \frac{2}{x^3}$, and

$$\frac{dy}{dx} = \frac{dy}{du}\frac{du}{dx} = \left[\frac{-1}{u^2}\right]\left[\frac{2}{x^3}\right]$$

When $x = \frac{1}{2}$, $u = 3 - \frac{1}{(1/2)^2} = 3 - 4 = -1$, and so

$$\frac{dy}{dx} = \left[\frac{-1}{(-1)^2}\right]\left[\frac{2}{(1/2)^3}\right] = \frac{-2}{1/8} = -16$$

17. If $f(x) = (2x+1)^4$, then

$$f'(x) = 4(2x+1)^3\frac{d}{dx}(2x+1) = 4(2x+1)^3(2) = 8(2x+1)^3$$

19. If $f(x) = (x^5 - 4x^3 - 7)^8$, then

$$f'(x) = 8(x^5 - 4x^3 - 7)^7\frac{d}{dx}(x^5 - 4x^3 - 7)$$

$$= 8(x^5 - 4x^3 - 7)^7(5x^4 - 12x^2) = 8x^2(x^5 - 4x^3 - 7)^7(5x^2 - 12)$$

21. If $f(x) = \frac{1}{5x^2 - 6x + 2} = (5x^2 - 6x + 2)^{-1}$, then

$$f'(x) = -(5x^2 - 6x + 2)^{-2}\frac{d}{dx}(5x^2 - 6x + 2) = -\frac{10x - 6}{(5x^2 - 6x + 2)^2}$$

23. If $f(x) = \frac{1}{\sqrt{4x^2+1}} = (4x^2 + 1)^{-1/2}$, then

$$f'(x) = -\frac{1}{2}(4x^2 + 1)^{-3/2}\frac{d}{dx}(4x^2 + 1) = -\frac{1}{2}(4x^2 + 1)^{-3/2}(8x) = \frac{-4x}{(4x^2 + 1)^{3/2}}$$

25. If $f(x) = \frac{3}{(1-x^2)^4} = 3(1 - x^2)^{-4}$, then

$$f'(x) = -12(1 - x^2)^{-5}\frac{d}{dx}(1 - x^2) = -12(1 - x^2)^{-5}(-2x) = \frac{24x}{(1 - x^2)^5}$$

27. If $f(x) = (1 + \sqrt{3x})^5$, then

$$f'(x) = 5(1 + \sqrt{3x})^4 \frac{d}{dx}(1 + \sqrt{3x})$$

$$= 5(1 + \sqrt{3x})^4 \frac{d}{dx}[1 + (3x)^{1/2}]$$

$$= 5(1 + \sqrt{3x})^4 \left(\frac{1}{2}\right)(3x)^{-1/2}\frac{d}{dx}(3x)$$

$$= 5(1 + \sqrt{3x})^4 \left(\frac{1}{2}\right)(3x)^{-1/2}(3)$$

$$= \frac{15}{2}\frac{(1 + \sqrt{3x})^4}{\sqrt{3x}}$$

29. If $f(x) = (x + 2)^3(2x - 1)^5$, then

$$f'(x) = (x + 2)^3 \frac{d}{dx}(2x - 1)^5 + (2x - 1)^5 \frac{d}{dx}(x + 2)^3$$

$$= (x + 2)^3[10(2x - 1)^4] + (2x - 1)^5[3(x + 2)^2]$$

$$= (2x - 1)^4(x + 2)^2[10(x + 2) + 3(2x - 1)]$$

$$= (2x - 1)^4(x + 2)^2(16x + 17)$$

31. If $f(x) = \sqrt{\frac{3x+1}{2x-1}} = \left[\frac{3x+1}{2x-1}\right]^{1/2}$, then

$$f'(x) = \frac{1}{2}\left[\frac{3x+1}{2x-1}\right]^{-1/2} \frac{d}{dx}\left[\frac{3x+1}{2x-1}\right]$$

$$= \frac{1}{2}\left[\frac{(3x+1)^{-1/2}}{(2x-1)^{-1/2}}\right]\left[\frac{(2x-1)(3) - (3x+1)(2)}{(2x-1)^2}\right]$$

$$= \frac{1}{2}\left[\frac{(2x-1)^{1/2}}{(3x+1)^{1/2}}\right]\left[\frac{-5}{(2x-1)^2}\right]$$

$$= \frac{-5}{2(3x+1)^{1/2}(2x-1)^{3/2}}$$

33. If $f(x) = \frac{(x+1)^5}{(1-x)^4}$, then

$$f'(x) = \frac{(1-x)^4\frac{d}{dx}(x+1)^5 - (x+1)^5\frac{d}{dx}(1-x)^4}{(1-x)^8}$$

$$= \frac{(1-x)^4(5)(x+1)^4(1) - (x+1)^5(4)(1-x)^3(-1)}{(1-x)^8}$$

$$= \frac{(1-x)^3(x+1)^4[5(1-x) + 4(x+1)]}{(1-x)^8}$$

$$= \frac{(x+1)^4(9-x)}{(1-x)^5}$$

35.   If $f(x) = \frac{3x+1}{\sqrt{1-4x}}$, then

$$f'(x) = \frac{\sqrt{1-4x}\,\frac{d}{dx}(3x+1) - (3x+1)\frac{d}{dx}[(1-4x)^{1/2}]}{(\sqrt{1-4x})^2}$$

$$= \frac{\sqrt{1-4x}\,\,3 - (3x+1)\left(\frac{1}{2}\right)(1-4x)^{-1/2}\frac{d}{dx}(1-4x)}{1-4x}$$

$$= \frac{3\sqrt{1-4x} - \frac{1}{2}(3x+1)(1-4x)^{-1/2}(-4)}{1-4x}$$

$$= \frac{3(1-4x)^{1/2} + 2(3x+1)(1-4x)^{-1/2}}{1-4x}$$

$$= \frac{(1-4x)^{-1/2}[3(1-4x) + 2(3x+1)]}{1-4x}$$

$$= \frac{3 - 12x + 6x + 2}{(1-4x)^{3/2}}$$

$$= \frac{5 - 6x}{(1-4x)^{3/2}}$$

37.   If $f(x) = (x^2 - 3)^5 (2x - 1)^3$, then

$$f'(x) = (x^2 - 3)^5 \frac{d}{dx}(2x-1)^3 + (2x-1)^3 \frac{d}{dx}(x^2-3)^5$$

$$= (x^2-3)^5(3)(2x-1)^2(2) + (2x-1)^3(5)(x^2-3)^4(2x)$$

$$= (x^2-3)^4(2x-1)^2[6(x^2-3) + 10x(2x-1)]$$

$$= (x^2-3)^4(2x-1)^2(26x^2 - 10x - 18)$$

the slope of the tangent line at the point $(2, f(2)) = (2, 27)$ is $f'(2) = 594$, and so the equation of the tangent line is

$$y - 27 = 594(x - 2) \quad \text{or} \quad y = 594x - 1{,}161$$

39.   If $f(x) = \left[\frac{x+1}{x-1}\right]^3$, then

$$f'(x) = 3\left[\frac{x+1}{x-1}\right]^2 \frac{d}{dx}\left[\frac{x+1}{x-1}\right]$$

$$= 3\left[\frac{x+1}{x-1}\right]^2 \left[\frac{(x-1)(1) - (x+1)(1)}{(x-1)^2}\right]$$

$$= \frac{3(x+1)^2}{(x-1)^2}\left[\frac{-2}{(x-1)^2}\right]$$

$$= -\frac{6(x+1)^2}{(x-1)^4}$$

The slope of the tangent at the point $(3, f(3)) = (3, 8)$ is $f'(3) = -6$, and so the equation of the tangent line is

$$y - 8 = -6(x - 3) \quad \text{or} \quad y = -6x + 26$$

41. (a) Since $f(t) = \sqrt{10t^2 + t + 236} = (10t^2 + t + 236)^{1/2}$ is the factory's gross annual earnings (in thousand-dollar units) t years after its formation in 1984, the rate at which the earnings are growing at that time is

$$f'(t) = \frac{1}{2}(10t^2 + t + 236)^{-1/2}(20t + 1)$$

$$= \frac{20t + 1}{2(10t^2 + t + 236)^{1/2}}$$

thousand dollars per year. The rate of growth in 1988 (when $t = 4$) is

$$f'(4) = \frac{20(4) + 1}{2(10(4)^2 + 4 + 236)^{1/2}}$$

$$= \frac{81}{2(400)^{1/2}} = \frac{81}{40} = 2.025$$

That is, in 1988 the gross annual earnings were increasing at the rate of \$2,025 per year.

(b) The percentage rate of increase of the earnings of 1988 was

$$100\left[\frac{f'(4)}{f(4)}\right] = \frac{100(2.025)}{\sqrt{10(4)^2 + 4 + 236}} = 10.125$$

percent per year.

43. Since the population t years from now is $p(t) = 20 - \frac{6}{t+1} = 20 - 6(t + 1)^{-1}$ thousand and the average daily level of carbon monoxide in the air is $c(p) = 0.5\sqrt{p^2 + p + 58} = 0.5(p^2 + p + 58)^{1/2}$ parts per million when the population is p thousand, the rate of change of carbon monoxide with respect to time is

$$\frac{dc}{dt} = \frac{dc}{dp}\frac{dp}{dt}$$

$$= \left[(0.5)\left(\frac{1}{2}\right)(p^2 + p + 58)^{-1/2}(2p + 1)\right]\left[\frac{6}{(t + 1)^2}\right]$$

$$= \left[\frac{2p + 1}{4(p^2 + p + 58)^{1/2}}\right]\left[\frac{6}{(t + 1)^2}\right]$$

parts per million per year. Two years from now, $t = 2$, $p(2) = 18$, and the rate of change will be

$$\frac{dc}{dt} = \left[\frac{2(18) + 1}{4((18)^2 + 18 + 58)^{1/2}}\right]\left[\frac{6}{(2 + 1)^2}\right] \simeq 0.31$$

parts per million per year.

45. Since $D(p) = \frac{4,374}{p^2} = 4,374p^{-2}$ pounds of coffee are sold per week if the price is p dollars per pound and the price is $p(t) = 0.02t^2 + 0.1t + 6$ dollars per pound,

$$\frac{dD}{dt} = \frac{dD}{dp}\frac{dp}{dt} = 4,374(-2p^{-3})(0.04t + 0.1)$$

$$= \left[-\frac{8,748}{p^3}\right](0.04t + 0.1) \quad \text{pounds per week}$$

Ten weeks from now, $t = 10$, $p(10) = 9$, and the rate of change of the demand with respect to time is

$$\frac{dD}{dt} = \left[-\frac{8,748}{(9)^3}\right][0.04(10) + 0.1] = -6 \quad \text{pounds per week}$$

47. The population is $p(t) = 12 - \frac{6}{t+1} = 12 - 6(t+1)^{-1}$ thousand, and the carbon monoxide level is $c(p) = 0.6\sqrt{p^2 + 2p + 24} = 0.6(p^2 + 2p + 24)^{1/2}$ units. The corresponding derivatives are

$$\frac{dp}{dt} = 6(t + 1)^{-2} = \frac{6}{(t + 1)^2}$$

and

$$\frac{dc}{dp} = 0.3(p^2 + 2p + 24)^{-1/2}(2p + 2) = \frac{0.6(p + 1)}{\sqrt{p^2 + 2p + 24}}$$

The rate of change of the carbon monoxide level with respect to time is

$$\frac{dc}{dt} = \frac{dc}{dp}\frac{dp}{dt} = \frac{0.6(p + 1)}{\sqrt{p^2 + 2p + 24}}\frac{6}{(t + 1)^2}$$

and the percentage rate of change is

$$\text{Percentage rate of change } = 100\frac{dc/dt}{c}$$

When $t = 2$,

$$p = p(2) = 12 - \frac{6}{3} = 10$$

and

$$c = c(10) = 0.6\sqrt{(10)^2 + 2(10) + 24} = 0.6\sqrt{144} = 7.2$$

Hence, when $t = 2$,

$$\text{Percentage rate of change } = 100\left[\frac{0.6(10 + 1)}{\sqrt{144}}\frac{6}{(2 + 1)^2}\right]\left[\frac{1}{7.2}\right]$$

$$\simeq 5.09 \quad \text{percent per year}$$

49. To prove that $\frac{d}{dx}[h(x)]^2 = 2h(x)h'(x)$, use the product rule to get:

$$\frac{d}{dx}[h(x)]^2 = \frac{d}{dx}[h(x)h(x)] = h(x)h'(x) + h(x)h'(x) = 2h(x)h'(x)$$

Chapter 2, Section 6

1.   If $x^2 + y^2 = 25$, then

$$2x + 2y\frac{dy}{dx} = 0 \quad \text{or} \quad \frac{dy}{dx} = -\frac{x}{y}$$

3.   If $x^3 + y^3 = xy$, then

$$3x^2 + 3y^2\frac{dy}{dx} = x\frac{dy}{dx} + y$$

$$(3y^2 - x)\frac{dy}{dx} = y - 3x^2$$

$$\frac{dy}{dx} = \frac{y - 3x^2}{3y^2 - x}$$

5.   If $y^2 + 2xy^2 - 3x + 1 = 0$, then (using the product rule for the second term),

$$2y\frac{dy}{dx} + \left[2x\left(2y\frac{dy}{dx}\right) + y^2(2)\right] - 3 = 0$$

$$(2y + 4xy)\frac{dy}{dx} = 3 - 2y^2$$

$$\frac{dy}{dx} = \frac{3 - 2y^2}{2y + 4xy} = \frac{3 - 2y^2}{2y(1 + 2x)}$$

7.   If $(2x + y)^3 = x$, then

$$3(2x + y)^2\left[2 + \frac{dy}{dx}\right] = 1$$

$$2 + \frac{dy}{dx} = \frac{1}{3(2x + y)^2}$$

$$\frac{dy}{dx} = \frac{1}{3(2x + y)^2} - 2$$

9.   If $(x^2 + 3y^2)^5 = 2xy$, then

$$5(x^2 + 3y^2)^4\frac{dy}{dx}(x^2 + 3y^2) = 2x\frac{dy}{dx} + 2y$$

$$5(x^2 + 3y^2)^4(2x + 6y\frac{dy}{dx}) = 2x\frac{dy}{dx} + 2y$$

$$10x(x^2 + 3y^2)^4 + 30y(x^2 + 3y^2)^4\frac{dy}{dx} = 2x\frac{dy}{dx} + 2y$$

$$[30y(x^2 + 3y^2)^4 - 2x]\frac{dy}{dx} = 2y - 10x(x^2 + 3y^2)^4$$

$$\frac{dy}{dx} = \frac{2y - 10x(x^2 + 3y^2)^4}{30y(x^2 + 3y^2)^4 - 2x}$$

11. If $x^2 = y^3$, then

$$2x = 3y^2 \frac{dy}{dx} \quad \text{or} \quad \frac{dy}{dx} = \frac{2x}{3y^2}$$

When $x = 8$, the original equation gives $8^2 = y^3$, $y^3 = 64$, or $y = 4$. Substituting in the equation for $\frac{dy}{dx}$ yields

$$\frac{dy}{dx} = \frac{2(8)}{3(4)^2} = \frac{1}{3}$$

that is, the slope of the tangent line at the point $(8, 4)$ is $\frac{1}{3}$.

13. If $xy = 2$, then

$$x \frac{dy}{dx} + y = 0 \quad \text{or} \quad \frac{dy}{dx} = -\frac{y}{x}$$

When $x = 2$, the original equation gives $2y = 2$ or $y = 1$. Substituting in the equation for $\frac{dy}{dx}$ yields $\frac{dy}{dx} = -\frac{1}{2}$. That is, the slope of the tangent line at the point $(2, 1)$ is $-\frac{1}{2}$.

15. If $(1 - x + y)^3 = x + 7$, then

$$3(1 - x + y)^2 \left[-1 + \frac{dy}{dx} \right] = 1$$

$$-1 + \frac{dy}{dx} = \frac{1}{3(1 - x + y)^2} \quad \text{or} \quad \frac{dy}{dx} = \frac{1}{3(1 - x + y)^2} + 1$$

When $x = 1$, the original equation gives $(1 - 1 + y)^3 = 1 + 7$, $y^3 = 8$, or $y = 2$. Substituting in the equation for $\frac{dy}{dx}$ yields

$$\frac{dy}{dx} = \frac{1}{3(1 - 1 + 2)^2} + 1 = \frac{1}{12} + 1 = \frac{13}{12}$$

That is, the slope of the tangent line at $(1, 2)$ is $\frac{13}{12}$.

17. If $(2xy^3 + 1)^3 = 2x - y^3$, then

$$3(2xy^3 + 1)^2 \frac{dy}{dx}(2xy^3 + 1) = 2 - 3y^2 \frac{dy}{dx}$$

$$3(2xy^3 + 1)^2 (6xy^2 \frac{dy}{dx} + 2y^3) = 2 - 3y^2 \frac{dy}{dx}$$

$$[18xy^2(2xy^3 + 1)^2 + 3y^2] \frac{dy}{dx} = 2 - 6y^3(2xy^3 + 1)^2$$

$$\frac{dy}{dx} = \frac{2 - 6y^3(2xy^3 + 1)^2}{18xy^2(2xy^3 + 1)^2 + 3y^2}$$

When $x = 0$, the original equation gives $1^3 = -y^3$ or $y = -1$. Hence,

$$\frac{dy}{dx} = \frac{2 - 6(-1)^3(0 + 1)^2}{0(0 + 1)^2 + 3(-1)^2} = \frac{8}{3}$$

19. <u>Explicit differentiation:</u> Solve $xy + 2y = x^2$ to get $y = \frac{x^2}{x+2}$ and differentiate using the quotient rule to get

$$\frac{dy}{dx} = \frac{(x+2)(2x) - x^2(1)}{(x+2)^2} = \frac{x^2 + 4x}{(x+2)^2} = \frac{x(x+4)}{(x+2)^2}$$

<u>Implicit differentiation:</u>

$$xy + 2y = x^2$$

$$\left[x\frac{dy}{dx} + y\right] + 2\frac{dy}{dx} = 2x, \quad (x+2)\frac{dy}{dx} = 2x - y, \quad \text{or} \quad \frac{dy}{dx} = \frac{2x - y}{x+2}$$

But, as shown previously, $y = \frac{x^2}{x+2}$, and so

$$\frac{dy}{dx} = \frac{2x - \left[\frac{x^2}{x+2}\right]}{x+2} = \frac{2x^2 + 4x - x^2}{(x+2)^2} = \frac{x(x+4)}{(x+2)^2}$$

21. <u>Explicit differentiation:</u> Solve $xy - x = y + 2$ to get $y = \frac{x+2}{x-1}$ and differentiate using the quotient rule to get

$$\frac{dy}{dx} = \frac{(x-1)(1) - (x+2)(1)}{(x-1)^2} = \frac{-3}{(x-1)^2}$$

<u>Implicit differentiation:</u>

$$xy - x = y + 2$$

$$\left[x\frac{dy}{dx} + y\right] - 1 = \frac{dy}{dx}, \quad (x-1)\frac{dy}{dx} = 1 - y, \quad \text{or} \quad \frac{dy}{dx} = \frac{1-y}{x-1}$$

But, as shown previously, $y = \frac{x+2}{x-1}$, and so

$$\frac{dy}{dx} = \frac{1 - \left[\frac{x+2}{x-1}\right]}{x-1} = \frac{(x-1) - (x+2)}{(x-1)^2} = \frac{-3}{(x-1)^2}$$

23. If output is to remain unchanged, the equation relating inputs x and y can be written as

$$C = 2x^2 + 3x^2y^2 + (1+y)^3$$

where C is a constant representing the current output. By implicit differentiation of this equation,

$$0 = 6x^2 + 6x^2y\frac{dy}{dx} + 6xy^2 + 3(1+y)^2\frac{dy}{dx}$$

$$- [6x^2y + 3(1+y)^2]\frac{dy}{dx} = 6x^2 + 6xy^2$$

$$\frac{dy}{dx} = -\frac{6x^2 + 6xy^2}{6x^2y + 3(1+y)^2}$$

Use the approximation formula

$$\text{Change in } y \simeq \frac{dy}{dx}\Delta x$$

from Section 4, with $x = 30$, $y = 20$, and $\Delta x = -0.8$ to get

$$\text{Change in } y \simeq \left[-\frac{6(30)^2 + 6(30)(20)^2}{6(30)^2(20) + 3(21)^2} \right] (-0.2) \simeq 0.57 \text{ unit}$$

That is, to maintain the current level of output, input y should be increased by approximately 0.57 unit to offset a decrease in input x of 0.8 unit.

25. If the output is to remain unchanged, the equation relating inputs x and y can be written as

$$C = .06x^2 + .14xy + .05y^2$$

where C is the constant representing the current output. By implicit differentiation of this equation,

$$0 = .12x + .14(x\frac{dy}{dx} + y) + .10y\frac{dy}{dx}$$

$$(.14x + .10y)\frac{dy}{dx} = -.12x - .14y$$

$$\frac{dy}{dx} = \frac{-.12x - .14y}{.14x + .10y} = \frac{-2(.06x + .07y)}{2(.07x + .05y)} = -\frac{.06x + .07y}{.07x + .05y}$$

Use the approximation formula

$$\text{Change in } y \approx \frac{dy}{dx}\Delta x$$

from Section 4, with $x = 60$, $y = 300$ and $\Delta x = -1$ to get

$$\text{Change in } y = -\left[\frac{(.06)(60) + (.07)(300)}{(.07)(60) + (.05)(300)} \right] (-1)$$

$$= 1.28125 \text{ hours of unskilled labor.}$$

That is, to maintain the current level of output, input y should be increased by 1.28125 hours to offset a decrease in input x of 1 hour.

Chapter 2, Section 7

1. If $f(x) = 5x^{10} - 6x^5 - 27x + 4$, then

$$f'(x) = 50x^9 - 30x^4 - 27 \quad \text{and} \quad f''(x) = 450x^8 - 120x^3$$

3. If $y = 5\sqrt{x} + \frac{3}{x^2} + \frac{1}{3\sqrt{x}} + \frac{1}{2} = 5x^{1/2} + 3x^{-2} + \frac{1}{3}x^{-1/2} + \frac{1}{2}$, then

$$\frac{dy}{dx} = \frac{5}{2}x^{-1/2} - 6x^{-3} - \frac{1}{6}x^{-3/2}$$

and

$$\frac{d^2y}{dx^2} = -\frac{5}{4}x^{-3/2} + 18x^{-4} + \frac{1}{4}x^{-5/2} = -\frac{5}{4x^{3/2}} + \frac{18}{x^4} + \frac{1}{4x^{5/2}}$$

5. If $f(x) = (3x + 1)^5$, then, by the chain rule for powers,

$$f'(x) = 5(3x + 1)^4(3) = 15(3x + 1)^4$$

and by the chain rule for powers again,

$$f''(x) = 60(3x + 1)^3(3) = 180(3x + 1)^3$$

7. If $y = (x^2 + 5)^8$, then, by the chain rule for powers,

$$\frac{dy}{dx} = 8(x^2 + 5)^7(2x) = 16x(x^2 + 5)^7$$

and, by the product rule,

$$\frac{d^2y}{dx^2} = 16x[7(x^2 + 5)^6(2x)] + (x^2 + 5)^7(16)$$

$$= 16(x^2 + 5)^6[14x^2 + (x^2 + 5)]$$

$$= 16(x^2 + 5)^6(15x^2 + 5)$$

$$= 80(x^2 + 5)^6(3x^2 + 1)$$

9. If $f(x) = \sqrt{1 + x^2} = (1 + x^2)^{1/2}$, then, by the chain rule for powers,

$$f'(x) = \frac{1}{2}(1 + x^2)^{-1/2}(2x) = \frac{x}{(1 + x^2)^{1/2}}$$

and, by the quotient rule,

$$f''(x) = \frac{(1 + x^2)^{1/2}(1) - x\left[\frac{1}{2}(1 + x^2)^{-1/2}(2x)\right]}{1 + x^2}$$

$$= \frac{(1 + x^2)^{-1/2}[(1 + x^2) - x^2]}{1 + x^2}$$

$$= \frac{1}{(1 + x^2)^{3/2}}$$

11.   If $y = \frac{2}{1+x^2} = 2(1 + x^2)^{-1}$, then, by the chain rule for powers,

$$\frac{dy}{dx} = -2(1 + x^2)^{-2}(2) = -\frac{4x}{(1 + x^2)^2}$$

and, by the quotient rule,

$$\begin{aligned} \frac{d^2y}{dx^2} &= -\frac{(1 + x^2)^2(4) - 4x[2(1 + x^2)(2x)]}{(1 + x^2)^4} \\ &= -\frac{4(1 + x^2)[(1 + x^2) - 4x^2]}{(1 + x^2)^4} \\ &= -\frac{4(1 - 3x^2)}{(1 + x^2)^3} = \frac{4(3x^2 - 1)}{(1 + x^2)^3} \end{aligned}$$

13.   If $f(x) = x(2x + 1)^4$, then, by the product rule,

$$\begin{aligned} f'(x) &= x(4)(2x + 1)^3(2) + (2x + 1)^4(1) \\ &= (2x + 1)^3[8x + (2x + 1)] = (2x + 1)^3(10x + 1) \end{aligned}$$

and, by the product rule again,

$$\begin{aligned} f''(x) &= (2x + 1)^3(10) + (10x + 1)(3)(2x + 1)^2(2) \\ &= 2(2x + 1)^2[5(2x + 1) + 3(10x + 1)] \\ &= 2(2x + 1)^2(10x + 5 + 30x + 3) \\ &= 2(2x + 1)^2(40x + 8) = 16(2x + 1)^2(5x + 1) \end{aligned}$$

15.   If $y = \left[\frac{x}{x+1}\right]^2$, then, by the chain rule for powers and the quotient rule,

$$\frac{dy}{dx} = 2\left[\frac{x}{x + 1}\right]\frac{(x + 1)(1) - x(1)}{(x + 1)^2} = \frac{2x}{(x + 1)^3}$$

and, by the quotient rule again,

$$\begin{aligned} \frac{d^2y}{dx^2} &= \frac{(x + 1)^3(2) - 2x(3)(x + 1)^2(1)}{(x + 1)^6} \\ &= \frac{(x + 1)^2(2x + 2 - 6x)}{(x + 1)^6} = \frac{2(1 - 2x)}{(x + 1)^4} \end{aligned}$$

17.   If $2y^2 - 5x^2 = 3$, then

$$4y\frac{dy}{dx} - 10x = 0 \quad \text{or} \quad \frac{dy}{dx} = \frac{10x}{4y} = \frac{5x}{2y}$$

By the quotient rule,

$$\frac{d^2y}{dx^2} = \frac{2y(5) - 5x\left[2\frac{dy}{dx}\right]}{(2y)^2}$$

$$= \frac{10y - 10x\frac{dy}{dx}}{4y^2} = \frac{5y - 5x\frac{dy}{dx}}{2y^2}$$

Since $\frac{dy}{dx} = \frac{5x}{2y}$,

$$\frac{d^2y}{dx^2} = \frac{5y - 5x\left[\frac{5x}{2y}\right]}{2y^2} = \frac{10y^2 - 25x^2}{4y^3} = \frac{5(2y^2 - 5x^2)}{4y^3}$$

From the original equation,

$$2y^2 - 5x^2 = 3$$

hence

$$\frac{d^2y}{dx^2} = \frac{5(3)}{4y^3} = \frac{15}{4y^3}$$

19. If $ax^2 + by^2 = 1$, where a and b are constants, then

$$2ax + 2by\frac{dy}{dx} = 0 \quad \text{or} \quad \frac{dy}{dx} = -\frac{2ax}{2by} = -\frac{ax}{by}$$

By the quotient rule,

$$\frac{d^2y}{dx^2} = -\frac{by(a) - ax\left[b\frac{dy}{dx}\right]}{(by)^2} = -\frac{aby - abx\frac{dy}{dx}}{b^2y^2}$$

Since $\frac{dy}{dx} = -\frac{ax}{by}$,

$$\frac{d^2y}{dx^2} = -\frac{aby - abx\left[-\frac{ax}{by}\right]}{b^2y^2}$$

$$= -\frac{ab^2y^2 + a^2bx^2}{b^3y^3} = -\frac{ab(by^2 + ax^2)}{b^3y^3} = -\frac{a(by^2 + ax^2)}{b^2y^3}$$

From the original equation,

$$by^2 + ax^2 = 1$$

and so

$$\frac{d^2y}{dx^2} = -\frac{a}{b^2y^3}$$

21. The price per unit t months from now will be $P(t) = -t^3 + 7t^2 + 200t + 300$ dollars.

(a)   The rate of change of the price with respect to time is the first derivative

$$P'(t) = -3t^2 + 14t + 200$$

which, 5 months from now, will be

$$P'(5) = -3(5)^2 + 14(5) + 200 = 195 \text{ dollars per month}$$

That is, the price will be increasing at the rate of $195 per month.

(b)   The rate of change of the rate of price increase is the second derivative

$$P''(t) = -6t + 14$$

which, 5 months from now, will be

$$P''(5) = -6(5) + 14 = -16 \text{ dollars per month per month}$$

That is, although the price is rising, the rate at which it does so is decreasing at the rate of $16 per month per month.

(c)   To estimate the change in the rate P' of price increase during the first half of the 6th month, apply the approximation formula

$$\text{Change in rate of price increase } = \Delta P' \simeq P''(t)\Delta t$$

with $t = 5$ and $\Delta t = 0.5$ to get

$$\text{Change in rate of price increase } \simeq P''(5)(0.5) = -16(0.5) = -8$$

dollars per month.

(d)   The actual change in the rate P' of price increase during the first half of the 6th month is

$$P'(5.5) - P'(5) \simeq 186.25 - 195 = -8.75 \text{ dollars per month}$$

23.   Since the distance function is $D(t) = t^3 - 12t^2 + 100t + 12$, the speed is

$$V(t) = D'(t) = 3t^2 - 24t + 100$$

and the acceleration is

$$A(t) = V'(t) = D''(t) = 6t - 24$$

After 3 seconds, the acceleration is $A(3) = -6$ meters per second per second, where the minus sign indicates that the speed is decreasing.

25.   (a)   Since the distance after t hours is $D(t) = 64t + \frac{10}{3}t^2 - \frac{2}{9}t^3$, the speed is

$$V(t) = D'(t) = 64 + \frac{20}{3}t - \frac{2}{3}t^2$$

and the acceleration is

$$A(t) = V'(t) = D''(t) = \frac{20}{3} - \frac{4}{3}t$$

(b) After 6 hours, the acceleration is $A(6) = -\frac{4}{3}$. That is, the speed is decreasing at the rate of $\frac{4}{3}$ kilometers per hour per hour.

(c) During the 7th hour, the actual change in speed is $V(7) - V(6) = -2$ kilometers per hour.

27. If $y = \sqrt{x} - \frac{1}{2x} + \frac{x}{\sqrt{2}} = x^{1/2} - \frac{1}{2}x^{-1} + \frac{1}{\sqrt{2}}x$, then

$$\frac{dy}{dx} = \frac{1}{2}x^{-1/2} + \frac{1}{2}x^{-2} + \frac{1}{\sqrt{2}}$$

$$\frac{d^2y}{dx^2} = -\frac{1}{4}x^{-3/2} - x^{-3} + 0$$

$$\frac{d^3y}{dx^3} = \frac{3}{8}x^{-5/2} + 3x^{-4} = \frac{3}{8x^{5/2}} + \frac{3}{x^4}$$

Chapter 2, Review Problems

1.   (a)   If $f(x) = x^2 - 3x + 1$, then

$$\frac{f(x + \Delta x) - f(x)}{\Delta x} = \frac{[(x + \Delta x)^2 - 3(x + \Delta x) + 1] - [x^2 - 3x + 1]}{\Delta x}$$

$$= \frac{x^2 + 2x(\Delta x) + (\Delta x)^2 - 3x - 3\Delta x + 1 - x^2 + 3x - 1}{\Delta x}$$

$$= \frac{2x(\Delta x) + (\Delta x)^2 - 3\Delta x}{\Delta x}$$

$$= 2x + \Delta x - 3$$

As $\Delta x \to 0$, this difference quotient approaches $2x - 3$, so $f'(x) = 2x - 3$.

(b)   If $f(x) = \frac{1}{x-2}$, then

$$\frac{f(x + \Delta x) - f(x)}{\Delta x} = \frac{\left[\frac{1}{(x + \Delta x) - 2}\right] - \left[\frac{1}{x - 2}\right]}{\Delta x}$$

$$= \left[\frac{\frac{1}{x + \Delta x - 2} - \frac{1}{x - 2}}{\Delta x}\right]\left[\frac{(x + \Delta x - 2)(x - 2)}{(x + \Delta x - 2)(x - 2)}\right]$$

$$= \frac{(x - 2) - (x + \Delta x - 2)}{\Delta x(x + \Delta x - 2)(x - 2)}$$

$$= \frac{-1}{(x + \Delta x - 2)(x - 2)}$$

As $\Delta x \to 0$, this difference quotient approaches $\frac{-1}{(x-2)^2}$, and so $f'(x) = \frac{-1}{(x-2)^2}$.

2.   (a)   If $f(x) = 6x^4 - 7x^3 + 2x + \sqrt{2}$, then

$$f'(x) = 24x^3 - 21x^2 + 2$$

(b)   If $f(x) = x^3 - \frac{1}{3x^5} + 2\sqrt{x} - \frac{3}{x} + \frac{1 - 2x}{x^3}$ rewrite $f(x)$ as

$$f(x) = x^3 - \frac{1}{3}x^{-5} + 2x^{1/2} - 3x^{-1} + x^{-3} - 2x^{-2}$$

and differentiate term by term to get

$$f'(x) = 3x^2 + \frac{5}{3}x^{-6} + x^{-1/2} + 3x^{-2} - 3x^{-4} + 4x^{-3}$$

$$= 3x^2 + \frac{5}{3x^6} + \frac{1}{\sqrt{x}} + \frac{3}{x^2} - \frac{3}{x^4} + \frac{4}{x^3}$$

(c)   If $y = \frac{2 - x^2}{3x^2 + 1}$, then

$$\frac{dy}{dx} = \frac{(3x^2 + 1)(-2x) - (2 - x^2)(6x)}{(3x^2 + 1)^2}$$

$$= \frac{2x(-3x^2 - 1 - 6 + 3x^2)}{(3x^2 + 1)^2} = \frac{-14x}{(3x^2 + 1)^2}$$

(d) If $y = (2x + 5)^3(x + 1)^2$, then

$$\frac{dy}{dx} = (2x + 5)^3(2)(x + 1)(1) + (x + 1)^2(3)(2x + 5)^2(2)$$

$$= 2(2x + 5)^2(x + 1)[(2x + 5) + 3(x + 1)]$$

$$= 2(2x + 5)^2(x + 1)(5x + 8)$$

(e) If $f(x) = (5x^4 - 3x^2 + 2x + 1)^{10}$, then

$$f'(x) = 10(5x^4 - 3x^2 + 2x + 1)^9(20x^3 - 6x + 2)$$

$$= 20(5x^4 - 3x^2 + 2x + 1)^9(10x^3 - 3x + 1)$$

(f) If $f(x) = \sqrt{x^2 + 1} = (x^2 + 1)^{1/2}$, then

$$f'(x) = \frac{1}{2}(x^2 + 1)^{-1/2}(2x) = \frac{x}{(x^2 + 1)^{1/2}}$$

(g) If $y = \left[x + \frac{1}{x}\right]^2 - \frac{5}{\sqrt{3x}} = (x + x^{-1})^2 - 5(3x)^{-1/2}$, then

$$\frac{dy}{dx} = 2(x + x^{-1})\frac{d}{dx}(x + x^{-1}) + \frac{5}{2}(3x)^{-3/2}\frac{d}{dx}(3x)$$

$$= 2(x + x^{-1})(1 - x^{-2}) + \frac{5}{2}(3x)^{-3/2}(3)$$

$$= 2\left[x + \frac{1}{x}\right]\left[1 - \frac{1}{x^2}\right] + \frac{15}{2(3x)^{3/2}}$$

(h) If $y = \left[\frac{x+1}{1-x}\right]^2$, then

$$\frac{dy}{dx} = 2\left[\frac{x+1}{1-x}\right]\frac{d}{dx}\left[\frac{x+1}{1-x}\right]$$

$$= 2\left[\frac{x+1}{1-x}\right]\left[\frac{(1-x)(1) - (x+1)(-1)}{(1-x)^2}\right]$$

$$= 2\left[\frac{x+1}{1-x}\right]\left[\frac{2}{(1-x)^2}\right]$$

$$= \frac{4(x+1)}{(1-x)^3}$$

(i) If $f(x) = (3x + 1)\sqrt{6x + 5} = (3x + 1)(6x + 5)^{1/2}$, then

$$f'(x) = (3x + 1)\left(\frac{1}{2}\right)(6x + 5)^{-1/2}(6) + (6x + 5)^{1/2}(3)$$

$$= \frac{3(3x + 1)}{(6x + 5)^{1/2}} + 3(6x + 5)^{1/2} = \frac{9(3x + 2)}{\sqrt{6x + 5}}$$

(j)   If $f(x) = \frac{(3x+1)^3}{(1-3x)^4}$, then

$$f'(x) = \frac{(1-3x)^4 \frac{d}{dx}(3x+1)^3 - (3x+1)^3 \frac{d}{dx}(1-3x)^4}{(1-3x)^8}$$

$$= \frac{(1-3x)^4[3(3x+1)^2(3)] - (3x+1)^3[4(1-3x)^3(-3)]}{(1-3x)^8}$$

$$= \frac{3(1-3x)^3(3x+1)^2[3(1-3x) + 4(3x+1)]}{(1-3x)^8}$$

$$= \frac{3(3x+1)^2(3 - 9x + 12x + 4)}{(1-3x)^5}$$

$$= \frac{3(3x+1)^2(3x+7)}{(1-3x)^5}$$

(k)   If $y = \sqrt{\frac{1-2x}{3x+2}} = \left[\frac{1-2x}{3x+2}\right]^{1/2}$, then

$$\frac{dy}{dx} = \frac{1}{2}\left[\frac{1-2x}{3x+2}\right]^{-1/2} \frac{d}{dx}\left[\frac{1-2x}{3x+2}\right]$$

$$= \frac{1}{2}\left[\frac{1-2x}{3x+2}\right]^{-1/2} \frac{(3x+2)\frac{d}{dx}(1-2x) - (1-2x)\frac{d}{dx}(3x+2)}{(3x+2)^2}$$

$$= \frac{1}{2}\left[\frac{1-2x}{3x+2}\right]^{-1/2} \frac{(3x+2)(-2) - (1-2x)(3)}{(3x+2)^2}$$

$$= \frac{1}{2}\frac{(1-2x)^{-1/2}}{(3x+2)^{-1/2}} \frac{-6x - 4 - 3 + 6x}{(3x+2)^2}$$

$$= \frac{-7}{2(1-2x)^{1/2}(3x+2)^{3/2}}$$

3.   (a)   If $f(x) = x^2 - 3x + 2$, then $f'(x) = 2x - 3$ and the slope of the tangent line at the point $(1, f(1)) = (1, 0)$ is $f'(1) = -1$. The equation of the tangent line is therefore

$$y - 0 = -1(x - 1) \quad \text{or} \quad y = -x + 1$$

(b)   If $f(x) = \frac{4}{x-3}$, then $f'(x) = \frac{-4}{(x-3)^2}$ and the slope of the tangent line at the point $(1, f(1)) = (1, -2)$ is $f'(1) = -1$. The equation of the tangent line is therefore

$$y - (-2) = -1(x - 1) \quad \text{or} \quad y = -x - 1$$

(c)   If $f(x) = \frac{x}{x^2+1}$, then

$$f'(x) = \frac{(x^2+1)(1) - 2(2x)}{(x^2+1)^2} = \frac{1 - x^2}{(x^2+1)^2}$$

and the slope of the tangent line at the point $(0, f(0)) = (0, 0)$ if $f'(0) = 1$. The equation of the tangent line is therefore

$$y - 0 = 1(x - 0) \quad \text{or} \quad y = x$$

(d) If $f(x) = \sqrt{x^2 + 5} = (x^2 + 5)^{1/2}$, then

$$f'(x) = \frac{1}{2}(x^2 + 5)^{-1/2}(2x) = \frac{x}{\sqrt{x^2 + 5}}$$

and the slope of the tangent line at the point $(-2, f(-2)) = (-2, 3)$ is $f'(-2) = -\frac{2}{3}$. The equation of the tangent line is therefore

$$y - 3 = -\frac{2}{3}(x - (-2)) \quad \text{or} \quad y = -\frac{2}{3}x + \frac{5}{3}$$

4. (a) The rate of change of $f(t) = t^3 - 4t^2 + 5t\sqrt{t} - 5 = t^3 - 4t^2 + 5t^{3/2} - 5$ is $f'(t) = 3t^2 - 8t + \frac{15}{2}t^{1/2}$ at any value of $t \geq 0$ and when $t = 4$, $f'(4) = 48 - 32 + \frac{15}{2}(2) = 31$.

(b) The rate of change of $f(t) = t^3(t^2 - 1) = t^5 - t^3$ is $f'(t) = 5t^4 - 3t^2$ at any value of t and when $t = 0$, $f'(0) = 0$.

5. (a) If $f(t) = t^2 - 3t + \sqrt{t} = t^2 - 3t + t^{1/2}$ then the percentage rate of change is

$$100\left(\frac{f'(t)}{f(t)}\right) = 100\left(\frac{2t - 3 + \frac{1}{2}t^{-1/2}}{t^2 - 3t + t^{1/2}}\right)$$

$$= 100\left(\frac{4t^{3/2} - 6t^{1/2} + 1}{2t^{5/2} - 6t^{3/2} + 2t}\right)$$

(b) If $f(t) = t^2(3 - 2t)^3$ then the percentage rate of change is

$$100\left(\frac{f'(t)}{f(t)}\right) = 100\left(\frac{(t^2)(3)(3 - 2t)^2(-2) + (3 - 2t)^3(2t)}{t^2(3 - 2t)^3}\right)$$

$$= 100(3 - 2t)^2\left(\frac{-6t^2 + 6t - 4t^2}{t^2(3 - 2t)^3}\right)$$

$$= 100\left(\frac{-10t^2 + 6t}{t^2(3 - 2t)}\right)$$

$$= 100\left(\frac{2t(3 - 5t)}{t^2(3 - 2t)}\right)$$

$$= 200\left(\frac{(3 - 5t)}{t(3 - 2t)}\right)$$

(c)   If $f(t) = \frac{1}{t+1} = (t+1)^{-1}$ then the percentage rate of change is

$$100 \left(\frac{f'(t)}{f(t)} \right) = 100 \left(\frac{-1(t+1)^{-2}}{(t+1)^{-1}} \right)$$

$$= -100 \left(\frac{1}{t+1} \right)$$

6.   (a)   Since $N(x) = 6x^3 + 500x + 8{,}000$ is the number of people using the system after x weeks, the rate at which use of the system is changing after x weeks is $N'(x) = 18x^2 + 500$ people per week and the rate after 8 weeks is $N'(8) = 1{,}652$ people per week.

     (b)   The actual increase in the use of the system during the 8th week is $N(8) - N(7) = 1{,}514$ people.

7.   (a)   Since $Q(x) = 50x^2 + 9{,}000x$ is the weekly output when x workers are employed, the marginal output is $Q'(x) = 100x + 9{,}000$. The change in output due to an increase from 30 to 31 workers is approximately $Q'(30) = 12{,}000$ units.

     (b)   The actual increase in output is $Q(31) - Q(30) = 12{,}050$ units.

8.   Since the population t months from now will be $P(t) = 3t + 5t^{3/2} + 6{,}000$, the rate of change of the population will be $P'(t) = 3 + \frac{15}{2}t^{1/2}$, and the percentage rate of change 4 months from now will be

$$100 \left[\frac{P'(4)}{P(4)} \right] = 100 \left[\frac{18}{6{,}052} \right] \simeq 0.30 \text{ percent per month}$$

9.   Since the daily output is $Q(L) = 20{,}000L^{1/2}$ units when L worker-hours are used, a change in the work force from 900 worker-hours to 885 worker-hours ($\Delta L = -15$) results in a change in the output of

$$\Delta Q = Q(900) - Q(885) \simeq Q'(900)(-15)$$

Since $Q'(L) = 10{,}000L^{-1/2}$ and $Q'(900) = 10{,}000(900)^{-1/2} = \frac{1{,}000}{3}$ it follows that $\Delta Q \simeq \frac{1{,}000}{3}(-15) = -5{,}000$, that is, a decrease in output of 5,000 units.

10.   The gross national product t years after 1984 is $N(t) = t^2 + 6t + 300$ billion dollars. The derivative is $N'(t) = 2t + 6$. At the beginning of the second quarter of 1988, $t = 4.25$. The change in t during this quarter is $\Delta t = 0.25$. Hence,

$$\text{Percentage change in } N \simeq 100 \frac{N'(4.25)\Delta t}{N(4.25)}$$

$$= 100 \frac{[2(4.25) + 6](0.25)}{(4.25)^2 + 6(4.25) + 300}$$

$$\simeq 1.055 \text{ percent}$$

11. Let A denote the level of air pollution and p the population. Then, $A = kp^2$, where k is a positive constant of proportionality. If the population increases by 5 percent, the change in the population is $\Delta p = 0.05p$. The corresponding increase in the level of air pollution is

$$\Delta A = A(p + 0.05p) - A(p) \simeq A'(p)(0.05p) = 2kp(0.05p) = 0.1kp^2 = 0.1A$$

That is, an increase of 5 percent in the population causes an increase of 10 percent in the level of pollution.

12. Output is $Q(L) = 600L^{2/3}$. The derivative is $Q'(L) = 400L^{-1/3}$. We are given that the percentage change in Q is 1 percent, and the goal is to find the percentage change in L, which can be represented as $100\frac{\Delta L}{L}$. Apply the formula

$$\text{Percentage change in } Q \simeq 100\frac{Q'(L)\Delta L}{Q(L)}$$

with 1 on the left-hand side and solve for $100\frac{\Delta L}{L}$ as follows:

$$1 \simeq 100\frac{400L^{-1/3}\Delta L}{600L^{2/3}} = 100\left(\frac{2}{3}\right)\left(\frac{\Delta L}{L}\right)$$
$$100\frac{\Delta L}{L} \simeq \frac{3}{2} = 1.5 \text{ percent}$$

That is, an increase in labor of approximately 1.5 percent is required to increase output by 1 percent.

13. (a) If $y = 5u^2 + u - 1$ and $u = 3x + 1$, then $\frac{dy}{du} = 10u + 1$, $\frac{du}{dx} = 3$, and

$$\frac{dy}{dx} = \frac{dy}{du}\frac{du}{dx} = (10u + 1)(3) = 3[10(3x + 1) + 1] = 3(30x + 11)$$

(b) If $y = \frac{1}{u^2}$ and $u = 2x + 3$, then $\frac{dy}{du} = \frac{-2}{u^3}$, $\frac{du}{dx} = 2$, and

$$\frac{dy}{dx} = \frac{dy}{du}\frac{du}{dx} = \left[\frac{-2}{u^3}\right](2) = \frac{-4}{(2x + 3)^3}$$

14. (a) If $y = u^3 - 4u^2 + 5u + 2$ and $u = x^2 + 1$, then $\frac{dy}{du} = 3u^2 - 8u + 5$, $\frac{du}{dx} = 2x$, and

$$\frac{dy}{dx} = \frac{dy}{du}\frac{du}{dx} = (3u^2 - 8u + 5)(2x)$$

When $x = 1$, $u = 2$ and so

$$\frac{dy}{dx} = [3(2)^2 - 8(2) + 5][2(1)] = 2$$

(b) If $y = \sqrt{u} = u^{1/2}$ and $u = x^2 + 2x - 4$, then $\frac{dy}{du} = \frac{1}{2u^{1/2}}$, $\frac{du}{dx} = 2x + 2$ and so

$$\frac{dy}{dx} = \frac{dy}{du}\frac{du}{dx} = \left[\frac{1}{2u^{1/2}}\right](2x + 2) = \frac{x + 1}{u^{1/2}}$$

When $x = 2$, $u = 4$ and so $\frac{dy}{dx} = \frac{2+1}{(4)^{1/2}} = \frac{3}{2}$.

15. Since $C(q) = 0.1q^2 + 10q + 400$ is the total cost of producing q units and $q(t) = t^2 + 50t$ is the number of units produced during the first t hours, then $\frac{dC}{dq} = 0.2q + 10$ (dollars per units) and $\frac{dq}{dt} = 2t + 50$ (units per hour), and the rate of change of cost with respect to time is

$$\frac{dC}{dt} = \frac{dC}{dq}\frac{dq}{dt} = (0.2q + 10)(2t + 50) \text{ dollars per hour}$$

After 2 hours, $t = 2$, $q(2) = 104$, and so

$$\frac{dC}{dt} = [0.2(104) + 10][2(2) + 50] = \$1,663.20 \text{ per hour}$$

16. The population is $p(t) = 10 - \frac{20}{(t+1)^2} = 10 - 20(t + 1)^{-2}$ and the carbon monoxide level is $c(p) = 0.8\sqrt{p^2 + p + 139} = 0.8(p^2 + p + 139)^{1/2}$ units. By the chain rule, the rate of change of the carbon monoxide level with respect to time is

$$\frac{dc}{dt} = \frac{dc}{dp}\frac{dp}{dt} = [0.4(p^2 + p + 139)^{-1/2}(2p + 1)][40(t + 1)^{-3}]$$

$$= \frac{0.4(2p + 1)}{\sqrt{p^2 + p + 139}}\frac{40}{(t + 1)^3}$$

When $t = 1$,

$$p = p(1) = 10 - \frac{20}{4} = 5 \quad \text{and} \quad c = c(5) = 0.8\sqrt{169} = 10.4$$

Hence, when $t = 1$,

$$\text{Percentage rate of change} = 100\frac{dc/dt}{c}$$

$$= \left[\frac{0.4(10 + 1)}{\sqrt{169}}\frac{40}{(1 + 1)^3}\right]\left[\frac{1}{10.4}\right]$$

$$\simeq 16.27 \text{ percent per year}$$

17. (a) If $5x + 3y = 12$, then

$$5(1) + 3\frac{dy}{dx} = 0 \quad \text{or} \quad \frac{dy}{dx} = -\frac{5}{3}$$

(b)   If $x^2 y = 1$, then (by the product rule)

$$x^2 \frac{dy}{dx} + y(2x) = 0 \quad \text{or} \quad \frac{dy}{dx} = \frac{2xy}{x^2} = -\frac{2y}{x}$$

(c)   If $(2x + 3y)^5 = x = 1$, then

$$5(2x + 3y)^4 \left(2 + 3\frac{dy}{dx} \right) = 1$$

$$10(2x + 3y)^4 + 15(2x + 3y)^4 \frac{dy}{dx} = 1$$

$$15(2x + 3y)^4 \frac{dy}{dx} = 1 - 10(2x + 3y)^4$$

$$\frac{dy}{dx} = \frac{1 - 10(2x + 3y)^4}{15(2x + 3y)^4}$$

(d)   If $(1 - 2xy^3)^5 = x + 4y$, then

$$5(1 - 2xy^3)^4 \frac{d}{dx}(1 - 2xy^3) = 1 + 4\frac{dy}{dx}$$

$$5(1 - 2xy^3)^4 \left(-6xy^2 \frac{dy}{dx} - 2y^3 \right) = 1 + 4\frac{dy}{dx}$$

$$-30xy^2(1 - 2xy^3)^4 \frac{dy}{dx} - 10y^3(1 - 2xy^3)^4 = 1 + 4\frac{dy}{dx}$$

$$[-4 - 30xy^2(1 - 2xy^3)^4]\frac{dy}{dx} = 1 + 10y^3(1 - 2xy^3)^4$$

$$\frac{dy}{dx} = -\frac{1 + 10y^3(1 - 2xy^3)^4}{4 + 30xy^2(1 - 2xy^3)^4}$$

18.   (a)   If $xy^3 = 8$, then (by the product rule)

$$x\left(3y^2 \frac{dy}{dx} \right) + y^3 = 0 \quad \text{or} \quad \frac{dy}{dx} = \frac{-y^3}{3xy^2} = -\frac{y}{3x}$$

When $x = 1$, the original equation gives $1(y^3) = 8$ or $y = 2$. To find the slope of the tangent at the point $(1, 2)$, substitute into the equation for $\frac{dy}{dx}$ getting

$$\text{Slope} \ = \frac{dy}{dx} = -\frac{2}{3(1)} = -\frac{2}{3}$$

(b)   If $x^2 y - 2xy^3 + 6 = 2x + 2y$, then

$$\frac{x^2 dy}{dx} + y(2x) - 2\left[x\left(3y^2 \frac{dy}{dx} \right) + y^3(1) \right] = 2 + 2\frac{dy}{dx}$$

$$x^2 \frac{dy}{dx} + 2xy - 6xy^2 \frac{dy}{dx} - 2y^3 = 2 + 2\frac{dy}{dx}$$

When $x = 0$, the original equation gives

$$(0)^2(y) - 2(0)(y^3) + 6 = 2(0) + 2y, \quad 6 = 2y, \quad \text{or } y = 3$$

To find the slope of the tangent at $(0, 3)$, substitute into the derivative equation and solve for $\frac{dy}{dx}$ getting

$$(0)^2\frac{dy}{dx} + 2(0)(3) - 6(0)(3)\frac{dy}{dx} - 2(3)^3 = 2 + 2\frac{dy}{dx}$$

$$-54 = 2 + 2\frac{dy}{dx} \quad \text{or} \quad \text{Slope} = \frac{dy}{dx} = -28$$

19. By the approximation formula from Section 4, $\Delta y \simeq \frac{dy}{dx}\Delta x$. To find $\frac{dy}{dx}$, differentiate the equation

$$C = x^3 + 2xy^2 = 2y^3$$

implicitly with respect to x, where C is a constant representing the current level of output. You get

$$0 = 3x^2 + 4xy\frac{dy}{dx} + 2y^2 + 6y^2\frac{d}{dx} \quad \text{or} \quad \frac{dy}{dx} = -\frac{3x^2 + 2y^2}{4xy + 6y^2}$$

When $x = 10$ and $y = 20$,

$$\frac{dy}{dx} = -\frac{3(10)^2 + 2(20)^2}{4(10)(20) + 6(20)^2} \simeq -0.344$$

Finally, apply the approximation formula with $\frac{dy}{dx} \simeq -0.344$ and $\Delta x = 0.5$ to get

$$\Delta y \simeq -0.344(0.5) = -0.172 \text{ unit}$$

That is, to maintain the current level of output, input y should be decreased by approximately 0.172 unit to offset a 0.5 unit increase in input x.

20. (a) If $f(x) = 6x^5 - 4x^3 + 5x^2 - 2x + \frac{1}{x}$, then

$$f'(x) = 3x^4 - 12x^2 + 10x - 2 - \frac{1}{x^2}$$

and

$$f''(x) = 120x^3 - 24x + 10 + \frac{2}{x^3}$$

(b) If $y = (3x^2 + 2)^4$, then

$$\frac{dy}{dx} = 4(3x^2 + 2)^3(6x) = 24x(3x^2 + 2)^3$$

and

$$\frac{d^2y}{dx^2} = 24x[3(3x^2+2)^2(6x)] + (3x^2+2)^3(24)$$
$$= 24(3x^2+2)^2[18x^2+(3x^2+2)]$$
$$= 24(3x^2+2)^2(21x^2+2)$$

(c)   If $f(x) = \frac{x-1}{(x+1)^2}$, then

$$f'(x) = \frac{(x+1)^2(1) - (x-1)(2)(x+1)(1)}{(x+1)^4}$$
$$= \frac{(x+1)[(x+1) - 2(x-1)]}{(x+1)^4}$$
$$= \frac{x+1-2x+2}{(x+1)^3} = \frac{3-x}{(x+1)^3}$$

and

$$f''(x) = \frac{(x+1)^3(-1) - (3-x)(3)(x+1)^2(1)}{(x+1)^6}$$
$$= \frac{(x+1)^2[-(x+1) - 3(3-x)]}{(x+1)^6}$$
$$= \frac{-x-1-9+3x}{(x+1)^4} = \frac{2x-10}{(x+1)^4} = \frac{2(x-5)}{(x+1)^4}$$

21.   If $3x^2 - 2y^2 = 6$, then

$$6x - 4y\frac{dy}{dx} = 0 \quad \text{or} \quad \frac{dy}{dx} = \frac{6x}{4y} = \frac{3x}{2y}$$

By the quotient rule,

$$\frac{d^2y}{dx^2} = \frac{2y(3) - 3x\left[2\frac{dy}{dx}\right]}{(2y)^2} = \frac{3y - 3x\frac{dy}{dx}}{2y^2}$$

Since $\frac{dy}{dx} = \frac{3x}{2y}$,

$$\frac{d^2y}{dx^2} = \frac{3y - 3x\left[\frac{3x}{2y}\right]}{2y^2} = \frac{6y^2 - 9x^2}{4y^3}$$

By the original equation,

$$6y^2 - 9x^2 = 3(2y^2 - 3x^2) = -3(3x^2 - 2y^2) = -3(6) = -18$$

and so

$$\frac{d^2y}{dx^2} = \frac{-18}{4y^3} = \frac{-9}{2y^3}$$

22. The production function is $Q(t) = -t^3 + 9t^2 + 12t$.

    (a) The rate of production is the derivative $Q'(t) = -3t^2 + 18t + 12$. At 9:00 A.M., $t = 1$ and the rate of production is $Q'(1) = 27$ units per hour.

    (b) The rate of change of the rate of production is the second derivative $Q''(t) = -6t + 18$. At 9:00 A.M., this rate is $Q''(1) = 12$ units per hour per hour.

    (c) The change in the rate of production between 9:00 A.M. and 9:06 A.M. is given by the approximation formula

    $$\text{Change in rate of production} \ = \Delta Q' \simeq Q''(t)\Delta t$$

    When $t = 1$ and $\Delta t = 0.1$ hour, this gives

    $$\text{Change in rate of production} \ \simeq Q''(1)\Delta t = 12(0.1) = 1.2$$

    units per hour.

    (d) The actual change in the worker's rate of production between 9:00 A.M. and 9:06 A.M. is
    $$Q'(1.1) - Q'(1) \simeq 1.17 \ \text{units per hour}$$

23. (a) If $y = 2x^5 + 5x^4 - 2x + \frac{1}{x}$, then

    $$\frac{dy}{dx} = 10x^4 + 20x^3 - 2 - \frac{1}{x^2}$$

    $$\frac{d^2y}{dx^2} = 40x^3 + 60x^2 + 0 + \frac{2}{x^3}$$

    $$\frac{d^3y}{dx^3} = 120x^2 + 120x - \frac{6}{x^4}$$

    $$\frac{d^4y}{dx^4} = 240x + 120 + \frac{24}{x^5}$$

    (b) If $f(x) = \sqrt{3x} + \frac{3}{2x^2} = \sqrt{3}(x^{1/2}) + \frac{3}{2}x^{-2}$, then

    $$f'(x) = \frac{1}{2}\sqrt{3}\ x^{-1/2} - 3x^{-3}$$

    $$f''(x) = -\frac{1}{4}\sqrt{3}\ x^{-3/2} + 9x^{-4}$$

    $$f'''(x) = \frac{3}{8}\sqrt{3}\ x^{-5/2} - 36x^{-5}$$

    $$f^{(4)}(x) = \frac{15}{16}\sqrt{3}\ x^{-7/2} + 180x^{-6} = \frac{15\sqrt{3}}{16x^{7/2}} + \frac{180}{x^6}$$

CHAPTER 3

DIFFERENTIATION: FURTHER TOPICS

---

1 INCREASE AND DECREASE; RELATIVE EXTREMA

**2 CURVE SKETCHING: CONCAVITY AND
THE SECOND-DERIVATIVE TEST**

3 ABSOLUTE MAXIMA AND MINIMA

4 PRACTICAL OPTIMIZATION PROBLEMS

5 APPLICATIONS TO BUSINESS AND ECONOMICS

REVIEW PROBLEMS

Chapter 3, Section 1

1.

Interval	Inc/dec	Sign of $f'(x)$
$x < -2$	decreasing	−
$-2 < x < 2$	increasing	+
$x > 2$	decreasing	−

3.

Interval	Inc/dec	Sign of $f'(x)$
$x < -4$	increasing	+
$-4 < x < -2$	decreasing	−
$-2 < x < 0$	decreasing	−
$0 < x < 2$	increasing	+
$x > 2$	decreasing	−

5.   Since $f(x) = x^3 + 3x^2 + 1$, the derivative is

$$f'(x) = 3x^2 + 6x = 3x(x + 2)$$

which is zero when $x = 0$ and $x = -2$. The corresponding critical points are

$(0, f(0)) = (0, 1)$ and $(-2, f(-2)) = (-2, 5)$.

Interval	$3x$	$x + 2$	$f'(x)$	Inc/dec
$x < -2$	$-$	$-$	$+$	increasing
$-2 < x < 2$	$-$	$+$	$-$	decreasing
$x > 2$	$+$	$+$	$+$	increasing

Notice that the critical point $(-2, 5)$ is a relative maximum and the critical point $(0, 1)$ is a relative minimum.

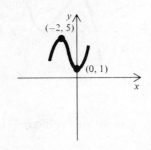

Graph for Problem 5.   Graph for Problem 7.

7.   Since $f(x) = \frac{1}{3}x^3 - 9x + 2$, the derivative is

$$f'(x) = x^2 - 9 = (x + 3)(x - 3)$$

which is zero when $x = -3$ and $x = 3$. The corresponding critical points are $(-3, f(-3)) = (-3, 20)$ and $(3, f(3)) = (3, -16)$.

Interval	$x + 3$	$x - 3$	$f'(x)$	Inc/dec
$x < -3$	$-$	$-$	$+$	increasing
$-3 < x < 3$	$+$	$-$	$-$	decreasing
$x > 3$	$+$	$+$	$+$	increasing

Notice that the critical point $(-3, 20)$ is a relative maximum and the critical point $(3, -16)$ is a relative minimum.

9.   Since $f(x) = 3x^5 - 5x^3$, the derivative is

$$f'(x) = 15x^4 - 15x^2 = 15x^2(x^2 - 1) = 15x^2(x + 1)(x - 1)$$

which is zero when $x = 0$, $x = -1$, and $x = 1$. The corresponding critical points are $(0, f(0)) = (0, 0), (-1, f(-1)) = (-1, 2)$, and $(1, f(1)) = (1, -2)$.

Interval	$15x^2$	$x + 1$	$x - 1$	$f'(x)$	Inc/dec
$x < -1$	$+$	$-$	$-$	$+$	increasing
$-1 < x < 0$	$+$	$+$	$-$	$-$	decreasing
$0 < x < 1$	$+$	$+$	$-$	$-$	decreasing
$x > 1$	$+$	$+$	$+$	$+$	increasing

Notice that the critical point $(-1, 2)$ is a relative maximum and the critical point $(1, -2)$ is a relative minimum.

11.   Since $f(x) = 324x - 72x^2 + 4x^3$, the derivative is

$$f'(x) = 324 - 144x + 12x^2 = 12(27 - 12x + x^2) = 12(x - 3)(x - 9)$$

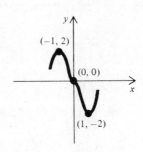

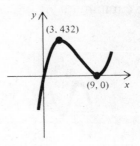

Graph for Problem 9.   Graph for Problem 11.

which is zero when $x = 3$ and $x = 9$. The corresponding critical points are $(3, f(3)) = (3, 432)$ and $(9, f(9)) = (9, 0)$.

Interval	$x - 3$	$x - 9$	$f'(x)$	Inc/dec
$x < 3$	$-$	$-$	$+$	increasing
$3 < x < 9$	$+$	$-$	$-$	decreasing
$x > 9$	$+$	$+$	$+$	increasing

Notice that the critical point $(3, 432)$ is a relative maximum and the critical point $(9, 0)$ is a relative minimum.

13.   Since $f(x) = 10x^6 + 24x^5 + 15x^4 + 3$, the derivative is

$$f'(x) = 60x^5 + 120x^4 + 60x^3 = 60x^3(x^2 + 2x + 1) = 60x^3(x + 1)^2$$

which is zero when $x = 0$ and $x = -1$. The corresponding critical points are $(0, f(0)) = (0, 3)$ and $(-1, f(-1)) = (-1, 4)$.

Interval	$60x^3$	$(x - 1)^2$	$f'(x)$	Inc/dec
$x < -1$	$-$	$+$	$-$	decreasing
$-1 < x < 0$	$-$	$+$	$-$	decreasing
$x > 0$	$+$	$+$	$+$	increasing

Notice that the critical point $(0, 3)$ is a relative minimum and that the critical point $(-1, 4)$ is not a relative extremum.

15.   Since $f(x) = 3 - (x + 1)^3$, the derivative is

$$f'(x) = -3(x + 1)^2$$

which is zero when $x = -1$. The corresponding critical point is
$(-1, f(-1)) = (-1, 3)$.

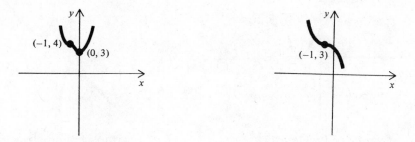

Graph for Problem 13.   Graph for Problem 15.

Interval	$-3(x + 1)^2$	$f'(x)$	Inc/dec
$x < -1$	$-$	$-$	decreasing
$x > -1$	$-$	$-$	decreasing

Notice that the critical point $(-1, 3)$ is not a relative extremum.

17.   Since $f(x) = (x^2 - 1)^4$, the derivative is

$$f'(x) = 4(x^2 - 1)^3(2x) = 8x(x^2 - 1)^3 = 8x(x + 1)^3(x - 1)^3$$

which is zero when $x = -1, x = 0$, and $x = 1$. The corresponding critical points are
$(-1, f(-1)) = (-1, 0), (0, f(0)) = (0, 1)$, and $(1, f(1)) = (1, 0)$.

Interval	$8x$	$(x + 1)^3$	$(x - 1)^3$	$f'(x)$	Inc/dec
$x < -1$	$-$	$-$	$-$	$-$	decreasing
$-1 < x < 0$	$-$	$+$	$-$	$+$	increasing
$0 < x < 1$	$+$	$+$	$-$	$-$	decreasing
$x > 1$	$+$	$+$	$+$	$+$	increasing

Notice that the critical point $(0, 1)$ is a relative maximum and the critical points
$(-1, 0)$ and $(1, 0)$ are relative minima.

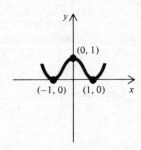

Graph for Problem 17.

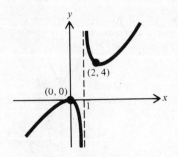

Graph for Problem 19.

19. Since $f(x) = \frac{x^2}{x-1}$, the derivative is

$$f'(x) = \frac{(x-1)(2x) - x^2(1)}{(x-1)^2} = \frac{x(x-2)}{(x-1)^2}$$

which is zero when $x = 0$ and $x = 2$. The corresponding critical points are $(0, f(0)) = (0, 0)$ and $(2, f(2)) = (2, 4)$. Note that the derivative is undefined when $x = 1$, which is not a critical point since $x = 1$ is not in the domain of f.

Interval	x	$x-2$	$(x-1)^2$	$f'(x)$	Inc/dec
$x < 0$	−	−	+	+	increasing
$0 < x < 1$	+	−	+	−	decreasing
$1 < x < 2$	+	−	+	−	decreasing
$x > 2$	+	+	+	+	increasing

Notice that the critical point $(0, 0)$ is a relative maximum and the critical point $(2, 4)$ is a relative minimum.

21. Since $f(x) = \frac{x^2 - 3x}{x+1}$, the derivative is

$$f'(x) = \frac{(x+1)(2x-3) - (x^2 - 3x)(1)}{(x+1)^2}$$

$$= \frac{x^2 + 2x - 3}{(x+1)^2} = \frac{(x+3)(x-1)}{(x+1)^2}$$

which is zero when $x = -3$ and $x = 1$. The corresponding critical points are $(-3, f(-3)) = (-3, -9)$ and $(1, f(1)) = (1, -1)$. Note that the derivative is undefined when $x = -1$, which is not a critical point since $x = -1$ is not in the domain of f.

Interval	$x+3$	$x-1$	$(x+1)^2$	$f'(x)$	Inc/dec
$x < -3$	−	−	+	+	increasing
$-3 < x < -1$	+	−	+	−	decreasing
$-1 < x < 1$	+	−	+	−	decreasing
$x > 1$	+	+	+	+	increasing

Notice that the critical point $(-3, -9)$ is a relative maximum and the critical point $(1, -1)$ is a relative minimum.

23. Since $f(x) = x + \frac{1}{x}$, the derivative is

$$f'(x) = 1 - \frac{1}{x^2} = \frac{x^2 - 1}{x^2} = \frac{(x + 1)(x - 1)}{x^2}$$

which is zero when $x = -1$ and $x = 1$. The corresponding critical points are $(-1, f(-1)) = (-1, -2)$ and $(1, f(1)) = (1, 2)$. Note that the derivative is undefined when $x = 0$, which is not a critical point since $x = 0$ is not in the domain of f.

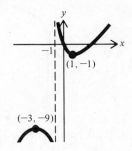

Graph for Problem 21.

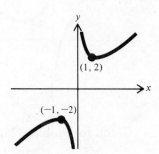

Graph for Problem 23.

Interval	$x + 1$	$x - 1$	x^2	$f'(x)$	Inc/dec
$x < -1$	−	−	+	+	increasing
$-1 < x < 0$	+	−	+	−	decreasing
$0 < x < 1$	+	−	+	−	decreasing
$x > 1$	+	+	+	+	increasing

Notice that the critical point $(-1, 2)$ is a relative maximum and the critical point $(1, 2)$ is a relative minimum.

25. Since $f(x) = 6x^2 + \frac{12,000}{x}$, the derivative is

$$f'(x) = 12x - \frac{12,000}{x^2} = \frac{12x^3 - 12,000}{x^2}$$

$$= \frac{12(x^3 - 1000)}{x^2} = \frac{12(x - 10)(x^2 + 10x + 100)}{x^2}$$

which is zero when $x = 10$. The corresponding critical point is $(10, f(10)) = (10, 1800)$. Note that the derivative is undefined when $x = 0$, which is not a critical point since $x = 0$ is not in the domain of f.

Interval	$(x - 10)$	$x^2 + 10x + 100$	$f'(x)$	Inc/dec
$x < 0$	−	+	−	decreasing
$0 < x < 10$	−	+	−	decreasing
$x > 10$	+	+	+	increasing

Notice that the critical point $(10, 1800)$ is a relative minimum.

27. Since $f(x) = x^{3/5}$, the derivative is

$$f'(x) = \frac{3}{5}x^{-2/5} = \frac{3}{5x^{2/5}}$$

which is never zero but is undefined when $x = 0$. The corresponding critical point is $(0, f(0)) = (0, 0)$.

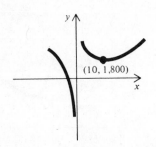

Graph for Problem 25.   Graph for Problem 27.

Interval	$5x^{2/5}$	$f'(x)$	Inc/dec
$x < 0$	+	+	increasing
$x > 0$	+	+	increasing

Notice that the critical point $(0, 0)$ is not a relative extremum.

29. Since $f(x) = x^3 - 2x^2 - 3x + 2$, the derivative is

$$f'(x) = 3x^2 - 4x - 3$$

which cannot be factored. By the quadratic formula, $f'(x) = 0$ when

$$x = \frac{-(-4) \pm \sqrt{(-4)^2 - 4(3)(-3)}}{2(3)}$$

$$= \frac{4 \pm \sqrt{52}}{6} = \frac{4 \pm 2\sqrt{13}}{6} = \frac{2 \pm \sqrt{13}}{3}$$

That is, $f'(x) = 0$ when

$$x_1 = \frac{2 + \sqrt{13}}{3} \simeq 1.87 \text{ and } x_2 = \frac{2 - \sqrt{13}}{3} \simeq -0.54$$

The corresponding critical points are $(1.87, f(1.87)) = (1.87, -4.06)$ and $(-0.54, f(-0.54)) = (-0.54, 2.88)$.

Interval	$x - x_1$	$x - x_2$	$f'(x)$	Inc/dec
$x < x_1$	−	−	+	increasing
$x_1 < x < x_2$	+	−	−	decreasing
$x > x_2$	+	+	+	increasing

Notice that the critical point $(-0.54, 2.88)$ is a relative maximum and the critical point $(1.87, -4.06)$ is a relative minimum.

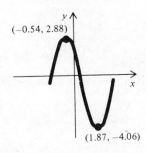

Graph for Problem 29.

31. (a) The fact that $f'(x) > 0$ when $x < -5$ and $x > 1$ means that the function $f(x)$ is increasing on these intervals.

    (b) The fact that $f'(x) < 0$ when $-5 < x < 1$ means that the function $f(x)$ is decreasing on this interval.

    (c) The facts that $f(-5) = 4$ and $f(1) = -1$ together with (a) and (b) imply that the function $f(x)$ has a relative maximum at $(-5, 4)$ and a relative minimum at $(1, -1)$.

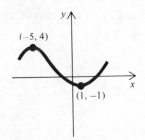

Graph for Problem 31.

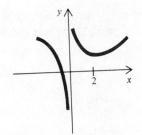

Graph for Problem 33.

33. (a) The fact that $f'(x) > 0$ when $x > 2$ means that the function $f(x)$ is increasing on this interval.

    (b) The fact that $f'(x) < 0$ when $x < 0$ and $0 < x < 2$ means that the function $f(x)$ is decreasing on these intervals. Facts (a) and (b) together imply that $f(x)$ has a relative minimum when $x = 2$.

    (c) The fact that $f(x)$ is undefined when $x = 0$ means that there is a break in the graph at $x = 0$.

35. The fact that $f(x) = ax^2 + bx + c$ crosses the y axis at $(0, 3)$ means $f(0) = 3$ or

$a(0)^2 + b(0) + c = 3$. Hence $c = 3$. The fact that $f(x) = ax^2 + bx + 3$ has a relative maximum at $(5, 12)$ implies that $f'(5) = 0$. Since $f'(x) = 2ax + b$, it follows that

$$0 = 2a(5) + b \text{ or } 10a + b = 0$$

Moreover, the fact that the point $(5, 12)$ is on the graph of f implies that $f(5) = 12$ or

$$12 = a(5)^2 + b(5) + 3 \text{ or } 25a + 5b = 9$$

From the first equation, $b = -10a$. Substitution in the second equation yields

$$25a + 5(-10a) = 9, \ 25a - 50a = 9, \ \text{or } a = -\frac{9}{25}$$

Finally, if $a = -\frac{9}{25}$,

$$b = -10\left(-\frac{9}{25}\right) = \frac{18}{5}$$

Hence the desired function is $f(x) = -\frac{9}{25}x^2 + \frac{18}{5}x + 3$.

37. Using the product rule to differentiate $f(x) = (x - p)(x - q)$ gives

$$f'(x) = (x - p)(1) + (x - q)(1) = 2x - p - q$$

The relative extremum of f is at the value of x for which $f'(x) = 0$, that is, for which

$$2x - p - q = 0, \ 2x = p + q, \ \text{or } x = \frac{p + q}{2}$$

which is the average value of the x intercepts p and q. Thus the relative extremum occurs midway between the x intercepts of f.

Chapter 3, Section 2

1. The graph is concave downward ($f'' < 0$) when $x < 2$ and concave upward ($f'' > 0$) when $x > 2$.

3. If $f(x) = \frac{1}{3}x^3 - 9x + 2$, the first derivative is

    $$f'(x) = x^2 - 9 = (x + 3)(x - 3)$$

    which is zero when $x = -3$ and $x = 3$. The corresponding first-order critical points are $(-3, 20)$ and $(3, -16)$. The second derivative is $f''(x) = 2x$, which is zero when $x = 0$, and the corresponding second-order critical point is $(0, 2)$.

Interval	$f'(x)$	$f''(x)$	Inc/dec	Concavity
$x < -3$	+	−	increasing	down
$-3 < x < 0$	−	−	decreasing	down
$0 < x < 3$	−	+	decreasing	up
$x > 3$	+	+	increasing	up

    Notice that the first-order critical point $(-3, 20)$ is a relative maximum, the first-order critical point $(3, -16)$ is a relative minimum, and the second-order critical point $(0, 2)$ is an inflection point.

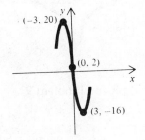

Graph for Problem 3.

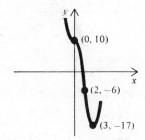

Graph for Problem 5.

5. If $f(x) = x^4 - 4x^3 + 10$, the first derivative is

    $$f'(x) = 4x^3 - 12x^2 = 4x^2(x - 3)$$

    which is zero when $x = 0$ and $x = 3$. The corresponding first-order critical points are $(0, 10)$ and $(3, -17)$. The second derivative is

    $$f''(x) = 12x^2 - 24x = 12x(x - 2)$$

    which is zero when $x = 0$ and $x = 2$. The corresponding second-order critical points are $(0, 10)$ and $(2, -6)$.

Interval	$f'(x)$	$f''(x)$	Inc/dec	Concavity
$x < 0$	−	+	decreasing	up
$0 < x < 2$	−	−	decreasing	down
$2 < x < 3$	−	+	decreasing	up
$x > 3$	+	+	increasing	up

Notice that the first-order critical point $(3, -17)$ is a relative minimum while the first-order critical point $(0, 10)$ is not a relative extremum and that both of the second-order critical points $(0, 10)$ and $(2, -6)$ are inflection points.

7.   If $f(x) = (x - 2)^3$, the first derivative is

$$f'(x) = 3(x - 2)^2$$

which is zero when $x = 2$, and the corresponding first-order critical point is $(2, 0)$. The second derivative is

$$f''(x) = 6(x - 2)$$

which is zero when $x = 2$, and the corresponding second-order critical point is $(2, 0)$.

Interval	$f'(x)$	$f''(x)$	Inc/dec	Concavity
$x < 2$	+	−	increasing	down
$x > 2$	+	+	increasing	up

Notice that the critical point is not a relative extremum but is an inflection point.

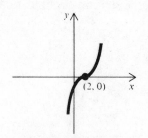

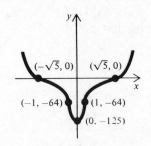

Graph for Problem 7.   Graph for Problem 9.

9.   If $f(x) = (x^2 - 5)^3$, the first derivative is

$$f'(x) = 3(x^2 - 5)^2(2x) = 6x(x^2 - 5)^2$$

which is zero when $x = 0, x = -\sqrt{5}$, and $x = \sqrt{5}$. The corresponding first-order critical points are $(0, -125), (-\sqrt{5}, 0)$, and $(\sqrt{5}, 0)$. The second derivative is

$$f''(x) = 6x[2(x^2 - 5)(2x)] + (x^2 - 5)^2(6) = 6(x^2 - 5)(5x^2 - 5)$$

$$= 30(x^2 - 5)(x + 1)(x - 1)$$

which is zero when $x = -\sqrt{5}$, $x = \sqrt{5}$, $x = -1$, and $x = 1$. The corresponding second-order critical points are $(-\sqrt{5}, 0)$, $(\sqrt{5}, 0)$, $(-1, -64)$, and $(1, -64)$.

Interval	$f'(x)$	$f''(x)$	Inc/dec	Concavity
$x < -\sqrt{5}$	−	+	decreasing	up
$-\sqrt{5} < x < -1$	−	−	decreasing	down
$-1 < x < 0$	−	+	decreasing	up
$0 < x < 1$	+	+	increasing	up
$1 < x < \sqrt{5}$	+	−	increasing	down
$x > \sqrt{5}$	+	+	increasing	up

Notice that the first-order critical point $(0, -125)$ is a relative minimum while the first-order critical points $(-\sqrt{5}, 0)$ and $(\sqrt{5}, 0)$ are not relative extrema but are inflection points along with $(-1, -64)$ and $(1, -64)$.

11.   If $f(x) = x + \frac{1}{x}$, th first derivative is

$$f'(x) = 1 - \frac{1}{x^2} = \frac{x^2 - 1}{x^2} = \frac{(x-1)(x+1)}{x^2}$$

which is zero when $x = 1$ and $x = -1$. The corresponding first-order critical points are $(1, 2)$ and $(-1, -2)$. The second derivative is

$$f''(x) = \frac{2}{x^3}$$

which is never zero. Note that f'' is undefined at $x = 0$, which is not a critical point since it is not in the domain of f.

Interval	$f'(x)$	$f''(x)$	Inc/dec	Concavity
$x < -1$	+	−	increasing	down
$-1 < x < 0$	−	−	decreasing	down
$0 < x < 1$	−	+	decreasing	up
$x > 1$	+	+	increasing	up

Notice that the first-order critical point $(-1, -2)$ is a relative maximum and the first-order critical point $(1, 2)$ is a relative minimum. Although $x = 0$ is not in the domain of the function, the graph changes concavity at $x = 0$.

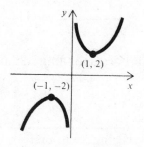

Graph for Problem 11.

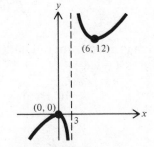

Graph for Problem 13.

13.   If $f(x) = \frac{x^2}{x-3}$, the first derivative is

$$f'(x) = \frac{(x-3)(2x) - x^2(1)}{(x-3)^2} = \frac{x^2 - 6x}{(x-3)^2} = \frac{x(x-6)}{(x-3)^2}$$

which is zero when $x = 0$ and $x = 6$. The corresponding first-order critical points are $(0, 0)$ and $(6, 12)$. Note that the derivative is undefined at $x = 3$, which is not a critical point since it is not in the domain of f. The second derivative is

$$f''(x) = \frac{(x-3)^2(2x-6) - (x^2 - 6x)[2(x-3)(1)]}{(x-3)^4}$$

$$= \frac{2(x-3)[(x-3)^2 - x(x-6)]}{(x-3)^4} = \frac{18}{(x-3)^3}$$

which is never zero. Note that f'' is undefined at $x = 3$, which is not a critical point because it is not in the domain of f.

Interval	$f'(x)$	$f''(x)$	Inc/dec	Concavity
$x < 0$	+	−	increasing	down
$0 < x < 3$	−	−	decreasing	down
$3 < x < 6$	−	+	decreasing	up
$x > 6$	+	+	increasing	up

Notice that the first-order critical point $(0,0)$ is a relative maximum and the first-order critical point $(6, 12)$ is a relative minimum. Although $x = 3$ is not in the domain of f, the graph changes concavity at $x = 3$.

15. If $f(x) = (x+1)^{1/3}$, the first derivative is

$$f'(x) = \frac{1}{3}(x+1)^{-2/3} = \frac{1}{3(x+1)^{2/3}}$$

which is never zero but is undefined at $x = -1$. The corresponding first-order critical point is $(-1, 0)$. The second derivative is

$$f''(x) = -\frac{2}{9}(x+1)^{-5/3} = \frac{-2}{9(x+1)^{5/3}}$$

which is never zero but undefined at $x = -1$. The corresponding second-order critical point is $(-1, 0)$.

Interval	$f'(x)$	$f''(x)$	Inc/dec	Concavity
$x < -1$	+	+	increasing	up
$x > -1$	+	−	increasing	down

Notice that the critical point $(-1, 0)$ is not a relative extremum but is an inflection point.

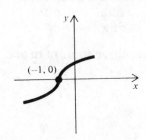

Graph for Problem 15.   Graph for Problem 17.

17. If $f(x) = (x+1)^{4/3}$, the first derivative is

$$f'(x) = \frac{4}{3}(x+1)^{1/3}$$

which is zero when $x = -1$, and the corresponding first-order critical point is $(-1, 0)$. The second derivative is

$$f''(x) = \frac{4}{9}(x+1)^{-2/3} = \frac{4}{9(x+1)^{2/3}}$$

which is never zero but is undefined at $x = -1$. The corresponding second-order critical point is $(-1, 0)$.

Interval	$f'(x)$	$f''(x)$	Inc/dec	Concavity
$x < -1$	$-$	$+$	decreasing	up
$x > -1$	$+$	$+$	increasing	up

Notice that the critical point $(-1, 0)$ is a relative minimum and that the graph does not change concavity.

19. If $f(x) = \sqrt{x^2 + 1} = (x^2 + 1)^{1/2}$, the first derivative is

$$f'(x) = \frac{1}{2}(x^2 + 1)^{-1/2}(2x) = x(x^2 + 1)^{-1/2} = \frac{x}{(x^2 + 1)^{1/2}}$$

which is zero when $x = 0$, and the corresponding first-order critical point is $(0, 1)$. The second derivative is

$$f''(x) = x\left[-\frac{1}{2}(x^2 + 1)^{-3/2}(2x)\right] + (x^2 + 1)^{-1/2}(1)$$

$$= \frac{-x^2}{(x^2 + 1)^{3/2}} + \frac{1}{(x^2 + 1)^{1/2}} = \frac{-x^2 + (x^2 + 1)}{(x^2 + 1)^{3/2}} = \frac{1}{(x^2 + 1)^{3/2}}$$

which is never zero or undefined.

Interval	$f'(x)$	$f''(x)$	Inc/dec	Concavity
$x < 0$	$-$	$+$	decreasing	up
$x > 0$	$+$	$+$	increasing	up

Notice that the first-order critical point $(0, 1)$ is a realtive minimum and that the graph never changes concavity.

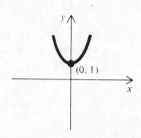

Graph for Problem 19.

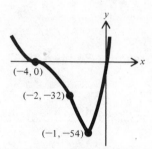

Graph for Problem 21.

21. If $f(x) = 2x(x + 4)^3$, the first derivative is

$$f'(x) = 2x[3(x + 4)^2(1)] + 2(x + 4)^3$$
$$= 2(x + 4)^2[3x + (x + 4)] = 8(x + 4)^2(x + 1)$$

which is zero when $x = -4$ and $x = -1$. The corresponding first-order critical points are $(-4, 0)$ and $(-1, -54)$. The second derivative is

$$f''(x) = 8[(x + 4)^2(1) + (x + 1)(2)(x + 4)]$$
$$= 8(x + 4)[(x + 4) + 2(x + 1)] = 24(x + 4)(x + 2)$$

which is zero when $x = -4$ and $x = -2$. The corresponding second-order critical points are $(-4, 0)$ and $(-2, -32)$.

Interval	$f'(x)$	$f''(x)$	Inc/dec	Concavity
$x < -4$	$-$	$+$	decreasing	up
$-4 < x < -2$	$-$	$-$	decreasing	down
$-2 < x < -1$	$-$	$+$	decreasing	up
$x > -1$	$+$	$+$	increasing	up

Notice that the first-order critical point $(-1, -54)$ is a relative minimum while the first- and second-order critical point $(-4, 0)$ is not a relative extremum but is an inflection point, as is the other second-order critical point $(-2, -32)$.

23. If $f(x) = \frac{2}{1+x^2} = 2(1 + x^2)^{-1}$, the first derivative is

$$f'(x) = -2(1 + x^2)^{-2}(2x) = \frac{-4x}{(1 + x^2)^2}$$

which is zero when $x = 0$. The corresponding first-order critical point is $(0, 2)$. The second derivative is

$$f''(x) = \frac{(1 + x^2)^2(-4) - (-4x)[2(1 + x^2)(2x)]}{(1 + x^2)^4}$$

$$= \frac{4(1 + x^2)[-(1 + x^2) + 4x^2]}{(1 + x^2)^4} = \frac{4(3x^2 - 1)}{(1 + x^2)^3}$$

which is zero when $x = \pm\frac{1}{\sqrt{3}}$. The corresponding second-order critical points are $(\frac{1}{\sqrt{3}}, \frac{3}{2})$ and $(-\frac{1}{\sqrt{3}}, \frac{3}{2})$.

Interval	$f'(x)$	$f''(x)$	Inc/dec	Concavity
$x < -\frac{1}{\sqrt{3}}$	+	+	increasing	up
$-\frac{1}{\sqrt{3}} < x < 0$	+	−	increasing	down
$0 < x < \frac{1}{\sqrt{3}}$	−	−	decreasing	down
$x > \frac{1}{\sqrt{3}}$	−	+	decreasing	up

Notice that the first-order critical point $(0, 2)$ is a relative maximum and that the second-order critical points are both inflection points. Notice also that the graph has no x intercepts since the original function is never zero.

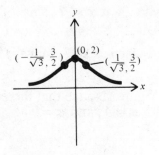

Graph for Problem 23.

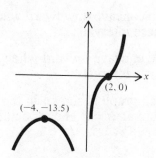

Graph for Problem 25.

25. If $f(x) = \frac{(x-2)^3}{x^2}$, then f has a discontinuity at $x = 0$. The first derivative is

$$f'(x) = \frac{x^2[3(x - 2)^2(x)] - (x - 2)^3(2x)}{x^4}$$

$$= \frac{x(x - 2)^2[3x - 2(x - 2)]}{x^4} = \frac{(x - 2)^2(x + 4)}{x^3}$$

which is zero when $x = 2$ and $x = -4$. The corresponding first-order critical points are $(2, 0)$ and $(-4, -13.5)$. The second derivative is

$$f''(x) = \frac{x^3[(x-2)^2(1) + (x+4)(2)(x-2)] - (x-2)^2(x+4)(3x^2)}{x^6}$$

$$= \frac{x^2(x-2)[x(x-2) + 2x(x+4) - 3(x-2)(x+4)]}{x^6}$$

$$= \frac{(x-2)[x^2 - 2x + 2x^2 + 8x - 3x^2 - 6x + 24]}{x^4} = \frac{24(x-2)}{x^4}$$

which is zero when $x = 2$. The corresponding second-order critical point is $(2, 0)$.

Interval	$f'(x)$	$f''(x)$	Inc/dec	Concavity
$x < -4$	+	−	increasing	down
$-4 < x < 0$	−	−	decreasing	down
$0 < x < 2$	+	−	increasing	down
$x > 2$	+	+	increasing	up

Notice that the first-order critical point $(-4, -13.5)$ is a relative maximum while the first- and second-order critical point $(2, 0)$ is not a relative extremum but is an inflection point.

27. There are many graphs that will have the given properties, but all should:

(a) Be increasing when $x < 2$ and when $2 < x < 5$, since $f'(x) > 0$ on these intervals;

(b) Be decreasing when $x > 5$, since $f'(x) < 0$ on this interval;

(c) Have a horizontal tangent at $x = 2$, since $f'(2) = 0$;

(d) Be concave downward when $x < 2$ and when $4 < x < 7$, since $f''(x) < 0$ on these intervals;

(e) Be concave upward when $2 < x < 4$ and when $x > 7$, since $f''(x) > 0$ on these intervals.

One such graph is shown in the accompanying figure. Notice that there is a relative maximum at $x = 5$ and inflection points at $x = 2$, $x = 4$, and $x = 7$.

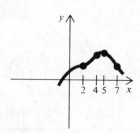

Graph for Problem 27.

29. (a) Since $f'(x) = x^2 - 4x = x(x - 4)$, the first-derivative is zero when $x = 0$ and

$x = 4$. Moreover, $f'(x) > 0$ (and f is increasing) when $x < 0$ and when $x > 4$, and $f'(x) < 0$ (and f is decreasing) when $0 < x < 4$.

(b)   The second derivative is $f''(x) = 2x - 4 = 2(x - 2)$, which is zero when $x = 2$. Moreover, $f''(x) < 0$ (and f is concave down) when $x < 2$, and $f''(x) > 0$ (and f is concave up) when $x > 2$.

(c)   From (a), f has a relative maximum when $x = 0$ and a relative minimum when $x = 4$, and from (b), f has an inflection point when $x = 2$.

31.   From the graph of $f'(x)$, $f'(x) < 0$ (and f is decreasing) when $x < 2$ and $f'(x) > 0$ (and f is increasing) when $x > 2$. Hence f must have a relative minimum at $x = 2$. The graph also shows that $f'(x)$ is increasing for all x, which implies that $f''(x) > 0$ and $f(x)$ is concave up for all x.

33.   From the graph of $f'(x)$, $f'(x) \leq 0$ (and f is decreasing) when $x < 2$ and $f'(x) > 0$ (and f is increasing) when $x > 2$. Hence f must have a relative minimum at $x = 2$. The graph also shows that $f'(x)$ is increasing when $x < -3$ and $x > -1$, which implies that $f''(x) > 0$ and $f(x)$ is concave up on these intervals. The graph also shows that $f'(x)$ is decreasing when $-3 < x < -1$, which implies that $f''(x) < 0$ and $f(x)$ is concave down on this interval. It follows that f must have inflection points at $x = -3$ and $x = -1$.

35.   If $f(x) = x^3 + 3x^2 + 1$, the first derivative is
$$f'(x) = 3x^2 + 6x = 3x(x + 2)$$
which is zero when $x = 0$ and $x = -2$. The corresponding first-order critical points are $(0, 1)$ and $(-2, 5)$. The second derivative is
$$f''(x) = 6x + 6 = 6(x + 1)$$
Since $f''(0) = 6 > 0, (0, 1)$ is a relative minimum, and since $f''(-2) = -6 < 0, (-2, 5)$ is a relative maximum.

37.   If $f(x) = (x^2 - 9)^2$, the first derivative is
$$f'(x) = 2(x^2 - 9)(2x) = 4x(x + 3)(x - 3)$$
which is zero when $x = 0, x = -3$, and $x = 3$. The corresponding first-order critical points are $(0, 81), (-3, 0)$, and $(3, 0)$. The first derivative can be rewritten as $f'(x) = 4x^3 - 36x$, and so the second derivative is
$$f''(x) = 12x^2 - 36 = 12(x^2 - 3)$$
Since $f''(0) = -36 < 0, (0, 81)$ is a relative maximum. Since $f''(-3) = 72 > 0, (-3, 0)$ is a relative minimum. And, since $f''(3) = 72 > 0, (3, 0)$ is a relative minimum.

39.   If $f(x) = 2x + 1 + \frac{18}{x}$, the first derivative is
$$f'(x) = 2 - \frac{18}{x^2} = \frac{2x^2 - 18}{x^2} = \frac{2(x + 3)(x - 3)}{x^2}$$
which is equal to zero when $x = -3$ and $x = 3$. The corresponding first-order critical points are $(-3, -11)$ and $(3, 13)$. The second derivative is $f''(x) = \frac{36}{x^3}$. Since $f''(-3) = -\frac{4}{3} < 0, (-3, -11)$ is a relative maximum. Since $f''(3) = \frac{4}{3} > 0$, $(3, 13)$ is a relative minimum.

Chapter 3, Section 3

1. Since $f(x) = x^2 + 4x + 5$ on $-3 \le x \le 1$, the derivative is

    $$f'(x) = 2x + 4 = 2(x + 2)$$

    which is zero when $x = -2$, which is in the interval. Comparing the function values at the critical point and endpoints

    $$f(-2) = 1, f(-3) = 2, \text{ and } f(1) = 10$$

    shows that the absolute maximum is $f(1) = 10$ and the absolute minimum is $f(-2) = 1$.

3. Since $f(x) = \frac{1}{3}x^3 - 9x + 2$ on $0 \le x \le 2$, the derivative is

    $$f'(x) = x^2 - 9 = (x + 3)(x - 3)$$

    which is zero when $x = -3$ and $x = 3$, neither of which is in the interval. Hence computing the function values at the endpoints

    $$f(0) = 2 \text{ and } f(2) = -\frac{40}{3}$$

    shows that the absolute maximum is $f(0) = 2$ and the absolute minimum is $f(2) = -\frac{40}{3}$.

5. Since $f(x) = 3x^5 - 5x^3$ on $-2 \le x \le 0$, the derivative is

    $$f'(x) = 15x^4 - 15x^2 = 15x^2(x^2 - 1) = 15x^2(x + 1)(x - 1)$$

    which is zero when $x = 0, x = -1$, and $x = 1$, of which only $x = 0$ and $x = -1$ are in the interval. Comparing the function values at the critical points and endpoints

    $$f(0) = 0, \ f(-1) = 2, \text{ and } f(-2) = -56$$

    shows that the absolute maximum is $f(-1) = 2$, and the absolute minimum is $f(-2) = -56$.

7. Since $f(x) = (x^2 - 4)^5$ on $-3 \le x \le 2$, the derivative is

    $$f'(x) = 5(x^2 - 4)^4(2x) = 10x(x + 2)(x - 2)$$

    which is zero when $x = 0, x = -2$, and $x = 2$, all of which are in the interval. Comparing the function values at the critical points and endpoints

    $$f(0) = -1,024, \ f(-2) = 0, \ f(2) = 0, \text{ and } f(-3) = 3,125$$

    shows that the absolute maximum is $f(-3) = 3,125$ and the absolute minimum is $f(0) = -1,024$.

9.   Since $f(x) = x + \frac{1}{x}$ on $\frac{1}{2} \le x \le 3$, the derivative is

$$f'(x) = 1 - \frac{1}{x^2} = \frac{x^2 - 1}{x^2} = \frac{(x+1)(x-1)}{x^2}$$

which is zero when $x = -1$ and $x = 1$, of which only $x = 1$ is in the interval. Note that $f'(x)$ is undefined when $x = 0$, which is not in the domain of f and hence is not a critical point. Comparing the function values at the critical points and endponts

$$f(1) = 2, \; f\left(\frac{1}{2}\right) = \frac{5}{2}, \text{ and } f(3) = \frac{10}{3}$$

shows that the absolute maximum is $f(3) = \frac{10}{3}$ and the absolute minimum is $f(1) = 2$.

11.   Since $f(x) = x + \frac{1}{x}$ on $x > 0$, the derivative is

$$f'(x) = 1 - \frac{1}{x^2} = \frac{x^2 - 1}{x^2} = \frac{(x+1)(x-1)}{x^2}$$

which is zero when $x = -1$ and $x = 1$, of which only $x = 1$ is in the interval. Note that $f'(x)$ is undefined when $x = 0$, which is not in the domain of f and hence is not a critical point. The function value at the only critical point in the interval is $f(1) = 2$, and there are no endpoints. Since $f'(x) < 0$ when $0 < x < 1$ (i.e., f is decreasing) and $f'(x) > 0$ when $x > 1$ (i.e., f is increasing), it follows that there is no absolute maximum and that the absoute minimum is $f(1) = 2$.

13.   Since $f(x) = \frac{1}{x}$ on $x > 0$, the derivative is $f'(x) = \frac{-1}{x^2}$. There are no endpoints and no critical points. Hence there is no absoute maximum and no absolute minimum.

15.   Since $f(x) = \frac{1}{x+1} = (x+1)^{-1}$ on $x \ge 0$, the derivative is

$$f'(x) = -(x+1)^{-2} = \frac{-1}{(x+1)^2}$$

which is less than zero for all $x \ge 0$. Hence the graph of f begins at $f(0) = 1$ and decreases for all $x > 0$. Thus the absolute maximum is $f(0) = 1$ and there is no absolute minimum.

17.   (a)   The membership of the association x years after 1975 is given by the function $f(x) = 100(2x^3 - 45x^2 + 264x)$. The period of time between 1975 and 1989 corresponds to the interval $0 \le x \le 14$. The derivative is

$$f'(x) = 100(6x^2 - 90x + 264)$$
$$= 600(x^2 - 15x + 44) = 600(x - 4)(x - 11)$$

which is zero when $x = 4$ and $x = 11$. Comparing the function values at the critical points and endpoints

$$f(11) = 12,100, \; f(4) = 46,400, \; f(0) = 0, \text{ and } f(14) = 36,400$$

shows that the absolute maximum on the interval is $f(4) = 46,400$. Hence the membership was greatest in 1979, four years after the founding of the association, when there were $46,400$ members.

(b)   The period of time between 1976 and 1989 corresponds to the interval $1 \leq x \leq 14$. Comparing the function values at the critical points and the endpoints

$$f(1) = 22,100, \ f(14) = 36,400, \ f(4) = 46,400, \text{ and } f(11) = 12,100$$

shows that the membership was smallest in 1986, eleven years after the founding of the association, when there were $12,100$ members.

19.   Let $P(x)$ denote the profit if the price is x dollars per radio. Then

$$P(x) = \text{(number of radios sold) (profit per radio)}$$
$$= (20 - x)(x - 5) = -100 + 25x - x^2 \text{ for } 0 \leq x \leq 20$$

The derivative is $P'(x) = 25 - 2x$, which is zero when $x = 12.5$. Comparing the function values at the critical point and the endpoints

$$P(12.5) = 56.25, \ P(0) = 0 \text{ and } P(20) = 0$$

shows that the profit is greatest when $x = 12.5$, that is, when the price is \$12.50 per radio.

21.   Let S denote the speed of the blood, R the radius of the artery, and r the distance from the central axis. Poiseuille's law states that $S(r) = c(R^2 - r^2)$, where c is a positive constant. The relevant interval is $0 \leq r \leq R$. The derivative is $S'(r) = -2cr$ which is zero when $r = 0$ (the left-hand endpoint of the interval). Comparing the function values at the endpoints

$$S(0) = cR^2 \text{ and } S(R) = 0$$

shows that the speed of the blood is greatest when $r = 0$, that is, at the central axis.

23.   Let R denote the rate at which the population changes, P the current population size, and B the upper bound for the population. Then,

$$R = kP(B - P) = kPB - kP^2$$

where k is a positive constant of proportionality. The derivative is

$$\frac{dR}{dP} = kB - 2kP$$

which is zero when $P = \frac{B}{2}$. Moreover, the derivative is positive (and the function is increasing) when $P < \frac{B}{2}$ and the derivative is negative (and the function is decreasing) when $P > \frac{B}{2}$. Hence R has an absolute maximum at the critical point $P = \frac{B}{2}$. That is, the rate of change is greatest when the population size P is $\frac{1}{2}B$ or 50 percent of its uppper bound.

25. Let s denote the speed of the truck. Then the wages are $\frac{k_1}{s}$, where k_1 is a positive constant, and the amount of gasoline used is $k_2 s$, where k_2, is another positive constant. The cost of gasoline is therefore $k_2 ps$, where p is the fixed price per gallon for gasoline. Thus the total cost $C(s)$ is given by

$$C(s) = \frac{k_1}{s} + k_2 ps$$

Setting the derivative equal to zero yields

$$C'(s) = -\frac{k_1}{s^2} + k_2 p = 0, \quad \frac{k_1}{s^2} = k_2 p, \quad \text{or} \quad \frac{k_1}{s} = k_2 ps$$

which says that wages $\left(\frac{k_1}{s}\right)$ are equal to the cost of gasoline $(k_2 ps)$.

27. (a) Since $C(q) = 3q^2 + q + 48$ is the total cost of producing q units, the average cost per unit is

$$A(q) = \frac{C(q)}{q} = \frac{3q^2 + q + 48}{q} = 3q + 1 + \frac{48}{q} \text{ for } q > 0$$

  (b) The derivative of the average cost is

$$A'(q) = 3 - \frac{48}{q^2} = \frac{3q^2 - 48}{q^2}$$

$$= \frac{3(q^2 - 16)}{q^2} = \frac{3(q+4)(q-4)}{q^2}$$

which is zero when $q = -4$ and $q = 4$, of which only $q = 4$ is in the interval $q > 0$. Moreover, $A'(q)$ is negative (and A is decreasing) for $0 < q < 4$ and $A'(q)$ is positive (and A is increasing) for $q > 4$. Hence the average cost is smallest when $q = 4$ units are manufactured.

  (c) The marginal cost is $C'(q) = 6q + 1$ and marginal cost equals average cost when

$$6q + 1 = 3q + 1 + \frac{48}{q}, \quad 3q - \frac{48}{q} = 0, \quad q^2 = 16, \quad \text{or } q = 4$$

that is, when 4 units are manufactured.

  (d) The graphs are shown below. TC stands for total cost, AC for average cost, and MC for marginal cost.

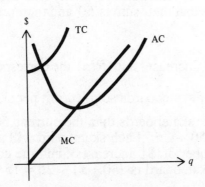

Total cost, average cost, and marginal cost for Problem 27.

Chapter 3, Section 4

1. Let x denote the number that exceeds its square, x^2, by the largest amount. Then

    $$f(x) = x - x^2$$

    is the function to be maximized. There is no restriction on the domain of $f(x)$.

    To find the critical points, set the deriative equal to zero and solve for x getting

    $$f'(x) = 1 - 2x = 0 \ \text{ or } \ x = \frac{1}{2}.$$

    The second derivative is
    $$f''(x) = -2 < 0$$

    so that $\left(\frac{1}{2}, f\left(\frac{1}{2}\right)\right) = \left(\frac{1}{2}, \frac{1}{4}\right)$ is a relative maximum. Since the graph of $f(x)$ is a parabola opening down, the relative maximum is an absolute maximum. Thus, $x = \frac{1}{2}$ is the number that maximizes $f(x) = x - x^2$.

3. Let x and y denote the two positive numbers. Then xy is their product and $x + y$ is their sum. The goal is to maximize $f = xy$, which is a function of two variables. Since

    $$x + y = 50 \ \text{ or } \ y = 50 - x,$$

    substituting for y in the formula for f yields

    $$f(x) = x(50 - x) = 50x - x^2,$$

    a function of one variable.

    To find the critical points, set the derivative equal to zero and solve for x getting

    $$f'(x) = 50 - 2x = 0 \ \text{ or } \ x = 25.$$

    The second derivative is
    $$f''(x) = -2 < 0$$

    so that $(25, f(25) = (25, 625)$ is a relative maximum. Since the graph of $f(x)$ is a parabola opening down, the relative maximum is an absolute maximum. Thus, $x = 25$ and $y = 25$ have the properties that their sum is 50 and their product, 625, is as large as possible.

5. Let x denote the number of $1 increases and $P(x)$ the corresponding profit function. Then
    $$P(x) = (\text{number of skateboards sold}) \ (\text{profit per skateboard}).$$

    For each $1 increase, 3 fewer skateboards than the current 50 will be sold, and so the total number sold will be $50 - 3x$. Each skateboard will sell for $40 + x$ dollars (the current price plus the number of $1 increases), and the cost of each skateboard is $25. Hence the profit per skateboard is $(40 + x) - 25 = 15 + x$ dollars. Putting it all together,
    $$P(x) = (50 - 3x)(15 + x) = 750 + 5x - 3x^2.$$

The relevant interval is $0 \leq x \leq 16$. If there are more than 16 reductions the total number sold will be a negative number. To find the critical points, set the derivative equal to zero and solve for x getting

$$P'(x) = 5 - 6x = 0 \text{ or } x = 5/6.$$

The second derivative is

$$P''(x) = -6 < 0$$

and so a relative maximum is achieved when $x = 5/6$. Since the graph of $P(x)$ is a parabola opening down, the relative maximum is an absolute maximum.

Since x needs to be a nonnegative integer, the profit is computed at $x = 0$ and $x = 1$;

$$P(0) = 750 \text{ and } P(1) = 752.$$

Also computing $P(x)$ at the right hand endpoint,

$$P(16) = 750 + 5(16) - (16)^2 = 574.$$

The greatest profit is $752 which is generated when there is only one $1 increase, that is, the skateboard sells for $41,

7. Let x denote the number of $1 reduction in the price and $P(x)$ the corresponding profit function. Then,

$$P(x) = (\text{number of books sold}) \, (\text{profit per book})$$

For each $1 reduction in price, 20 more books than the current 200 will be sold, and so the total number sold will be $200 + 20x$. Each book will sell for $15 - x$ dollars (the current price minus the number of $1 reductions), and the cost of each book is $3. Hence the profit per book is $(15 - x) - 3 = 12 - x$ dollars. Putting it all together

$$P(x) = (200 + 20x)(12 - x) = 20(10 + x)(12 - x) = 20(120 + 2x - x^2)$$

The relevant interval is $0 \leq x \leq 12$. To find the critical points, set the derivative equal to zero and solve for x getting

$$P'(x) = 20(2 - 2x) = 40(1 - x) = 0 \text{ or } x = 1$$

Comparing the function values at the critical point and endpoints

$$P(1) = 2,420, \ P(0) = 2,400 \text{ and } P(12) = 0$$

shows that the greatest possible profit is $2,420, which is generated when there is only one $1 reduction, that is, when the book sells for $14.

9. Let x denote the number of additional trees to be planted and $N(x)$ the corresponding yield. Then,

$$N(x) = (\text{number of oranges per tree}) \, (\text{number of trees})$$

Since there are 60 trees to begin with and x additional trees are planted, the total number of trees is $60 + x$. For each additional tree, the average yield of 400 oranges per tree is decreased by 4. Thus, for x additional trees, the average yield per tree is $400 - 4x$. Putting it all together,

$$N(x) = (400 - 4x)(60 + x) = 4(100 - x)(60 + x) = 4(6,000 + 40x - x^2)$$

The relevant interval is $0 \leq x \leq 100$. To find the critical points, set the derivative equal to zero and solve for x getting

$$N'(x) = 4(40 - 2x) = 8(20 - x) = 0 \text{ or } x = 20$$

Comparing the function values at the critical point and endpoints

$$N(20) = 25,600, \ N(0) = 2,400, \text{ and } N(100) = 0$$

shows that the greatest possible yield is $25,600$ oranges, which is generated by planting 20 additional trees, that is, when 80 trees are planted.

11. If x is the number of additional days and $R(x)$ the corresponding revenue,

$$R(x) = (\text{number of pounds collected})(\text{price per pound})$$

Over of period of 80 days, $24,000$ pounds have been collected at a rate of 300 pounds per day, and so for each day over 80, an additional 300 pounds will be collected. Thus, the total number of pounds collected and sold is $24,000 + 300x$. Currently, the recycling center pays 1 cent per pound. For each additional day, it reduces the price it pays by 1 cent per 100 pounds, that is, by $\frac{1}{100}$ cents per pound. Hence, after x additional days, the price per pound will be $1 - \frac{x}{100}$ cents. Putting it all together,

$$R(x) = (24,000 + 300x)\left[1 - \frac{x}{100}\right] = 24,000 + 60x - 3x^2$$

The relevant interval is $0 \leq x \leq 100$. To find the critical points, set the derivative equal to zero and solve for x getting

$$R'(x) = 60 - 6x = 0 \text{ or } x = 10$$

Comparing the function at the critical point and endpoints

$$R(10) = 24,300, \ R(0) = 24,000, \text{ and } R(100) = 0$$

shows that the most profitable time to conclude the project will be 10 days from now.

13. Let x denote the dimensions of the rectangle, A the area, and P the (fixed) perimeter. The goal is to maximize the area $A = xy$, which is a function of two variables. The fact that the perimeter is to be P gives

$$2x + 2y = P \text{ or } y = \frac{P}{2} - x$$

Substuting for y in the formula for A yields

$$A(x) = x\left[\frac{P}{2} - x\right] = \frac{P}{2}x - x^2 \text{ (where P is a constant)}$$

The relevant interval is $x > 0$. To find the critical points, set the derivative equal to zero and solve for x getting

$$A'(x) = \frac{P}{2} - 2x = 0 \text{ or } x = \frac{P}{4}$$

Since this is the only critical point in the interval, apply the second-derivative test for absolute extrema. The second derivative is $A''(x) = -2$, which is negative. Hence the area has an absolute maximum when $x = \frac{P}{4}$. The corresponding value of y is $y = \frac{P}{2} - \frac{P}{4} = \frac{P}{4}$. Thus, the area is greatest when the rectangle is a square.

15. Let x and y denote the dimensions of the rectangle, A the (fixed) area, and P the perimeter. The goal is to minimize the perimeter $P = 2x + 2y$, which is a function of two variables. The fact that the area is to be A gives

$$A = xy \text{ or } y = \frac{A}{x}$$

Substituting for y in the formula for P gives

$$P(x) = 2x + \frac{2A}{x} \qquad \text{(where A is a positive constant)}$$

The relevant interval is $x > 0$. To find the critical points, set the derivative equal to zero and solve for x getting

$$P'(x) = 2 - \frac{2A}{x^2} = 0 \qquad x^2 = A \text{ or } x = \pm\sqrt{A}$$

Only the positive root is in the interval $x > 0$, and this is the only critical point in the interval. Since the second derivative $P''(x) = \frac{4A}{x^3}$ is positive for $x > 0$, it follows that the perimeter is minimal when $x = \sqrt{A}$. The corresponding value of y is $y = \frac{A}{x} = \frac{A}{\sqrt{A}} = \sqrt{A}$, making the rectangle a square.

17. Label the sides of the open box as indicated and let V denote the volume. The goal is to maximize $V = x^2 y$, which is a function of two variables.

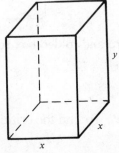

Open box for Problem 17.

The cost of the sides is \$3 per square meter and since the area of the four sides is $4xy$, the total cost of the sides is $3(4xy) = 12xy$ dollars. The cost of the bottom is

$4 per square meter, and since the area of the bottom is x^2, the cost of the bottom is $4x^2$ dollars. The fact that the total cost is to be $48 implies

$$12xy + 4x^2 = 48, \text{ or } y = \frac{12 - x^2}{3x} = \frac{4}{x} - \frac{x}{3}$$

Substituting for y in the formula for V yields

$$V(x) = x^2 \left[\frac{4}{x} - \frac{x}{3} \right] = 4x - \frac{x^3}{3}$$

The relevant interval is $x > 0$. To find the crical points, set the derivative equal to zero and solve for x getting

$$V'(x) = 4 - x^2 = 0 \text{ or } x = 2$$

Since this is the only critical point in the interval, apply the second-derivative test for absolute extrema. Since $V''(x) = -2x$ is negative when $x > 0$, it follows that V has an absolute maximum when $x = 2$. The corresponding value of y is $y = \frac{4}{2} - \frac{2}{3} = \frac{4}{3}$. Hence, the box of greatest volume is 2 meters by 2 meters by $\frac{4}{3}$ meters.

19. Label the square piece of cardboard as indicated in the figure and let V denote the volume of the box. Then,

$$V(x) = (18 - 2x)^2(x) = 324x - 72x^2 + 4x^3$$

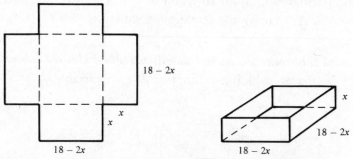

Square piece of cardboard and folded box for Problem 19.

The relevant interval is $0 \le x \le 9$. To find the critical points, set the derivative equal to zero and solve for x getting

$$V'(x) = 324 - 144x + 12x^2 = 12(27 - 12x + x^2) = 12(x - 9)(x - 3) = 0$$

or $x = 9$ and $x = 3$. Comparing the function values at the critical points and endpoints

$$V(0) = 0, \ V(3) = 432, \text{ and } V(9) = 0$$

shows that the volume will be greatest when $x = 3$. When $x = 3$, the sides of the base of the box are $18 - 2(3) = 12$ inches long. Hence the dimensions of the box of greatest volume are 12 inches by 12 inches by 3 inches.

21. Let x be the distance indicated in the accompanying figure.

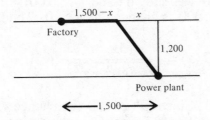

Figure for Problem 21.

The goal is to minimize the total cost

$$C = \text{cost under water} + \text{cost over land}$$

Since the distance over land is $1,500 - x$ and the distance under water is $\sqrt{x^2 + (1,200)^2}$ (by the pythagorean theorem), the total cost is

$$C(x) = 25\sqrt{x^2 + (1,200)^2} + 20(1,500 - x)$$

The relevant interval is $0 \leq x \leq 1,500$. The derivative is

$$C'(x) = 25\left(\frac{1}{2}\right)\left[x^2 + (1,200)^2\right]^{-1/2}(2x) - 20$$

$$= \frac{25x}{\sqrt{x^2 + (1,200)^2}} - 20$$

which is zero when

$$25x = 20\sqrt{x^2 + (1,200)^2} \text{ or } 5x = 4\sqrt{x^2 + (1,200)^2}$$

$$25x^2 = 16\left[x^2 + (1,200)^2\right]$$

$$9x^2 = 16(1,200)^2 \quad x^2 = \frac{15(1,200)^2}{9} \text{ or } x = \pm\frac{4,(1,200)}{3} = \pm 1,600$$

neither of which is in the interval $0 \leq x \leq 1,500$. Hence, the absolute minimum must occur at an endpoint. Comparing

$$C(0) = 25(1,200) + 20(1,500) = 60,000$$

and

$$C(1,500) = 25\sqrt{(1,500)^2 + (1,200)^2} + 0 \simeq 48,023$$

shows that the minimal cost is $48,023$ and will occur if $x = 1,500$, that is, if the cable is run entirely under water.

23. In Example 4.3, the minimum value of the cost function

$$C(x) = 5\sqrt{(900)^2 + x^2} + 4(3,000 - x)$$

on the interval $0 \leq x \leq 3,000$ was shown to be $C(1,200) = 14,700$. Suppose now that the restriction $0 \leq x \leq 3,000$ is changed to $x > 0$. Then, as in Example 4.3,

$$C'(x) = \frac{5x}{\sqrt{(900)^2 + x^2}} - 4$$

which is zero when $x = 1,200$. Since $C'(x) < 0$ (and hence C is decreasing) when $0 < x < 1,200$ and $C'(x) > 0$ (and hence C is increasing) when $x > 1,200$, it follows that C has an absolute minimum on the inteval $x > 0$ when $x = 1,200$. Thus, no matter how far downstream (beyond the critical point at $1,200$ meters) the factory is located, the most economical location of the cable on the opposite bank is $1,200$ meters downstream from the power plant.

25. Label the sides of the rectangular poster as indicated in the figure and let A denote the area of the poster. The goal is to minimize

$$A = (x + 4)(y + 8)$$

which is a function of two variables.

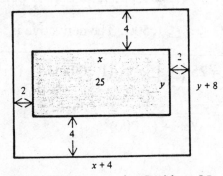

Rectangular poster for Problem 25.

The fact that the printed area is to be 25 square centimeters implies

$$xy = 25 \quad \text{or} \quad y = \frac{25}{x}$$

Substituting for y in the equation for A yields

$$A(x) = (x + 4)\left(\frac{25}{x} + 8\right) = 8x + \frac{100}{x} + 57$$

The relevant interval is $x > 0$. To find the critical points, set the derivative equal to zero and solve for x getting

$$A'(x) = 8 - \frac{100}{x^2} = \frac{8x^2 - 100}{x^2} = 0$$

$$8x^2 - 100 = 0, \ x^2 = \frac{100}{8} = \frac{25}{2}, \ \text{or} \ x = \sqrt{\frac{25}{2}} = \frac{5\sqrt{2}}{2}$$

Since this is the only critcal point in the interval, apply the second-derivative test for absolute extrema. The second derivative is $A''(x) = \frac{200}{x^3}$, which is positive when $x > 0$. Hence $A(x)$ has an absolute minimum at $x = \frac{5\sqrt{2}}{2}$. When $x = \frac{5\sqrt{2}}{2}, y = 25 \left[\frac{1}{x} \right] = 25 \left[\frac{2}{5\sqrt{2}} \right] = 5\sqrt{2}$. Hence the area of the poster will be minimal when one side is $4 + \frac{5\sqrt{2}}{2}$ centimeters long and the other is $8 + 5\sqrt{2}$ centimeters long.

27. Let r be the radius of the can and h its height. Let k denote the cost per square inch of constructing the cardboard side. Then $2k$ is the cost per square inch of constructing the top and bottom. The area of the side is $2\pi rh$, and so the cost of constructing the side is $k(2\pi rh)$. The combined area of the top and bottom is $2\pi r^2$, and so the cost of constructing that part of the can is $2k(2\pi r^2)$. The total construction cost is therefore

$$C = 2k\pi rh + 4k\pi r^2$$

which is a function of the two variables r and h. The fact that the volume is to be 4π cubic inches implies

$$4\pi = \pi r^2 h \ \text{or} \ h = \frac{4}{r^2}$$

Substituting for h in the formula for C yields

$$C(r) = 2k\pi r \left[\frac{4}{r^2} \right] + 4k\pi r^2 = \frac{8k\pi}{r} + 4k\pi r^2$$

The relevant interval is $r > 0$. To find the critical points, set the derivative equal to zero and solve for r getting

$$C'(r) = -\frac{8k\pi}{r^2} + 8k\pi r = 8k\pi \left[r - \frac{1}{r^2} \right] = 0$$

$$r - \frac{1}{r^2} = 0, \ r^3 - 1 = 0, \ \text{or} \ r = 1$$

Since this is the only critical point in the interval, apply the second-derivative test for absolute extrema. The second derivative is $C''(r) = \frac{16k r}{r^3} + 8k\pi$, which is positive when $r > 0$ (since k is positive). Hence, C has an absolute minimum when $r = 1$. When $r = 1, h = \frac{4}{(1)^2} = 4$. Hence the least expensive can has a radius of 1 inch and a height of 4 inches.

29. Let r denote the radius, h the height, A the (fixed) area, and V the volume. The goal is to maximize the volume $V = \pi r^2 h$, which is a function of the two variables r and h. To write h in terms of r, use the fact that the area is to be A. That is,

$$A = \text{area of top} + \text{area of bottom} + \text{area of side}$$
$$= \pi r^2 + \pi r^2 + 2\pi r h = 2\pi r^2 + 2\pi r h$$

and solve for h to get

$$2\pi r h = A - 2\pi r^2 \ \text{ or } \ h = \frac{A - 2\pi r^2}{2\pi r}$$

(where A is a constant). Now substitute for h in the volume formula and simplify to get

$$V(r) = \pi r^2 \left[\frac{A - 2\pi r^2}{2\pi r}\right] = \left[\frac{A}{2}\right] r - \pi r^3$$

The relevant interval is $r > 0$. The derivative is

$$V'(r) = \frac{A}{2} - 3\pi r^2 \quad \text{(Remember that A is a constant.)}$$

which is zero when

$$\frac{A}{2} = 3\pi r^2 \ \text{ or } \ A = 6\pi r^2$$

To get h back in the picture, replace A by $2\pi r^2 + 2\pi r h$ and simplify as follows:

$$2\pi r^2 + 2\pi r h = 6\pi r^2 \quad 2\pi r h = 4\pi r^2 \ \text{ or } \ h = 2r$$

Since $V(r)$ has only one critical point in the interval $r > 0$ and since the second derivative is $V''(r) = -6\pi r$ which is negative, it follows that the volume will be greatest at this critical point, that is, when the height is twice the radius.

31. (a) Let x denote the number of machines used and $C(x)$ the corresponding total cost. Then,

$$C(x) = \text{set–up cost} + \text{operating cost}$$
$$= 20\,(\text{number of machines}) + 4.80\,(\text{number of hours})$$

Since each machine produces 30 kickboards per hour, x machines produce $30x$ kickboards per hours and the number of hours required to produce $8,000$ kickboards is $\frac{8,000}{30x}$. Hence,

$$C(x) = 20x + 4.8 \left[\frac{8,000}{30x}\right] = 20x + \frac{1,280}{x}$$

Since the firm owns only 10 machines, the relevant interval is $0 \le x \le 10$. To find the critical points, set the deriative equal to zero and solve for x getting

$$C'(x) = 20 - \frac{1,280}{x^2} = \frac{20(x^2 - 64)}{x^2} = 0 \ \text{or} \ x = 8$$

Since this is the only critical point in the interal, apply the second-derivative test for absolute extrema. The second derivative is $C''(x) = \frac{2,560}{x^3}$, which is positive on the interval. Hence the total cost C has an absolute minimum when $x = 8$, that is, when 8 machines are used.

(b)   When $x = 8$, the supervisor earns $\frac{1,280}{8} = \$160$.

(c)   The cost of setting up 8 machines is $20(8) = \$160$.

33.   The goal is to optimize the rate of output on the interval $0 \leq t \leq 4$. Since $Q(t) = -t^3 + 6t^2 + 15t$ is the output, the rate of output is

$$R(t) = Q'(t) = -3t^2 + 12t + 15$$

To maximize $R(t)$ on $0 \leq t \leq 4$, find the critical points by setting $R'(t)$ equal to zero and solving for t:

$$R'(t) = Q''(t) = -6t + 12 = -6(t - 2) = 0 \text{ or } t = 2$$

Comparing the values of R at the critical point and endpoints

$$R(0) = 15, \ R(2) = 27, \ \text{and} \ R(4) = 15$$

shows that $R(t)$ has an absolute maximum when $t = 2$ and an absolute minimum when $t = 0$ and $t = 4$. That is, the worker is performing most efficiently at 10:00 A.M. and least efficiently at 8:00 A.M. and at 12:00 noon.

35.   The slope of the tangent is given by the derivative. Hence, to find the point on the curve $y = 2x^3 - 3x^2 + 6x$ at which the tangent has smallest slope, minimize the derivative function

$$m(x) = y'(x) = 6x^2 - 6x + 6$$

The derivative of $m(x)$ is

$$m'(x) = y''(x) = 12x - 6$$

which is zero when $x = \frac{1}{2}$. When $x = \frac{1}{2}$, the corresponding y value comes from the original function, that is, $y = 2\left(\frac{1}{2}\right)^3 - 3\left(\frac{1}{2}\right)^2 + 6\left(\frac{1}{2}\right) = \frac{5}{2}$. Since the second derivative $m''(x) = 12$ is positive, it follows from the second-derivative test that the slope is minimal at $\left(\frac{1}{2}, \frac{5}{2}\right)$. The value of the slope at this point is $m\left(\frac{1}{2}\right) = \frac{9}{2}$.

37.   The rate of population growth is the derivative

$$R(t) = P'(t) = -3t^2 + 18t + 48$$

The goal is to optimize the function $R(t)$ on the interval $0 \leq t \leq 5$. The derivative of R is

$$R'(t) = P''(t) = -6t + 18$$

which is zero when $t = 3$. Comparing the values

$$R(0) = 48 \ \ R(3) = 75 \ \text{and} \ R(5) = 63$$

you can conclude that the rate of growth is greatest 3 years from now (when it is 75 thousand people per year) and that the rate of growth is smallest now (at $t = 0$ when it is 48 thousand people per year).

39. Let n denote the number of floors and $A(n)$ the corresponding average cost. Since the total cost is $C(n) = 2n^2 + 500n + 600$ (thousand dollars),

$$A(n) = \frac{C(n)}{n} = 2n + 500 + \frac{600}{n}$$

The relevant interval is $n > 0$. To find the critical points, set the derivative equal to zero and solve for n getting

$$A'(n) = 2 - \frac{600}{n^2} = \frac{2(n^2 - 300)}{n^2} = 0 \text{ or } n = \sqrt{300} \simeq 17.32$$

Since this is the only critical point in the interval $n > 0$, apply the second-derivative test for absolute extrema. The second derivative is $A''(n) = \frac{1,200}{n^3}$, which is positive when $n > 0$. Hence, $A(n)$ has an absolute minimum when $n = \sqrt{300}$. In the context of this practical problem, the optimal value of n (which denotes the number of floors) must be an integer. Since

$$A(17) = 569.294 \text{ and } A(18) = 569.333$$

it follows that to minimize the average cost per floor, 17 floors should be built.

41. Let x denote the number of hours after 8:00 A.M. at which the 15 minute coffee break begins. Then $f(x)$ is the number of radios assembled before the break. After the break, $4 - x$ hours remain until lunchtime at 12:15 (4 hours 15 minutes minus x hours before the break minus 15 minutes for the break). Thus, the number of radios assembled during the $4 - x$ hours after the break is $g(4 - x)$, and the total number of radios assembled between 8:00 A.M. and 12:15 P.M. is

$$N(x) = f(x) + g(4 - x)$$
$$= -x^3 + 6x^2 + 15x - \frac{1}{3}(4 - x)^3 + (4 - x)^2 + 23(4 - x)$$

The relevant interval is $0 \le x \le 4$. To find the critical points, first use the chain rule to find the derivative

$$N'(x) = -3x^2 + 12x + 15 - (4 - x)^2(-1) + 2(4 - x)(-1) - 23$$
$$= -3x^2 + 12x + 15 + 16 - 8x + x^2 - 8 + 2x - 23$$
$$= -2x^2 + 6x = -2x(x - 3)$$

which is zero when $x = 0$ and $x = 3$. Comparing the function values at the critical points and endpoints

$$N(0) = 84.67, \quad n(3) = 95.67, \quad \text{and } N(4) = 92$$

shows that the worker will assemble the maximum number of radios by lunch time if the coffee break begins when $x = 3$, that is, at 11:00 A.M.

Chapter 3, Section 5

1. Let x denote the number of transistors in each shipment and $C(x)$ the corresponding (variable) cost. Then

$$C(x) = \text{(storage cost)} + \text{(ordering cost)}$$

The storage cost is $\left(\frac{x}{2}\right)(0.9) = 0.45x$. (See example 5.1 in the text for an explanation of the storage cost.) Since 600 transistors are used each year, $\frac{600}{x}$ is the number of shipments, and so the ordering cost is $30\left[\frac{600}{x}\right] = \frac{18,000}{x}$. Thus,

$$C(x) = 0.45x + \frac{18,000}{x}$$

The relevant interval is $0 < x \le 600$. The derivative is

$$C'(x) = 0.45 - \frac{18,000}{x^2} = \frac{0.45x^2 - 18,000}{x^2} = \frac{0.45(x^2 - 40,000)}{x^2}$$

which is zero when $x^2 = 40,000$ or $x = 200$. Since this is the only critical point in the interval, apply the second-derivative test for absolute extrema. The second derivative is $C''(x) = -\frac{36,000}{x^3}$, which is negative on the interval. Hence, C has an absolute minimum when $x = 200$, that is, when there are 200 cases per shipment and the transistors are ordered $\frac{600}{200} = 3$ times per year.

3. Let x denote the number of maps per batch and $C(x)$ the corresponding cost. Then,

$$C(x) = \text{(storage cost)} + \text{(production cost)} + \text{(set-up cost)}$$

The storage cost is $\left(\frac{x}{2}\right)(0.20) = 0.1x$. (see Example 5.1 in the text for an explanation of the storage cost.) The production cost is $0.06(16,000) = 960$. Since x maps are produced per batch and 16,000 maps are needed, the number of batches is $\frac{16,000}{x}$, and the set-up cost is $100\left[\frac{16,000}{x}\right] = \frac{1,600,000}{x}$. Putting it all together,

$$C(x) = 0.1x + 960 + \frac{1,600,000}{x}$$

The relevant interval is $0 < x \le 16,000$. The derivative is

$$C'(x) = 0.1 - \frac{1,600,000}{x^2} = \frac{0.1x^2 - 1,600,000}{x^2}$$

$$= \frac{0.1(x^2 - 16,000,000)}{x^2} = \frac{0.1(x - 4,000)(x + 4,000)}{x^2}$$

which is zero when $x = 4,000$. Since this is the only critical point in the relevant interval, apply the second-derivative test for absolute extrema. The second derivative is $C''(x) = \frac{3,200,000}{x^3}$, which is positive on the interval. Hence, C has an absolute minimum when $x = 4,000$ maps per batch.

5.   (a)   If total cost is $C(q) = 3q^2 + 5q + 75$, then the average cost per unit is

$$A(q) = \frac{C(q)}{q} = \frac{3q^2 + 5q + 75}{q} = 3q + 5 + \frac{75}{q}$$

The goal is to find the absolute minimum of $A(q)$ on the interval $q > 0$. The first derivate is

$$A'(q) = 3 - \frac{75}{q^2} = \frac{3q^2 - 75}{q^2} = \frac{3(q + 5)(q - 5)}{q^2}$$

which is zero on the interval $q > 0$ only when $q = 5$. Since the second derivative $A''(q) = \frac{150}{q^3}$ is positive when $q > 0$, it follows from the second-derivative test that the average cost is minimal on $q > 0$ when $q = 5$, that is, when 5 units are produced.

(b)   The marginal cost is the derivative $C'(q) = 6q + 5$ of the total cost function and equals average cost when

$$6q + 5 = 3q + 5 + \frac{75}{q} \qquad 3q^2 = 75 \qquad q^2 = 25 \text{ or } q = 5$$

which is the same level of production as that in part (a) for which average cost is minimal.

(c)   The marginal cost $C'(q) = 6q + 5$ is a linear function and its graph is a straight line with slope 6 and vertical intecept 5. To graph the average cost function $A(q)$, observe from part (a) that its derivative

$$A'(q) = \frac{3(q + 5)(q - 5)}{q^2}$$

is negative for $0 < q < 5$ and positive for $q > 5$. Hence $A(q)$ is decreasing for $0 < q < 5$, increasing for $q > 5$, and has a relative minimum at $q = 5$. The graphs are shown below.

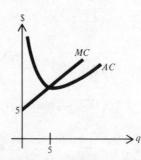

Graphs for Problem 5.

7.   (a)   If total revenue is $R(q) = -2q^2 + 68q - 128$, the average revenue per unit is

$$A(q) = \frac{R(q)}{q} = -2q + 68 - \frac{128}{q}$$

The marginal revenue is $R'(q) = -4q + 68$ and is equal to the average revenue when

$$-4q + 60 = -2q + 68 - \frac{128}{q} \qquad 2q = \frac{128}{q} \qquad q^2 = 64 \text{ or } q = 8$$

(b) The derivative of the average revenue is

$$A'(q) = -2 + \frac{128}{q^2} = \frac{-2q^2 + 128}{q^2} = \frac{-2(q+8)(q-8)}{q^2}$$

If $0 < q < 8, A'(q) > 0$ and $A(q)$ is increasing. If $q > 8, A'(q) < 0$ and $A(q)$ is decreasing.

(c) The graphs of the average and marginal revenue functions are sketched below. Notice that the average revenue has a maximum at $q = 8$ and q intercepts at $q = 2$ and $q = 32$.

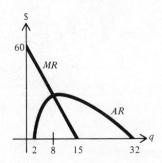

Graphs for Problem 7.

9.   (a) If the demand equation is $q = 60 - 0.1p$ (for $0 \le p \le 600$), the elasticity of demand is

$$\eta = \frac{p}{q}\frac{dq}{dp} = \frac{p}{60 - 0.1p}(-0.1) = -\frac{0.1p}{60 - 0.1p}$$

(b) When $p = 200$, the elasticity of demand is

$$\eta = -\frac{0.1(200)}{60 - 0.1(200)} = 0.5$$

That is, when the price is $p = 200$, a 1-percent increase in price will produce a decrease in demand of approximately 0.5 percent.

(c) The elasticity of demand will be -1 when

$$-1 = -\frac{0.1p}{60 - 0.1p} \qquad 60 - 0.1p = 0.1p$$

$$0.2p = 60 \quad \text{or} \quad p = \frac{60}{0.2} = 300$$

11. (a) If the demand equation is $q = 500 - 2p$ (for $0 \le p \le 250$), the elasticity of demand is

$$\eta = \frac{p}{q}\frac{dq}{dp} = \frac{p}{500 - 2p}(-2) = -\frac{p}{250 - p}$$

The demand is of unit elasticity when $|\eta| = 1$, that is, when

$$\frac{p}{250 - p} = 1 \qquad p = 250 - p \text{ or } 2p = 250 \quad p = 125$$

If $0 \le p < 125$,

$$|\eta| = \frac{p}{250 - p} < \frac{125}{250 - 125} = 1$$

and so the demand is inelastic. If $125 < p \le 250$,

$$|\eta| = \frac{p}{250 - p} > \frac{125}{250 - 125} = 1$$

and so the demand is elastic.

(b) The total revenue increases for $0 \le p < 125$ (where the demand is inelastic), decreases for $125 < p \le 250$ (where the demand is elastic), and has a maximum at $p = 125$ (where the demand is of unit elasticity).

(c) The revenue function is

$$R = pq = p(500 - 2p) = 500p - 2p^2$$

Its derivative is

$$R'(p) = 500 - 4p$$

which is zero when $p = 125$. On the interal $0 \le p < 125$, $R'(p)$ is positive and so $R(p)$ is increasing. On the interval $125 < p \le 250$, $R'(p)$ is negative and so $R(p)$ is decreasing. $R(p)$ has a relative maximum at the critical point at $p = 125$.

(d) The graphs of the demand and revenue fucntions are drawn below. Notice that for this linear demand function, the price of unit elasticity and maximum revenue is the midpoint $p = 125$ of the relevant interval $0 \le p \le 250$.

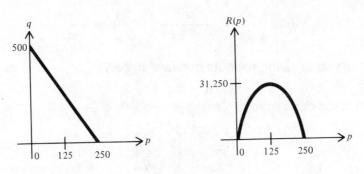

Demand and revenue graphs for Problem 11.

13.   If the demand is $q = \frac{a}{p^m} = ap^{-m}$, the elasticity of demand is

$$\eta = \frac{p}{q}\frac{dq}{dp} = \frac{p}{a/p^m}[-am^{-m-1}] = \frac{p^{m+1}}{a}\left[-\frac{am}{p^{m+1}}\right] = -m$$

a constant which is independent of the price p. Thus, at any price, a 1-percent increase in price will produce a decrease in demand of approximately m percent.

15.   Suppose the demand equation is $p = 60 - 2q$ (for $0 \le q \le 30$).

(a)   The elasticity of demand is

$$\eta = \frac{p/q}{dp/dq} = \frac{(60 - 2q)/q}{-2} = -\frac{30 - q}{q}$$

(b)   When $q = 10$, the elasticity of demand is

$$\eta = \frac{30 - 10}{10} = -2$$

That is, when demand is $q = 10$, a 1-percent icnrease in price will produce a decrease in demand of approximately 2 percent.

(c)   To express η in terms of p, solve the demand equation for q

$$p = 60 - 2q \quad 2q = 60 - p \quad q = 30 - \frac{p}{2}$$

and substitute for q in the formula in part (a) getting

$$\eta = -\frac{30 - (30 - p/2)}{30 - p/2} = -\frac{p}{60 - p}$$

(d)   Start with the demand equation in the form

$$q = 30 - \frac{p}{2}$$

and apply the original definition of elasticity of demand to get

$$\eta = \frac{p}{q}\frac{dq}{dp} = \frac{p}{30 - p/2}\left(-\frac{1}{2}\right) = -\frac{p}{60 - p}$$

which is the same as the formula obtained in part (c).

17. From Problem 8, the derivative of R with respect to q is

$$\frac{dR}{dq} = p\left[1 + \frac{1}{\eta}\right]$$

If $|\eta| > 1$, then $\eta < -1$ (since η is negative) and

$$\frac{dR}{dq} = p\left[1 + \frac{1}{\eta}\right] > p(1 - 1) = 0$$

and hence R is an increasing function of q. If $|\eta| < 1$, then $\eta > -1$ (since η is negative) and

$$\frac{dR}{dq} = p\left[1 + \frac{1}{\eta}\right] < p(1 - 1) = 0$$

and hence R is a decreasing function of q.

Chapter 3, Review Problems

1. (a) If $f(x) = -2x^3 + 3x^2 + 12x - 5$, the derivative is

$$f'(x) = -6x^2 + 6x + 12 = -6(x^2 - x - 2) = -6(x + 1)(x - 2)$$

which is zero when $x = -1$ and $x = 2$. The corresponding critical points are $(-1, f(-1)) = (-1, -12)$ and $(2, f(2)) = (2, 15)$.

Interval	$-6(x+1)$	$x - 2$	$f'(x)$	Inc/dec
$x < -1$	+	−	−	decreasing
$-1 < x < 2$	−	−	+	increasing
$x > 2$	−	+	−	decreasing

Notice that the critical point $(2, 15)$ is a relative maximum and the critical point $(-1, -12)$ is a relative minimum.

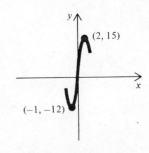

Graph for Problem 1a.

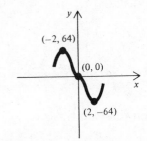

Graph for Problem 1b.

(b) If $f(x) = 3x^5 - 20x^3$, the derivative is

$$f'(x) = 15x^4 - 60x^2 = 15x^2(x^2 - 4) = 15x^2(x + 2)(x - 2)$$

which is zero when $x = 0, x = -2$, and $x = 2$. The corresponding critical points are $(-2, f(-2)) = (-2, 64), (0, f(0)) = (0, 0)$, and $(2, f(2)) = (2, -64)$.

Interval	$15x^2$	$x + 2$	$x - 2$	$f'(x)$	Inc/dec
$x < -2$	+	−	−	+	increasing
$-2 < x < 0$	+	+	−	−	decreasing
$0 < x < 2$	+	+	−	−	decreasing
$x > 2$	+	+	+	+	increasing

Notice that the critical point $(-2, 64)$ is a relative maximum, that the critical point $(2, -64)$ is a relative minimum, and that the critical point $(0, 0)$ is not a relative extremum.

(c) If $f(x) = \frac{x^2}{x+1}$, the derivative is

$$f'(x) = \frac{(x + 1)(2x) - (x^2)(1)}{(x + 1)^2} = \frac{x(x + 2)}{(x + 1)^2}$$

which is zero when $x = 0$ and $x = -2$. The corresponding critical points are $(0, f(0)) = (0, 0)$ and $(-2, f(-2)) = (-2, -4)$. Note that the derivative is undefined when $x = -1$, which is not a critical point since it is not in the domain of f.

Interval	x	$x + 2$	$(x + 1)^2$	$f'(x)$	Inc/dec
$x < -2$	$-$	$-$	$+$	$+$	increasing
$-2 < x < -1$	$-$	$+$	$+$	$-$	decreasing
$-1 < x < 0$	$-$	$+$	$+$	$-$	decreasing
$x > 0$	$+$	$+$	$+$	$+$	increasing

Notice that the critical point $(-2, -4)$ is a relative maximum and the critical point $(0, 0)$ is a relative minimum.

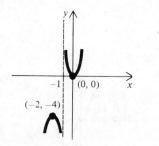

Graph for Problem 1c.

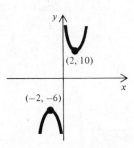

Graph for Problem 1d.

(d)   If $f(x) = 2x + \frac{8}{x} + 2$, the derivative is

$$f'(x) = 2 - \frac{8}{x^2} = \frac{2x^2 - 8}{x^2} = \frac{2(x^2 - 4)}{x^2} = \frac{2(x + 2)(x - 2)}{x^2}$$

which is zero when $x = -2$ and $x = 2$. The corresponding critical points are $(-2, f(-2)) = (-2, -6)$ and $(2, f(2)) = (2, 10)$. Note that the derivative is undefined when $x = 0$, which is not a critical point since it is not in the domain of f.

Interval	$x + 2$	$x - 2$	x^2	$f'(x)$	Inc/dec
$x < -2$	$-$	$-$	$+$	$+$	increasing
$-2 < x < 0$	$+$	$-$	$+$	$-$	decreasing
$0 < x < 2$	$+$	$-$	$+$	$-$	decreasing
$x > 2$	$+$	$+$	$+$	$+$	increasing

Notice that the critical point $(-2, -6)$ is a relative maximum and the critical point $(2, 10)$ is a relative minimum.

2.   There are many graphs with these properties, but all should be:

(a)   increasing for $x < 0$ and $x > 5$, since $f'(x) > 0$ on these intervals;

(b)   decreasing for $0 < x < 5$, since $f'(x) < 0$ on this interval;

(c)   concave upward for $-6 < x < -3$ and $x > 2$, since $f''(x) > 0$ on these intervals;

(d)   concave downward for $x < -6$ and for $-3 < x < 2$, since $f''(x) < 0$ on these intervals.

A graph will all these properties is shown in the accompanying figure.

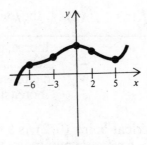

Graph for Problem 2.

3.   (a)   If $f(x) = x^2 - 6x + 1$, the first derivative is

$$f'(x) = 2x - 6 = 2(x - 3)$$

which is zero when $x = 3$, and the corresponding first-order critical point is $(3, -8)$. The second derivative is

$$f''(x) = 2$$

which is always positive

Interval	$f'(x)$	$f''(x)$	Inc/dec	Concavity
$x < 3$	−	+	decreasing	up
$x > 3$	+	+	increasing	up

Notice that the critical point $(3, -8)$ is a relative minimum.

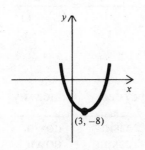

Graph for Problem 3a.

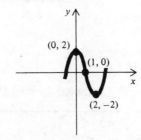

Graph for Problem 3b.

(b)   If $f(x) = x^3 - 3x^2 + 2$, the first derivative is

$$f'(x) = 3x^2 - 6x = 3x(x - 2)$$

which is zero when $x = 0$ and $x = 2$. The corresponding first-order critical points are $(0, 2)$ and $(2, -2)$. The second derivative is

$$f''(x) = 6x - 6 = 6(x - 1)$$

which is zero when $x = 1$, and the corresponding second-order critical point is $(1, 0)$.

Interval	$f'(x)$	$f''(x)$	Inc/dec	Concavity
$x < 0$	+	−	increasing	down
$0 < x < 1$	−	−	decreasing	down
$1 < x < 2$	−	+	decreasing	up
$x > 2$	+	+	increasing	up

Notice that the first-order critical point $(0, 2)$ is a relative maximum, the first-order critical point $(2, -2)$ is a relative minimum, and the second-order critical point $(1, 0)$ is an inflection point.

(c) If $f(x) = \frac{x^2+3}{x-1}$, the first derivative is

$$f'(x) = \frac{(x - 1)(2x) - (x^2 + 3)(1)}{(x - 1)^2} = \frac{x^2 - 2x - 3}{(x - 1)^2} = \frac{(x + 1)(x - 3)}{(x - 1)^2}$$

which is zero when $x = -1$ and $x = 3$. The corresponding first-order critical points are $(-1, -2)$ and $(3, 6)$. Note that $f'(x)$ is undefined when $x = 1$, which is not a critical point since it is not in the domain of f. The second derivative is

$$f''(x) = \frac{(x - 1)^2(2x - 2) - (x^2 - 2x - 3)(2)(x - 1)}{(x - 1)^4}$$

$$= \frac{2(x - 1)[(x - 1)^2 - (x^2 - 2x - 3)]}{(x - 1)^4} = \frac{8}{(x - 1)^3}$$

which is never zero.

Interval	$f'(x)$	$f''(x)$	Inc/dec	Concavity
$x < -1$	+	−	increasing	down
$-1 < x < 1$	−	−	decreasing	down
$1 < x < 3$	−	+	decreasing	up
$x > 3$	+	+	increasing	up

Notice that the first-order critical point $(-1, -2)$ is a relative maximum, the first-order critical point $(3, 6)$ is a relative minimum, and even though $x = 1$ is not in the domain of f, the graph changes concavity at $x = 1$.

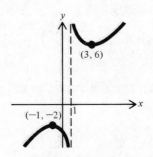

Graph for Problem 3c.

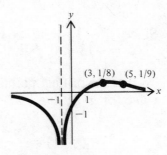

Graph for Problem 3d.

(d) If $f(x) = \frac{x-1}{(x+1)^2}$, the graph has a discontinuity at $x = -1$. The first derivative is

$$f'(x) = \frac{(x+1)^2(1) - (x-1)[2(x+1)(1)]}{(x+1)^4}$$

$$= \frac{(x+1)[(x+1) - 2(x-1)]}{(x+1)^4} = \frac{3-x}{(x+1)^3}$$

which is zero when $x = 3$. The corresponding first-order critical point is $\left(3, \frac{1}{8}\right)$. The second derivative is

$$f''(x) = \frac{(x+1)^3(-1) - (3-x)[3(x+1)^2(1)]}{(x+1)^6}$$

$$= \frac{(x+1)^2[-(x+1) - 3(3-x)]}{(x+1)^6}$$

$$= \frac{-x-1-9+3x}{(x+1)^4} = \frac{2x-10}{(x+1)^4} = \frac{2(x-5)}{(x+1)^4}$$

which is zero when $x = 5$. The corresponding second-order critical point is $\left(5, \frac{1}{9}\right)$.

Interval	$f'(x)$	$f''(x)$	Inc/dec	Concavity
$x < -1$	−	−	decreasing	down
$-1 < x < 3$	+	−	increasing	down
$3 < x < 5$	−	−	decreasing	down
$x > 5$	−	+	decreasing	up

Notice that the first-order critical point $\left(3, \frac{1}{8}\right)$ is a relative maximum and that the second-order critical point $\left(5, \frac{1}{9}\right)$ is an inflection point.

4. (a) To find the critical points of $f(x) = -2x^3 + 3x^2 + 12x - 5$, set the derivative equal to zero and solve for x getting

$$f'(x) = -6x^2 + 6x + 12 = -6(x+1)(x-2) = 0 \quad \text{or} \quad x = -1 \text{ and } x = 2$$

The corresponding first-order critical points are $(-1, -12)$ and $(2, 15)$. The second derivative is

$$f''(x) = -12x + 6 = -6(2x - 1)$$

Since $f''(-1) = 18 > 0$, f has a relative minimum at $(-1, -12)$ and since $f''(2) = -18 < 0$, f has a relative maximum at $(2, 15)$.

(b)   To find the critical points of $f(x) = \frac{x^2}{x+1}$, set the derivative equal to zero and solve for x getting

$$f'(x) = \frac{(x+1)(2x) - (x^2)}{(x+1)^2} = \frac{x^2 + 2x}{(x+1)^2} = \frac{x(x+2)}{(x+1)^2} = 0$$

or $x = 0$ and $x = -2$. The corresponding first-order critical points are $(0, 0)$ and $(-2, -4)$. The second derivative is

$$f''(x) = \frac{(x+1)^2(2x+2) - (x^2 + 2x)(2)(x+1)(1)}{(x+1)^4} = \frac{2}{(x+1)^3}$$

Since $f''(0) = 2 > 0$, f has a relative minimum at $(0, 0)$, and since $f''(-2) = -2 < 0$, f has a relative maximum at $(-2, -4)$.

(c)   To find the critical points of $f(x) = 2x + \frac{8}{x} + 2$, set the derivative equal to zero and solve for x getting

$$f'(x) = 2 - \frac{8}{x^2} = \frac{2(x+2)(x-2)}{x^2} = 0 \text{ or } x = -2 \text{ and } x = 2$$

The corresponding first-order critical points are $(-2, -6)$ and $(2, 10)$. The second derivative $f''(x) = \frac{16}{x^3}$. Since $f''(-2) = -2 < 0$, f has a relative maximum at $(-2, -6)$, and since $f''(2) = 2 > 0$, f has a relative minimum $(2, 10)$.

5.   (a)   If $f(x) = -2x^3 + 3x^2 + 12x - 5$, the derivative is

$$f'(x) = -6x^2 + 6x + 12 = -6(x - 2)(x + 1)$$

which is zero when $x = -1$ and $x = 2$, both of which are in the interval $-3 \leq x \leq 3$. Comparing the function values at the critical points and endpoints

$$f(-1) = -12, \ f(2) = 15, \ f(-3) = 40, \text{ and } f(3) = 4$$

shows that the absolute maximum is $f(-3) = 40$ and the absolute minimum is $f(-1) = -12$.

(b)   If $f(x) = -3x^4 + 8x^3 - 10$, the derivative is

$$f'(x) = -12x^3 + 24x^2 = -12x^2(x - 2)$$

which is zero when $x = 0$ and $x = 2$, both of which are in the interval $0 \leq x \leq 3$. Comparing function values at critical points and endpoints

$$f(0) = -10, \ f(2) = 6, \text{ and } f(3) = -37$$

shows that the absolute maximum is $f(2) = 6$ and the absoute minimum is $f(3) = -37$.

(c)   If $f(x) = \frac{x^2}{x+1}$, the derivative is

$$f'(x) = \frac{(x+1)(2x) - x^2(1)}{(x+1)^2} = \frac{x(x+2)}{(x+1)^2}$$

which is zero when $x = 0$ and $x = -2$, of which only $x = 0$ is in the interval $-\frac{1}{2} \le x \le 1$. Note that the derivative is undefined when $x = -1$, which is not a critical point since it is not in the domain of f. Comparing the function values at the critical point in the interval and at the endpoints

$$f(0) = 0, \quad f\left(-\frac{1}{2}\right) = \frac{1}{2}, \quad \text{and} \quad f(1) = \frac{1}{2}$$

shows that the absolute maximum is $f\left(-\frac{1}{2}\right) = f(1) = \frac{1}{2}$ and the absolute minimum is $f(0) = 0$.

(d)   If $f(x) = 2x + \frac{8}{x} + 2$, the derivative is

$$f'(x) = 2 - \frac{8}{x^2} = \frac{2x^2 - 8}{x^2} = \frac{2(x+2)(x-2)}{x^2}$$

which is zero when $x = -2$ and $x = 2$, of which only $x = 2$ is in the interval $x > 0$. There are no endpoints, so the only possible absolute extremum is $f(2) = 10$. Since $f'(x) < 0$ (and f is decreasing) when $0 < x < 2$ and $f'(x) > 0$ (and f is increasing) when $x > 2$, it follows that $f(2) = 10$ is the absoute minimum and that there is no absolute maximum.

6.   The goal is to find the absolute maximum and absolute minimum of the function $S(t) = t^3 - 9t^2 + 15t + 45$ on the interval $0 \le t \le 7$. The derivative is

$$S'(t) = 3t^2 - 18t + 15 = 3(t^2 - 6t + 5) = 3(t-1)(t-5)$$

which is zero when $t = 1$ and $t = 5$. Comparing the function values at the critical points and endpoints

$$S(0) = 45 \quad S(1) = 52 \quad S(5) = 20 \quad S(7) = 52$$

shows that the traffic is moving fastest at 1:00 P.M. and 7:00 P.M. when its speed is 52 miles per hours and slowest at 5:00 P.M. when its speed is 20 miles per hour.

7.   Let R denote the rate at which the rumor is spreading, Q the number of people who have heard the rumor, and P the total population of the community. Then,

$$R(Q) = kQ(P - Q)$$

where k is a positive constant of proportionality. The goal is to find the absolute maximum of $R(Q)$ for $0 \le Q \le P$. By the product rule, the derivative is

$$R'(Q) = kQ(-1) + (P - Q)(k) = Pk - 2kQ$$

which is zero when

$$2kQ = Pk \text{ or } Q = \frac{P}{2}$$

that is, when one-half of the population has heard the rumor. Since there is only one critical point in the interval, you can apply the second-derivative test for absolute extrema. The second derivative is $R''(Q) = -2k$, which is negative (since k is positive). Hence, the function $R(Q)$ has an absolute maximum at its critical point $Q = \frac{P}{2}$.

8. Let x denote the number of \$5 reductions in the price and $P(x)$ the corresponding profit. Since 40 cameras can be sold at the price of \$80 per camera and 10 more cameras will be sold for each \$5 reduction, the total number of cameras sold is $40 + 10x$. Moreover, the decrease in price is $5x$ (5 times the number of \$5 reductions), and so the price is $80 - 5x$. The cost is \$50, and so the profit per camera is $(80 - 5x) - 50 = 30 - 5x$. Thus,

$$P(x) = (\text{number of cameras sold}) (\text{profit per camera})$$
$$= (40 + 10x)(30 - 5x) = 50(4 + x)6 - x) = 50(24 + 2x - x^2)$$

The relevant interval is $0 \le x \le 6$. To find the critical points, set the derivative equal to zero and solve for x getting

$$P'(x) = 50(2 - 2x) = 100(1 - x) = 0 \text{ or } x = 1$$

Since $x = 1$ is the only critical point in the interal, you may apply the second-derivative text for absolute extrema. Since the second derivative $P''(x) = -100$ is negative, it follows that $P(x)$ has an absolute maximum when $x = 1$. That is, to maximize profit, there should be one \$5 reduction, and so the selling price should be \$75.

9. Label the sides of the rectangular plot as indicated in the figure and let A denote the area. Then

$$A = xy + xy = 2xy$$

which is a function of two variables. The fact that 300 meters of fencing are to be used implies

$$4x + 3y = 300 \text{ or } y = \frac{300 - 4x}{3} = 100 - \frac{4}{3}x$$

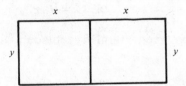

Rectangular plot for Problem 9.

Substituing for y in the equation for A yields

$$A(x) = 2x\left[100 - \frac{4}{3}x\right] = 200x - \frac{8}{3}x^2$$

The relevant interval is $x > 0$. To find the critical points, set the derivative equal to zero and solve for x getting

$$A'(x) = 200 - \frac{16}{3}x = 0 \text{ or } x = 37.5$$

Since $x = 37.5$ is the only critical point in the interval, apply the second-derivative test for absolute extrema. The second derivative is $A''(x) = -\frac{16}{3}$, which is negative. Hence, $A(x)$ has an absolute maximum when $x = 37.5$. The corresponding value of y is $y = 100 - \frac{4}{3}(37.5) = 50$. Thus, each plot should be 37.5 meters by 50 meters.

10. Let r denote the radius, h the height, C the (fixed) cost (in cents), and V the volume. The goal is to maximize the volume

$$V = \pi r^2 h$$

which is a function of the two variables r and h. To write h in terms of r, use the fact that the cost is to be C cents. That is,

$$C = \text{cost of bottom} + \text{cost of side}$$
$$= 3(\text{area of bottom}) + 2(\text{area of side})$$

or

$$C = 3\pi r^2 + 2(2\pi rh) = 3\pi r^2 + 4\pi rh$$

Solve for h to get

$$4\pi rh = C - 3\pi r^2 \text{ or } h = \frac{C - 3\pi r^2}{4\pi r}$$

where C is a constant. Now substitute this expression for h into the volume formula and simplify to get

$$V(r) = \pi r^2 \left[\frac{C - 3\pi r^2}{4\pi r}\right] = \frac{r(C - 3\pi r^2)}{4} = \left[\frac{C}{4}\right]r - \frac{3}{4}\pi r^3$$

Since the radius can be any positive number, the goal is to find the absolute maximum of $V(r)$ on the interval $r > 0$. The derivative is

$$V'(r) = \frac{C}{4} - \frac{9}{4}\pi r^2 \qquad \text{(Remember that } C \text{ is a constant.)}$$

which is zero when

$$\frac{C}{4} = \frac{9}{4}\pi r^2 \quad \text{or} \quad C = 9\pi r^2$$

Since we want a relationship between r and h, replace C by its formula $3\pi r^2 + 4\pi rh$ and simplify to get

$$3\pi r^2 + 4\pi rh = 9\pi r^2 \qquad 4\pi rh = 6\pi r^2 \quad \text{or} \quad h = \frac{3}{2}r$$

That is, assuming that the volume is maximized when its derivative is zero, the height of the container of largest volume is $\frac{3}{2}$ times the radius. To verify that this really does correspond to the absolute maximum, observe that the equation $C = 9\pi r^2$ gives $r = \pm\frac{1}{3}\sqrt{\frac{C}{\pi}}$ of which only the positive value is in the interval $r > 0$. Since this is the only first-order critical point in the interval, you can apply the second-derivative test for absolute extrema. The second derivative is

$$V''(r) = -\frac{9}{2}\pi r$$

which is negative when $r > 0$. Hence the critical point that gave the relationship $h = \frac{3}{2}r$ does indeed give the absolute maximum of the volume function on the interval.

11.   Let x be as indicated in the accompanying figure. Then

$$\text{Distance along bank} = 1 - x$$

and, by the pythagorean theorem,

$$\text{Distance across water} = \sqrt{1 + x^2}$$

The goal is to minimize time T which is given by

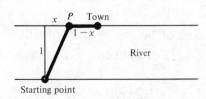

Figure for Problem 11.

$$\begin{aligned}
T &= \text{time in water} + \text{time on land} \\
&= \frac{\text{distance in water}}{\text{speed in water}} + \frac{\text{distance on land}}{\text{speed on land}}
\end{aligned}$$

and so

$$T(x) = \frac{\sqrt{1+x^2}}{4} + \frac{1-x}{5} = \frac{1}{4}\sqrt{1+x^2} + \frac{1}{5}(1-x)$$

The relevant interval is $0 \le x \le 1$. The deriative is

$$T'(x) = \frac{1}{4}\left(\frac{1}{2}\right)(1+x^2)^{-1/2}(2x) - \frac{1}{5} = \frac{x}{4\sqrt{1+x^2}} - \frac{1}{5}$$

which is zero if

$$\frac{x}{4\sqrt{1+x^2}} = \frac{1}{5}$$
$$5x = 4\sqrt{1+x^2}$$
$$25x^2 = 16(1+x^2) = 16 + 16x^2$$
$$9x^2 = 16 \quad x^2 = \frac{16}{9} \text{ or } x = \pm\frac{4}{3}$$

Neither of these critical values lies in the interval $0 \le x \le 1$. Hence, the absolute minimum must occur at an endpoint. Since

$$T(0) = \frac{1}{4} + \frac{1}{5} = 0.45 \text{ hour}$$
and
$$T(1) = \frac{\sqrt{2}}{4} \simeq 0.354 \text{ hours}$$

it follows that time is minimized if $x = 1$, that is, if you row all the way to the town.

12.   Since $f(t) = -t^3 + 7t^2 + 200t$ is the number of letters the clerk can sort in t hours, the clerk's rate of output is

$$R(t) = f'(t) = -3t^2 + 14t + 200 \text{ letters per hour}$$

The goal is to maximize $R(t)$ for $0 \le t \le 4$. To find the critical points, set the derivative equal to zero and solve for t getting

$$R'(t) = f''(t) = -6t + 14 = 0 \text{ or } t = \frac{7}{3}$$

Comparing the values of $R(t)$ at the critical point and endpoints

$$R\left(\frac{7}{3}\right) = 216.33, \quad R(0) = 200, \quad \text{and } R(4) = 208$$

shows that the rate of output is greatest when $t = \frac{7}{3}$ hours, that is, after 2 hours and 20 minutes.

13.   Let x denote the number of machines used and $C(x)$ the corresponding cost of producing the $400,000$ medals. Then

$$C(x) = \text{set–up cost } + \text{ operating cost}$$
$$= 80 \text{ (number of machines)} + 5.76 \text{ (number of hours)}$$

Each machine can produce 200 medals per hour, so x machines can produce $200x$ medals per hours, and it will take $\frac{400,000}{200x}$ hours to produce the 400,000 medals. Hence,

$$C(x) = 80x + 5.76 \left[\frac{400,000}{200x} \right] = 80x + \frac{11,520}{x}$$

To find the critical points, set the derivative equal to zero and solve for x getting

$$C'(x) = 80 - \frac{11,520}{x^2} = \frac{80(x^2 - 144)}{x^2} = \frac{80(x - 12)(x + 12)}{x^2}$$

or $x = 12$. For $0 < x < 12$, $C'(x) < 0$ and $C(x)$ is decreasing, and for $x > 12$, $C'(x) > 0$ and $C(x)$ is increasing. Thus $C(x)$ has an absolute minimum when $x = 12$ machines.

14. Let x denote the number of additional trees planted and $f(x)$ the corresponding total yield. The total number of trees planted is $60 + x$. The average yield per tree of 475 lemons when 60 trees are planted will decrease by 5 lemons for each additional tree. Thus, the average yield per tree is $475 - 5x$. Putting it all together,

$$f(x) = \text{(number of trees) (yield per tree)}$$
$$= (60 + x)(475 - 5x) = 5(50 + x)(95 - x) = 5(5,700 + 35x - x^2)$$

To find the critical points, set the deivative equal to zero and solve for x getting

$$f'(x) = 5(35 - 2x) = 0 \text{ or } x = 17.5$$

For $x < 17.5$, $f'(x) > 0$ and $f(x)$ is increasing, and for $x > 17.5$, $f'(x) < 0$ and $f'(x)$ is decreasing. Hence $f(x)$ has an absolute maximum when $x = 17.5$. Since the number of additional trees planted must be a whole number, compare $f(17) = 30,030$ and $f(18) = 30,030$ to conclude that planting either 17 or 18 additional trees (for a total of either 77 or 78 trees) will maximize the total yield.

15. (a) The graph of the linear demand function $D(p) = mp + b$ is shown below. Notice that the vertical intercept is $D(0) = b$ and the horizontal is $D\left(-\frac{b}{m}\right) = 0$. The fact that the vertical intercept b is positve reflects the fact that there will be demand for the product when the selling price is zero. The fact that the slope is negative means that $D(p)$ is a decreasing function, that is, demand decreases as the price increases.

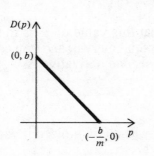 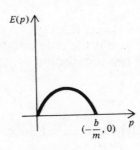

Demand and expenditure curves for Problem 15.

(b)   Let E denote the total monthly expenditure. Then

$$E(p) = \text{(price per unit) (number of units sold)}$$
$$= p(mp + b) = mp^2 + pb$$

Notice that $E(0) = 0$ and $E\left(-\frac{b}{m}\right) = 0$. Since $E'(p) = 2mp + b$, which is zero when $p = -\frac{b}{2m}$ and $E''(p) = 2m$, which is always negative (since $m < 0$), the expenditure function has a relative maximum at $p = -\frac{b}{2m}$, as shown in the figure.

(c)   From part (b), the optimal price is $p = -\frac{b}{2m}$, which is midway between the two p intercepts of $E(p)$.

16.   Let x denote the number of units per shipment and $C(x)$ the corresponding total cost. Then,

$$\text{Total cost} = \text{ordering cost} + \text{storage cost} + \text{purchase cost}$$

where

$$\text{Ordering cost} = \text{(ordering cost per shipment) (number of shipments)}$$
$$= b\left[\frac{q}{x}\right] = \frac{bq}{x}$$
$$\text{Storage cost} = \frac{x}{2} \text{ (cost of storing 1 unit 1 year)} = \frac{x}{2}(s) = \left[\frac{s}{2}\right] x$$

and

$$\text{Purchase cost} = \text{(number of units ordered) (cost per unit)} = qp$$

Putting it all together, the total cost is

$$C(x) = \frac{bq}{x} + \left[\frac{s}{2}\right] x + qp$$

The goal is to find the absolute minimum of $C(x)$ on the interval $0 < x \leq q$. The derivative is

$$C'(x) = \frac{bq}{x^2} + \frac{s}{2}$$

which is zero when

$$\frac{bq}{x^2} = \frac{s}{2} \qquad x^2 = \frac{2bq}{s} \quad \text{or} \quad x = \pm\sqrt{\frac{2bq}{s}}$$

Only the positive value is relevant. Assuming that this value of x is in the relevant interval $0 < x \le q$, it is the only critical point in this interval, and you can apply the second-derivative test for absolute extrema. The second derivative is

$$C''(x) = \frac{2bq}{x^3}$$

which is positive when $x > 0$. Hence, total cost will be minimized if each shipment contains $\sqrt{\frac{2bq}{s}}$ units.

17. If the total cost is $C(q) = aq^2 + bq + c$, the marginal cost is $C'(q) = 2aq + b$, and the average cost is $A(q) = \frac{C(q)}{q} = aq + b + \frac{c}{q}$.

(a) Average cost equals marginal cost when

$$aq + b + \frac{c}{q} = 2aq + b \qquad aq = \frac{c}{q} \qquad q^2 = \frac{c}{a} \qquad q = \sqrt{\frac{c}{a}}$$

(b) The first derivative of $A(q)$ is

$$A'(q) = a - \frac{c}{q^2}$$

If $0 < q < \sqrt{\frac{c}{a}}$, then

$$q^2 < \frac{c}{a} \quad \text{or} \quad \frac{c}{q^2} > a \quad \text{(since } a \text{ and } c \text{ are positive)}$$

and so

$$A'(q) = a - \frac{c}{q^2} < a - a = 0$$

which implies that $A(q)$ is decreasing. Similarly, if $q > \sqrt{\frac{c}{a}}$, then

$$q^2 > \frac{c}{a} \quad \text{or} \quad \frac{c}{q^2} < a \quad \text{since } a \text{ and } c \text{ are positive)}$$

and so

$$A'(q) = a - \frac{c}{q^2} > a - a = 0$$

which implies that $A(q)$ is increasing.

18. (a) If the demand equation is $q = 27 - 0.01p^2$ for $0 \le p \le \sqrt{2,700}$, the elasticity of demand is

$$\eta = \frac{p}{q}\frac{dq}{dp} = \frac{p}{27 - 0.01p^2}(-0.02p) = -\frac{2p^2}{2,700 - p^2}$$

The demand is of unit elasticity when $|\eta| = 1$, that is, when

$$\frac{2p^2}{2,700 - p^2} = 1 \qquad 2p^2 = 2,700 - p^2$$

$$3p^2 = 2,700 \qquad p^2 = 900 \text{ or } p = \pm 30$$

of which only $p = 30$ is in the relevant interval. If $0 \le p < 30$,

$$|\eta| = \frac{2p^2}{2,700 - p^2} < \frac{2(30)^2}{2,700 - (30)^2} = \frac{1,800}{1,800} = 1$$

and hence the demand is inelastic. If $30 < p \le \sqrt{2,700}$, then

$$|\eta| = \frac{2p^2}{2,700 - p^2} > \frac{2(30)^2}{2,700 - (30)^2} = 1$$

and hence the demand is elastic.

(b) The total revenue is increasing for $0 \le p < 30$ (when demand is inelastic), decreasing for $30 < p \le \sqrt{2,700}$ (when demand is elastic), and has a relative maximum when $p = 30$ (where demand is of unit elasticity.)

(c) The total revenue function is

$$R = pq = p(27 - 0.01p^2) = 27p - 0.01p^3$$

Its derivative is

$$R'(p) = 27 - 0.03p^2 = 0.03(900 - p^2) = 0.03(30 + p)(30 - p)$$

which is zero when $p = \pm 30$, of which only $p = 30$ is in the relevant interval. If $0 \le p < 30$, $R'(p)$ is positive and so $R(p)$ is increasing. If $30 < p \le \sqrt{2,700}, R'(p)$ is negative and so $R(p)$ is decreasing. $R(p)$ has a relative maximum at $p = 30$.

(d) The graphs of the demand and revenue functions are sketched below.

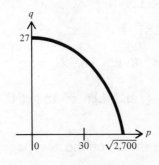
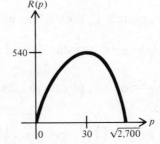

Graphs for Problem 18.

CHAPTER 4

EXPONENTIAL AND LOGARITHMIC FUNCTIONS

---

1 EXPONENTIAL FUNCTIONS

2 EXPONENTIAL MODELS

3 THE NATURAL LOGARITHM

4 DIFFERENTIATION OF LOGARITHMIC AND EXPONENTIAL FUNCTIONS

5 COMPOUND INTEREST

REVIEW PROBLEMS

Chapter 4, Section 1

1. The following calculations are for a calculator with an Algebraic Operating System (AOS) and an e^x key.

   For e^2, punch 2, then e^x to get 7.389.

   For e^{-2}, punch 2, then $+/-$, and then e^x to get 0.135.

   For $e^{0.05}$, punch 0.05, then e^x to get 1.051.

   For $e^{-0.05}$, punch 0.05, then $+/-$, and then e^x to get 1.

   For e^0, punch 0, then e^x to get 0.951.

   For e, punch 1, then e^x to get 2.718.

   For $\frac{1}{\sqrt{e}} = e^{1/2}$, punch $.5$, and then e^x to get 1.649.

   For $\frac{1}{\sqrt{e}} = e^{-1/2}$, punch $.5$, then $+/-$, and then e^x to get 0.607.

3. The graphs of $y = \left(\frac{1}{3}\right)^x$ and $y = \left(\frac{1}{4}\right)^x$ are shown on the next page.

5. If $f(x) = e^{-x}$, then $f(x)$ approaches zero as x increases without bound, and $f(x)$ increases without bound as x decreases without bound. The graph is shown on the next page.

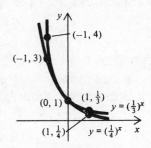

Graphs of $y = (\frac{1}{3})^x$ and $y = (\frac{1}{4})^x$ for Problem 3.

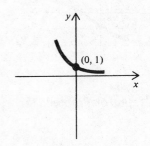

Graph for Problem 5.

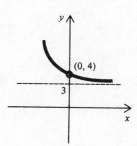

Graph for Problem 7.

7.   If $f(x) = 3 + e^{-x}$, then $f(x)$ approaches 3 as x increases without bound, and $f(x)$ increases without bound as x decreases without bound. The y intercept is $f(0) = 4$. Notice that the graph is the graph of e^{-x} moved up three units.

9.   If $f(x) = 3 - 2e^x$, then $f(x)$ decreases without bound as x increases without bound, and $f(x)$ approaches 3 as x decreases without bound. The y intercept is $f(0) = 1$.

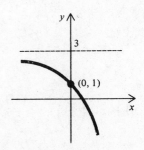

Graph for Problem 9.

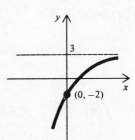

Graph for Problem 11.

11. If $f(x) = 3 - 5e^{-x}$, then $f(x)$ approaches 3 as x increases without bound, and $f(x)$ decreases without bound as x decreases without bound. The y intercept is $f(0) = -2$.

13. If $f(x) = \frac{2}{1+3e^{2x}}$, the y intercept is $f(0) = \frac{2}{1+3e^0} = \frac{2}{4} = \frac{1}{2}$. As x increases without bound, e^{2x} increases without bound. Hence, the denominator $1 + 3e^{2x}$ increases without bound and $f(x)$ approaches zero. As x decreases without bound, e^{2x} approaches zero. Hence, the denominator $1 + 3e^{2x}$ approaches $1 + 0 = 1$ and $f(x)$ approaches $\frac{2}{1} = 2$.

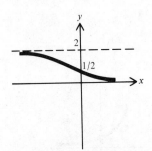

Graph for Problem 13.

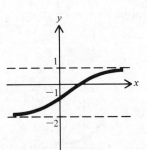

Graph for Problem 15.

15. If $f(x) = 1 - \frac{6}{2+e^{3x}}$, the y intercept is $f(0) = 1 - \frac{6}{2+e^0} = 1 - \frac{6}{2+1} = 1 - 2 = -1$. As x increases without bound, e^{3x} increases without bound. Hence, $\frac{6}{2+e^{3x}}$ approaches zero and $f(x)$ approaches $1 - 0 = 1$. As x decreases without bound, e^{3x} approaches zero. Hence, $\frac{6}{2+e^{3x}}$ approaches $\frac{6}{2+0} = 3$ and $f(x)$ approaches $1 - 3 = -2$.

17. If $f(x) = e^{kx}$ and $f(3) = e^{3k} = 2$, then $f(9) = e^{9k} = (e^{3k})^3 = 2^3 = 8$.

19. If $f(x) = 50 - Ae^{-kx}$ and $f(0) = 20$, it follows that

$$20 = 50 - A \quad \text{or} \quad A = 30$$

Hence, $f(x) = 50 - 30e^{-kx}$. Since $f(2) = 30$,

$$30 = 50 - 30e^{-2k}, \quad -20 = -30e^{-2k}, \quad \text{or} \quad e^{-2k} = \frac{2}{3}$$

Hence,

$$f(4) = 50 - 30e^{4k} = 50 - 30(e^{2k})^2 = 50 - 30\left(\frac{2}{3}\right)^2 = \frac{110}{3}$$

21. If P dollars is invested at an annual interest rate r and interest is compounded k times per year, the balance after t years will be

$$B(t) = P\left(1 + \frac{r}{k}\right)^{kt} \quad \text{dollars}$$

and if interest is compounded continously, the balance will be

$$B(t) = Pe^{rt} \text{ dollars}$$

(a)   If $P = 1,000, r = 0.07, t = 10$, and $k = 1$, then

$$B(10) = 1,000 \left[1 + \frac{0.07}{4} \right]^{10} = 1,000(1.07)^{10} \simeq \$1,967.15$$

(b)   If $P = 1,000, r = 0.07, t = 10$, and $k = 4$, then

$$B(10) = 1,000 \left[1 + \frac{0.07}{1} \right]^{40} \simeq \$2,001.60$$

(c)   If $P = 1,000, r = 0.07, t = 10$, and $k = 12$, then

$$B(10) = 1,000 \left[1 + \frac{0.07}{12} \right]^{120} \simeq \$2,009.66$$

(d)   If $P = 1,000, r = 0.07, t = 10$, and interest is compounded continously, then

$$B(10) = 1,000e^{0.7} \simeq \$2,013.75$$

23.   (a)   Since $B = Pe^{rt}$, then $P = Be^{-rt}$.

(b)   If $r = 0.06, t = 10$, and $B = 10,000$, then, from part (a),

$$P = 10,000e^{-0.6} \simeq \$5,488.12$$

25.   The following BASIC program evaluates $\left(1 + \frac{1}{n}\right)^{n}$ for $n = 1000, 2000, ...,$
50,000, and presents the output in tablular form.

```
10 PRINT "N", "(1 + 1/N) ↑ N"
20 PRINT" – – – – – – – – – – – – – – –"
30 FOR N = 1000 TO 50000 STEP 1000
40 PRINT N, (1 + 1/N) ↑ N
50 NEXT N
60 END
```

Chapter 4, Section 2

1.   The population in t years will be $P(t) = 50e^{0.02t}$ million.

   (a)   The current population is $P(0) = 50e^0 = 50$ million.

   (b)   The population 30 years from now will be

$$P(30) = 50e^{0.02(30)} = 50e^{0.6} \simeq 91.11 \text{ million}$$

3.   Let $P(t)$ denote the population (in millions) t years after 1984. Since the population grows exponentially, $P(t) = P_0 e^{kt}$, where P_0 is the intitial population (in 1984), which was 60 million. Hence $P(t) = 60e^{kt}$. Moreover, since the population was 90 million in 1989 (when $t = 5$),

$$90 = P(5) = 60e^{5k} \text{ or } e^{5k} = \frac{3}{2}$$

The population in 1999 will be

$$P(15) = 60e^{15k} = 60(e^{5k})^3 = 60\left(\frac{3}{2}\right)^3 = 202.5 \text{ million}$$

5.   Let $G(t)$ denote the gross national product (GNP) in billions t years after 1975. Since the GNP grows exponentially and was 100 billion in 1975 (when $t = 0$), $G(t) = 100e^{kt}$. Moreover, since the GNP was 180 billion in 1985 (when $t = 10$),

$$180 = G(10) = 100e^{10k} \text{ or } e^{10k} = \frac{180}{100} = \frac{9}{5}$$

The GNP in 1995 (when $t = 20$) is

$$G(20) = 100e^{20k} = 100(e^{10k})^2 = 100\left(\frac{9}{5}\right)^2 = 324 \text{ billion dollars}$$

7.   The population density x miles from the center of the city is $D(x) = 12e^{-0.07x}$ thousand people per square mile.

   (a)   At the center of the city, the density is $D(0) = 12$ thousand people per square mile

   (b)   Ten miles from the center, the density is

$$D(10) = 12e^{-0.07(10)} = 12e^{-0.7} \simeq 5.959$$

that is, 5,959 people per square mile.

9. Let $Q(t)$ denote the amount of the radioactive substance present after t years. Since the decay is exponential and 500 grams were present initially, $Q(t) = 500e^{-kt}$. Moreover, since 400 grams are present 50 years later,

$$400 = Q(50) = 500e^{-50k} \quad \text{or} \quad e^{-50k} = \frac{4}{5}$$

The amount present after 200 years will be

$$Q(200) = 500e^{-200k} = 500(e^{-50k})^4 = 500 \left(\frac{4}{5}\right)^4 \simeq 204.8 \text{ grams}$$

11. (a) The reliability function is $f(t) = 1 - e^{-0.03t}$. As t increases without bound, $e^{-0.03t}$ approaches zero and so $f(t)$ approaches 1. Furthermore, $f(0) = 0$. The graph, shown below, is like that of a learning curve.

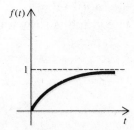

Reliability function for Problem 11.

(b) The fraction of tankers that sink in fewer than 10 days is

$$f(10) = 1 - e^{-0.03(10)} = 1 - e^{-0.3}$$

The fraction that remain afloat for at least 10 days is therefore

$$1 - f(10) = 1 - (1 - e^{-0.3}) = e^{-0.3} \simeq 0.7408$$

(c) The fraction of tankers that can be expected between the 15th and 20th days is

$$f(20) - f(15) = (1 - e^{-0.6}) - (1 - e^{-0.45})$$
$$= -e^{-0.6} + e^{-0.45} \simeq -0.5488 + 0.6373 = 0.0888$$

13. (a) The number of facts recalled after t minutes is $Q(t) = A(1 - e^{-kt})$, where k is a positive constant. As t increses without bound, e^{-kt} approaches zero and so $Q(t)$ approaches A. Moreover, $Q(0) = 0$. The graph, shown below is like that of a learning curve.

(b)   As t increases without bound, $Q(t)$ approaches A, the total number of relevant facts in the person's memory.

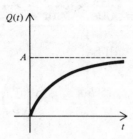

Recall function for Problem 13.

15.   (a)   The resale value of the machine when it is t years old is $V(t) = 4,800e^{-t/5} + 400$ dollars. As t increases without bound, $e^{-t/5}$ approaches zero and $V(t)$ approaches 400 dollars.
Moreover, $V(0) = 5,200$ dollars. The graph is shown below.

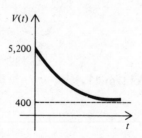

Resale value function for Problem 15.

(b)   When the machine was new, its value was $V(0) = \$5,200$.

(c)   After 10 years, the value of the machine is

$$V(10) = 4,800e^{-10/5} + 400 = 4,800e^{-2} + 400 \simeq \$1,049.61$$

17.   The temperature of the drink t minutes after leaving the refrigerator is $f(t) = 30 - Ae^{-kt}$. Since the temperature of the drink when it left the refrigerator was 10 degrees Celsius,

$$10 = f(0) = 30 - A \text{ or } A = 20$$

Thus, $f(t) = 30 - 20e^{-kt}$. Since the temperature of the drink was 15 degrees Celsius 20 minutes later,

$$15 = f(20) = 30 - 20e^{-20k} \text{ or } e^{-20k} = \frac{-15}{-20} = \frac{3}{4}$$

The temperature of the drink after 40 minutes is therefore

$$f(40) = 30 - 20e^{-40k} = 30 - 20(e^{-20k})^2 = 30 - 20\left(\frac{3}{4}\right)^2 = 18.75 \text{ degrees}$$

19.   (a)   The population t years from now is

$$P(t) = \frac{20}{2 + 3e^{-0.06t}} \text{ million}$$

The vertical intercept of this population function is

$$P(0) = \frac{20}{2 + 3e^0} = \frac{20}{5} = 4 \text{ million}$$

As t increases without bound, $e^{-0.06t}$ approaches zero. Hence,

$$\lim_{t \to \infty} P(t) = \frac{20}{2 + 0} = 10 \text{ million}$$

As t decreases without bound, $e^{-0.06t}$ increases without bound. Hence, the denominator $2 + 3e^{-0.06t}$ increases without bound and so

$$\lim_{t \to -\infty} P(t) = 0$$

The graph is shown below.

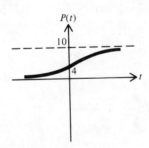

Graph for Problem 19.

(b)   The current population is $P(0) = 4$ million

(c)   Fifty years from now, the population wil be

$$P(50) = \frac{29}{2 + 3e^{-3}} \simeq 9.31 \text{ million}$$

(d)   In the long run, (as t increases without bound), $e^{-0.06t}$ approaches zero, and so $P(t)$ approaches $\frac{20}{2} = 10$ million.

21.   The number of residents who heard about the accident t hours later is

$$f(t) = \frac{B}{1 + Ce^{-kt}}$$

where B is the population of the town. Since $\frac{1}{10}$ of the residents $\left(\frac{B}{10}\right)$ witnessed the accident (when $t = 0$),

$$\frac{B}{10} = f(0) = \frac{B}{1 + C} \quad \text{or} \quad C = 9$$

Hence $f(t) = \frac{B}{1 + 9e^{-kt}}$. Since $\frac{1}{4}$ of the residents had heard about the accident after 2 hours,

$$\frac{B}{4} = f(2) = \frac{B}{1 + 9e^{-2k}}, \quad 4 = 1 + 9e^{-2k}, \quad \text{or} \quad e^{-2k} = \frac{1}{3}$$

One half of the residents heard the news when $f(t) = \frac{B}{2}$. That is, when

$$\frac{B}{2} = \frac{B}{1 + 9e^{-kt}} \qquad 2 = 1 + 9e^{-kt}$$

or

$$e^{-kt} = \frac{1}{9} = \left(\frac{1}{3}\right)^2 = (e^{-2k})^2 = e^{-4k}$$

which implies that $t = 4$ hours.

Chapter 4, Section 3

1. To use a calcualtor with Algebraic Operating System (AOS) to find $\ln a$, enter the number a and then press the $\ln x$ key. This gives $\ln 1 = 0$, $\ln 2 = 0.6931472$, $\ln 5 = 1.6094379$, $\ln \frac{1}{5} = \ln 0.2 = -1.5094379$. To find $\ln e^n$, enter the number n, then press the e^x key, and then the $\ln x$ key. This gives $\ln e = \ln e^1 = 1$ and $\ln e^2 = 2$ as expected since $\ln x$ and e^x are inverse functions and $\ln e^n = n$. For $\ln 0$ and $\ln(-2)$, the calculator displays an error signal (such as the letter E or a flashing light). This is because $\ln x$ is defined only for $x > 0$.

3. $\ln \sqrt{e} = \ln e^{1/2} = \frac{1}{2}$

5. $e^{2 \ln 3} = e^{\ln 3^2} = e^{\ln 9} = 9$

7. $\ln \frac{e^3 \sqrt{e}}{e^{1/3}} = \ln \frac{e^3 e^{1/2}}{e^{1/3}} = \ln e^{3+1/2-1/3} = \ln e^{19/6} = \frac{19}{6}$

9. If $\frac{1}{2}Q_0 = Q_0 e^{-1.2x}$, then $\frac{1}{2} = e^{-1.2x}$, and taking the natural logarithm of each side of the equation yields

$$\ln \frac{1}{2} = \ln e^{-1.2x}, \quad \ln 1 - \ln 2 = -1.2x, \quad \text{or } x = \frac{\ln 2}{1.2} \simeq 0.58$$

11. If $-2 \ln x = b$, then $\ln x = -\frac{b}{2}$, and applying the exponetial function to each side of the equation yields

$$e^{\ln x} = e^{-b/2} \quad \text{or } x = e^{-b/2}$$

13. If $5 = 3 \ln x - \frac{1}{2} \ln x$, then

$$5 = \ln x^3 - \ln x^{1/2} = \ln \frac{x^3}{x^{1/2}} = \ln x^{5/2}$$

and applying the exponential function to each side of the equation yields

$$e^5 = e^{\ln x^5/2} = x^{5/2}$$

Raising each side of this equation to the power $\frac{2}{5}$ gives

$$(x^{5/2})^{2/5} = (e^5)^{2/5} \quad \text{or } x = e^2 \simeq 7.39$$

15. If $\ln x = 2(\ln 3 - \ln 5)$, then

$$\ln x = 2(\ln \frac{3}{5}) = \ln(\frac{3}{5})^2 = \ln \frac{9}{25} \quad \text{or } x = \frac{9}{25}$$

17. If $a^k = e^{kx}$, taking the natural logarithm of each side of the equation yields

$$\ln a^k = \ln e^{kx}, \ k \ln a = kx, \ \text{or } x = \ln a$$

19. If $x^{\ln x} = e$, taking the natural logarithm of each side of the equation yields

$$\ln(x^{\ln x}) = \ln e, \ (\ln x)(\ln x) = 1, \ (\ln x)^2 = 1$$

$$\ln x = 1 \text{ and } \ln x = -1, \text{ or } x = e^1 = e \text{ and } x = e^{-1} = \frac{1}{e}$$

21. If $\ln b = 6$ and $\ln c = -2$, then

$$\frac{1}{a} \ln \left[\frac{\sqrt{b}}{c} \right]^a = \frac{a}{a} \ln \frac{b^{1/2}}{c} = \ln b^{1/2} - \ln c$$

$$= \frac{1}{2} \ln b - \ln c = \frac{1}{2}(6) - (-2) = 5$$

23. The balance after t years is $B(t) = Pe^{rt}$, where P is the intitial investment and r is the interest rate compounded continuously. Since money doubles in 13 years,

$$2P = B(13) = Pe^{13r}, \ 2 = e^{13r}, \ \ln 2 = 13r, \ \text{ or } r = \frac{\ln 2}{13} \simeq 0.0533$$

That is, the annual interest rate is 5.33 percent.

25. Let t denote the number of years after 1960. If the population P (measured in billions) is growing exponentially and was 3 billion in 1960 (when $t = 0$), then $P(t) = 3e^{kt}$. Since the population in 1975 (when $t = 15$) was 4 billion,

$$4 = P(15) = 3e^{15k}, \ e^{15k} = \frac{4}{3}, \ 15k = \ln \frac{4}{3}, \ \text{or } k = \frac{1}{15} \ln \frac{4}{3}$$

To find t for which $P(t) = 40$,

$$3e^{kt} = 40, \ e^{kt} = \frac{40}{3}, \ kt = \ln \frac{40}{3}, \ \text{or } t = \frac{\ln 40/3}{k}$$

Since $k = \frac{1}{15} \ln \frac{4}{3}$, it follows that

$$t = \frac{15 \ln 40/3}{\ln 4/3} \simeq 135$$

that is, the population will reach 40 billion in 2095 (135 years after 1960).

27. Since radium decays exponentially, the amount present after t years is $Q(t) = Q_0 e^{-kt}$, where Q_0 is the amount initially present. Since the half-life of radium is $1,690$ years,

$$\frac{1}{2}Q_0 = Q(1,690) = Q_0 e^{-1,690k}$$

$$e^{-1,690k} = \frac{1}{2}, \; -1,690k = \ln\frac{1}{2} = \ln 1 - \ln 2 = -\ln 2, \; \text{ or } \; k = \frac{\ln 2}{1,690}$$

Given $Q_0 = 50$, the goal is to find t so that $Q(t) = 5$. That is,

$$5 = 50e^{-kt}, \; e^{-kt} = \frac{1}{10}, \; -kt = \ln\frac{1}{10} = -\ln 10, \; \text{ or } \; t = \frac{\ln 10}{k}$$

Since $k = \frac{\ln 2}{1,690}$, it follows that $t = \frac{1,690 \, \ln 10}{\ln 2} \simeq 5,614.06$ years .

29. If λ is the half-life of a radioactive substance, then (from Problem 28) the amount $Q(t)$ of the substance remaining after t years is

$$Q(t) = Q_0 e^{-(\ln 2/\lambda)t}$$

where Q_0 is the amount initially present. Since

$$e^{-(\ln 2/\lambda)t} = e^{(-\ln 2)(t/\lambda)} = e^{(\ln 1/2)(t/\lambda)} = (e^{\ln 1/2})^{t/\lambda} = \left(\frac{1}{2}\right)^{t/\lambda}$$

it follows that $Q(t) = Q_0 \left(\frac{1}{2}\right)^{t/\lambda}$.

31. The number of bacteria is $Q(t) = Q_0 e^{kt}$. Since $6,000$ bacteria were present initially, $Q_0 = 6,000$ and so $Q(t) = 6,000 e^{kt}$. Since $9,000$ bacteria were present after 20 minutes,

$$9,000 = Q(20) = 6,000 e^{20k}, \; e^{20k} = \frac{3}{2}, \; 20k = \ln\frac{3}{2}, \; \text{ or } \; k = \frac{1}{20}\ln\frac{3}{2} \simeq 0.02$$

Hence the desired function is $Q(t) = 6,000 e^{0.02t}$.

33. Since the output after t months is $Q(t) = 500 - Ae^{-kt}$ and the output of a worker with no experience is 300, it follows that

$$300 = Q(0) = 500 - Ae^0 = 500 - A \; \text{ or } \; A = 200$$

Thus, $Q(t) = 500 - 200e^{-kt}$. Since $Q(6) = 410$,

$$410 = 500 - 200e^{-6k}, \; -200e^{-6k} = -90,$$

$$e^{-6k} = \frac{9}{20}, \; -6k = \ln\frac{9}{20}, \; \text{ or } \; k = -\frac{1}{6}\ln\frac{9}{20} \simeq 0.133$$

Thus the desired function is $Q(t) = 500 - 200e^{-0.133t}$.

35. The age of the fossil is the value of t for which $R(t) = \frac{1}{3}R_0$, that is, for which

$$\frac{1}{3}R_0 = R_0 e^{-kt}, \frac{1}{3} = e^{-kt}, -kt = \ln\frac{1}{3} = -\ln 3, \text{ or } t = \frac{\ln 3}{k}$$

Since $k = \frac{\ln 2}{5,730}$ (from page 242 of the text), it follows that

$$t = \frac{\ln 3}{k} = \frac{5,730 \ln 3}{\ln 2} \simeq 9,082 \text{ years.}$$

37. Since e^x and $\ln x$ are inverses,

$$e^{\ln uv} = uv, \ e^{\ln u} = u, \text{ and } e^{\ln v} = v$$

Hence,

$$e^{\ln(u/v)} = \frac{u}{v} = \frac{e^{\ln u}}{e^{\ln v}} = e^{(\ln u - \ln v)}$$

and taking the natural logarithm gives $\ln\frac{u}{v} = \ln u - \ln v$.

39. Let $u = \log_a x$. Then, by definition of $\log_a x, a^u = x$. Taking the natural logarithm of each side of this equation yields

$$u \ln a = \ln x, u = \frac{\ln x}{\ln a}, \text{ or } \log_a x = \frac{\ln x}{\ln a}$$

Chapter 4, Section 4

1. If $f(x) = e^{5x}$, then $f'(x) = e^{5x}\frac{d}{dx}(5x) = 5e^{5x})$.

3. If $f(x) = e^{x^2+2x-1}$, then

$$f'(x) = e^{x^2+2x-1}\frac{d}{dx}(x^2 + 2x - 1) = (2x + 2)e^{x^2+2x-1}$$

5. If $f(x) = 30 + 10e^{-0.05x}$, then

$$f'(x) = 0 + 10e^{-0.05x}\frac{d}{dx}(-0.05x) = 10e^{-0.05x}(-0.05) = -0.5e^{-0.05x}$$

7. If $f(x) = (x^2 + 3x + 5)e^{6x}$, then

$$f'(x) = (x^2 + 3x + 5)\frac{d}{dx}(e^{6x}) + e^{6x}\frac{d}{dx}(x^2 + 3x + 5)$$
$$= (x^2 + 3x + 5)(e^{6x})(6) + e^{6x}(2x + 3) = (6x^2 + 20x + 33)e^{6x}$$

9. If $f(x) = \frac{x}{e^x} = xe^{-x}$, then

$$f'(x) = x\frac{d}{dx}(e^{-x}) + e^{-x}\frac{d}{dx}(x) = -xe^{-x} + e^{-x} = \frac{1-x}{e^x}$$

11. If $f(x) = (1 - 3e^x)^2$, then

$$f'(x) = 2(1 - 3e^x)\frac{d}{dx}(1 - 3e^x) = 2(1 - 3e^x)(-3e^x) = -6e^x(1 - 3e^x)$$

13. If $f(x) = e^{\sqrt{3x}} = e^{(3x)^{1/2}}$, then

$$f'(x) = e^{\sqrt{3x}}\frac{d}{dx}(3x)^{1/2} = e^{\sqrt{3x}}\left[\frac{1}{2}(3x)^{-1/2}\frac{d}{dx}(3x)\right]$$
$$= e^{\sqrt{3x}}\left[\frac{1}{2}(3x)^{-1/2}(3)\right] = \frac{3}{2\sqrt{3x}}e^{\sqrt{3x}}$$

15. If $f(x) = \ln x^3$, then

$$f'(x) = \frac{1}{x^3}\frac{d}{dx}(x^3) = \frac{3x^2}{x^3} = \frac{3}{x}$$

An alternate solution is to rewrite the function as $f(x) = 3 \ln x$. Then,

$$f'(x) = 3\left(\frac{1}{x}\right) = \frac{3}{x}$$

17. If $f(x) = \ln(x^2 + 5x - 2)$, then

$$f'(x) = \frac{1}{x^2 + 5x - 2}\frac{d}{dx}(x^2 + 5x - 2) = \frac{2x + 5}{x^2 + 5x - 2}$$

19. If $f(x) = x^2 \ln x$, then

$$f'(x) = x^2 \frac{d}{dx}(\ln x) + (\ln x)\frac{d}{dx}(x^2)$$

$$= x^2\left(\frac{1}{x}\right) + (\ln x)(2x) = x + 2x \ln x = x(1 + 2\ln x)$$

21. If $f(x) = \frac{\ln x}{x}$, then

$$f'(x) = \frac{x\frac{d}{dx}(\ln x) - (\ln x)\frac{d}{dx}(x)}{x^2} = \frac{x(1/x) - (\ln x)(1)}{x^2} = \frac{1 - \ln x}{x^2}$$

23. If $f(x) = \ln\left(\frac{x+1}{x-1}\right)$, then

$$f'(x) = \frac{x-1}{x+1}\frac{d}{dx}\left[\frac{x+1}{x-1}\right] = \frac{x-1}{x+1}\left[\frac{(x-1)(1) - (x+1)(1)}{(x-1)^2}\right]$$

$$= \frac{x-1}{x+1}\left[\frac{-2}{(x-1)^2}\right] = \frac{-2}{(x+1)(x-1)} = \frac{-2}{x^2-1}$$

25. If $f(x) = \ln e^{2x}$, then

$$f'(x) = \frac{1}{e^{2x}}\frac{d}{dx}(e^{2x}) = \frac{1}{e^{2x}}(2e^{2x}) = 2$$

An alternate (and easier) solution is to use the inverse properties of the logarithm and exponential function to rewrite the function as $f(x) = 2x$, from which it is clear that $f'(x) = 2$.

27. Differentiate both sides of the equation $e^{xy} = xy^2$ with respect to x to get

$$\left[\frac{d}{dx}(xy)\right]e^{xy} = y^3 + 3xy^2\frac{dy}{dx}$$

$$\left[x\frac{dy}{dx} + y\right]e^{xy} = y^3 + 3xy^2\frac{dy}{dx}$$

Solve for $\frac{dy}{dx}$ to get

$$xe^{xy}\frac{dy}{dx} + ye^{xy} = y^3 + 3xy^2\frac{dy}{dx}$$

$$(xe^{xy} - 3xy^2)\frac{dy}{dx} = y^3 - ye^{xy}$$

$$\frac{dy}{dx} = \frac{y^3 - ye^{xy}}{xe^{xy} - 3xy^2} = \frac{y(y^2 - e^{xy})}{x(e^{xy} - 3y^2)}$$

To simplify the answer further, substitute $e^{xy} = xy^3$ (from the original equation) to get

$$\frac{dy}{dx} = \frac{y(y^2 - xy^3)}{x(xy^3 - 3y^2)} = \frac{y^3(1 - xy)}{xy^2(xy - 3)} = \frac{y(1 - xy)}{x(xy - 3)}$$

29.   To simplify the calculation, first write the equation $\ln \frac{y}{x} = x^2y^3$ as

$$\ln y - \ln x = x^2y^3$$

Then differentiate both sides with respect to x to get

$$\frac{1}{y}\frac{dy}{dx} - \frac{1}{x} = 2xy^3 + 3x^2y^2\frac{dy}{dx}$$

and solve for $\frac{dy}{dx}$ as follows:

$$x\frac{dy}{dx} - y = 2x^2y^4 + 3x^3y^3\frac{dy}{dx}$$

$$(x - 3x^3y^3)\frac{dy}{dx} = 2x^2y^4 + y \quad \text{or} \quad \frac{dy}{dx} = \frac{y(2x^2y^3 + 1)}{x(1 - 3x^2y^3)}$$

31.   (a)   The population t years from now will be $P(t) = 50e^{0.02t}$ million. Hence the rate of change of the population t years from now will be

$$P'(t) = 50e^{0.02t}(0.02) = e^{0.02t}$$

and the rate of change 10 years from now will be $P'(10) = e^{0.2} \simeq 1.22$ million per year.

(b)   The percentage rate of change t years from now will be

$$100\left[\frac{P'(t)}{P(t)}\right] = 100\left[\frac{e^{0.02t}}{50e^{0.02t}}\right] = 100\left(\frac{1}{50}\right) = 2$$

percent per year, which is a constant, independent of time.

33. (a) The value of the machine after t years is $Q(t) = 20,000e^{-0.4t}$ dollars. Hence, the rate of depreciation after t years is

$$Q'(t) = 20,000e^{-0.4t}(-0.4) = -8,000e^{-0.4t}$$

and the rate after 5 years is $Q'(5) = -8,000e^{-2} \simeq \$1,082.68$ per year.

(b) The percentage rate of change t years from now will be

$$100\left[\frac{Q'(t)}{Q(t)}\right] = 100\left[\frac{-8,000e^{-0.4t}}{20,000e^{-0.4t}}\right] = 100\frac{-8,000}{20,000} = -40$$

percent per year, which is a constant, independent of time.

35. If Q decreases exponentially, then $Q(t) = Ae^{-kt}$, where A and k are constants and $k > 0$. Then, $Q'(t) = -kAe^{-kt}$, and the percentage rate of change is

$$100\left[\frac{Q'(t)}{Q(t)}\right] = 100\left[\frac{-kAe^{-kt}}{Ae^{-kt}}\right] = -100k$$

which is constant, independent of time.

37. (a) If the world's population in billions t years after 1960 is

$$P(t) = \frac{40}{1 + 12e^{-0.08t}}$$

then the rate of change of the population will be

$$P'(t) = \frac{(1 + 12e^{-0.08t})(0) - 40[12e^{-0.08t}(-0.08)]}{(1 + 12e^{-0.08t})^2}$$
$$= \frac{38.4e^{-0.08t}}{(1 + 12e^{-0.08t})^2}$$

and the rate of change in 1995 is

$$P'(35) = \frac{38.4e^{-2.8}}{(1 + 12e^{-2.8})^2} \simeq 0.7805 \text{ billions per year}$$

(b) The percentage rate of change in 1995 is

$$100\left[\frac{P'(35)}{P(35)}\right] = 100\left[\frac{0.7805}{P(35)}\right]$$

Since $P(35) = \frac{40}{1+12e^{-2.8}} \simeq 23.1251$, it follows that

$$100\left[\frac{P'(35)}{P(35)}\right] = 100\frac{0.7805}{23.1251} \simeq 3.375 \text{ percent per year}$$

39. (a) First year sales of the text will be $f(x) = 20 - 15e^{-0.2x}$ thousand copies when x thousand complementary copies are distributed. If the number of complementary copies distributed is increased from $10,000(x = 10)$ by $1,000(\Delta x = 1)$, the approximate change in sales is $\Delta f \simeq f'(10)\Delta x$. Since $f'(x) = 3e^{-0.2x}$ and $\Delta x = 1$, it follows that

$$\Delta f \simeq f'(10) = 3e^{-2} \simeq 0.406 \text{ thousand}$$

or 406 copies.

(b) The actual change in sales is

$$\Delta f = f(11) - f(10) = [20 - 15e^{-2.2}] - [20 - 15e^{-2}] \simeq 0.368 \text{ thousand}$$

or 368 copies. Thus the estimate in part (a) is not a particularly good one.

41. Let $P(x)$ denote the profit. Then

$$P(x) = \text{(number of radios sold) (Profit per radio)} = (1,000e^{-0.1x})(x - 5)$$

To find the critical points of $P(x)$, use the product rule to get the derivative

$$P'(x) = (1,000e^{-0.1x})(1) + (x - 5)(1,000e^{-0.1x})(-0.1)$$
$$= 1,000e^{-0.1x}[1 - 0.1(x - 5)] = 100e^{-0.1x}(15 - x)$$

Since $100e^{-0.1x} > 0$, $P'(x)$ is zero only when $15 - x = 0$ or $x = 15$. Since $P'(x) > 0$ (and P is increasing) for $x < 15$, and $P'(x) < 0$ (and P is decreasing) for $x > 15$, it follows that $P(x)$ has an absolute maximum when $x = \$15$.

43. If $f(x) = xe^{-x}$, the first derivative is

$$f'(x) = -xe^{-x} + e^{-x} = e^{-x}(1 - x)$$

which is zero only when $x = 1$ (since e^{-x} is always positive). The corresponding first-order critical point is $(1, e^{-1}) = (1, \frac{1}{e})$. The second derivative is

$$f''(x) = -e^{-x} + (1 - x)e^{-x}(-1) = xe^{-x} - 2e^{-x} = e^{-x}(x - 2)$$

which is zero when $x = 2$. The corresponding second-order critical point is $(2, 2e^{-2}) = (2, \frac{2}{e^2})$.

Interval	$f'(x)$	$f''(x)$	Inc/dec	Concavity
$x < 1$	+	−	increasing	down
$1 < x < 2$	−	−	decreasing	down
$x > 2$	−	+	decreasing	up

Notice that the first-order critical point $\left(1, \frac{1}{e}\right)$ is a relative maximum and the second-order critical point $\left(2, \frac{2}{e^2}\right)$ is an inflection point.

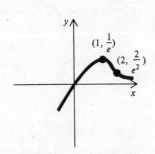

Graph for Problem 43.

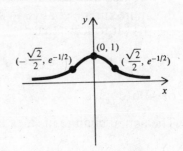

Graph for Problem 45.

45. If $f(x) = e^{-x^2}$, the first derivative is $f'(x) = -2xe^{-x^2}$, which is zero only when $x = 0$ (since e^{-x^2} is always positive). The corresponding first-order critical point is $(0, 1)$. The second derivative is

$$f''(x) = (-2x)(e^{-x^2})(-2x) + (e^{-x^2})(-2) = 2e^{-x^2}(2x^2 - 1)$$

which is zero when $x = \pm\frac{\sqrt{2}}{2}$. The corresponding second-order critical points are $(\pm\frac{\sqrt{2}}{2}, e^{-1/2}) = (\pm\frac{\sqrt{2}}{2}, \frac{1}{e^{1/2}})$.

Interval	$f'(x)$	$f''(x)$	Inc/dec	Concavity
$x < -\frac{\sqrt{2}}{2}$	+	+	increasing	up
$-\frac{\sqrt{2}}{2} < x < 0$	+	−	increasing	down
$0 < x < \frac{\sqrt{2}}{2}$	−	−	decreasing	down
$x > -\frac{\sqrt{2}}{2}$	−	+	decreasing	up

Notice that the first-order critical point $(0, 1)$ is a realtive maximum and that the second-order critical points are inflection points.

47. If $f(x) = e^x + e^{-x}$, the first derivative is

$$f'(x) = e^x - e^{-x}$$

which is zero when $x = 0$. The corresponding first-order critical point is $(0, 2)$. The second derivative is

$$f''(x) = e^x + e^{-x}$$

which is never zero. Hence there are no second-order critical points.

Interval	$f'(x)$	$f''(x)$	Inc/dec	Concavity
$x < 0$	−	+	decreasing	up
$x > 0$	+	+	increasing	up

Notice that the first-order critical point $(0, 2)$ is a relative minimum and that the graph is always concave up.

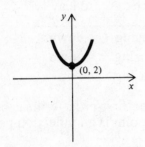

Graph for Problem 47.

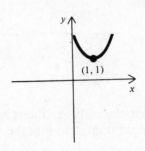

Graph for Problem 49.

49. If $f(x) = x - \ln x$ (for $x > 0$), the first derivative is

$$f'(x) = 1 - \frac{1}{x} = \frac{x-1}{x}$$

which is zero when $x = 1$. The corresponding first-order critical point is $(1, 1)$. The second derivative is

$$f''(x) = \frac{1}{x^2}$$

which is never zero. Hence there are no second-order critical points.

Interval	$f'(x)$	$f''(x)$	Inc/dec	Concavity
$0 < x < 1$	$-$	$+$	decreasing	up
$x > 1$	$+$	$+$	increasing	up

Notice that the first-order critical point $(1, 1)$ is a relative minimum and that the graph is always concave up.

51. If $f(x) = \frac{\ln x}{x}$ (for $x > 0$), the first derivative is (by the quotient rule)

$$f'(x) = \frac{x(\frac{1}{x}) - \ln x}{x^2} = \frac{1 - \ln x}{x^2}$$

which is zero when

$$1 - \ln x = 0 \quad \ln x = 1 \quad \text{or} \quad x = e$$

The corresponding point $(e, e^{-1}) \simeq (e, 0.37)$ is the only first-order critical point. The second derivative is

$$f''(x) = \frac{x^2(-\frac{1}{x}) - (1 - \ln x)(2x)}{x^4} = \frac{-x - 2x + 2x \ln x}{x^4} = \frac{2 \ln x - 3}{x^3}$$

which is zero when

$$2 \ln x - 3 = 0 \quad \ln x = \frac{3}{2} \quad \text{or} \quad x = e^{3/2}$$

The corresponding point $(e^{3/2}, \frac{3}{2}e^{-3/2}) \simeq (e^{3/2}, 0.33)$ is the only second-order critical point.

Interval	$f'(x)$	$f''(x)$	Inc/dec	Concavity
$0 < x < e$	+	−	increasing	down
$e < x < e^{2/3}$	−	−	decreasing	down
$x > e^{2/3}$	−	+	decreasing	up

Notice that the only x intercept is $(1,0)$, that is the first-order critical point is a relative maximum, and that the second-order critical point is an inflection point.

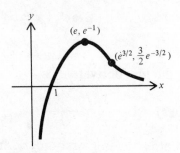

Graph for Problem 51.

53. Taking the logarithm of each side of the equation $f(x) = \frac{(x+2)^5}{(3x-5)^6}$ yields

$$\ln f(x) = \ln(x+2)^5 - \ln(3x-5)^6 = 5\ln(x+2) - 6\ln(3x-5)$$

Differentiating both sides of this equation yields

$$\frac{f'(x)}{f(x)} = \frac{5}{x+2} - \frac{6}{3x-5}(3) = \frac{5}{x+2} - \frac{18}{3x-5}$$

$$f'(x) = f(x)\left[\frac{5}{x+2} - \frac{18}{3x-5}\right]$$

$$= \left[\frac{(x+2)^5}{(3x-5)^6}\right]\left[\frac{5}{x+2} - \frac{18}{3x-5}\right]$$

$$= \left[\frac{(x+2)^5}{(3x-5)^6}\right]\left[\frac{15x - 25 - 18x - 36}{(x+2)(3x-5)}\right]$$

$$= \frac{(x+2)^4(-3x-61)}{(3x-5)^7} = -\frac{(3x+61)(x+2)^4}{(3x-5)^7}$$

55. Taking the natural logarithm of each side of the equation $f(x) = 2^x$ yields

$$\ln f(x) = x \ln 2$$

Differentiating both sides of this equation yields

$$\frac{f'(x)}{f(x)} = \ln\ 2, \quad \text{or} \quad f'(x) = f(x)\ln\ 2 = 2^x\ \ln\ 2$$

57. Taking the natural logarithm of each side of the equation $f(x) = x^x$ gives

$$\ln\ f(x) = x\ \ln\ x$$

Differentiating both sides of this equation (using the product rule) gives

$$\frac{f'(x)}{f(x)} = x\left(\frac{1}{x}\right) + \ln\ x, \quad \text{or} \quad f'(x) = f(x)(1 + \ln\ x) = x^x(1 + \ln\ x)\ .$$

59. Taking the natural logarithm of each side of the equation $f(x) = (2^x)^{\ln\ x}$ gives

$$\ln\ f(x) = \ln x\ (\ln\ 2^x) = \ln x\ (x\ \ln\ 2) = (\ln\ 2)(x\ \ln\ x)$$

Differentiating both sides of this equation (using the product rule) gives

$$\frac{f'(x)}{f(x)} = (\ln\ 2)\left[x\left(\frac{1}{x}\right) + \ln\ x\right] = (\ln\ 2)(1 + \ln\ x)$$

and so

$$f'(x) = f(x)(\ln\ 2)(1 + \ln\ x) = (2^x)^{\ln\ x}(\ln\ 2)(1 + \ln\ x)$$

61. The population of the country is

$$P(t) = \frac{160}{1 + 8e^{-0.01t}}$$

The rate of change of the population is

$$P'(t) = \frac{(1 + 8e^{-0.01t})(0) - 160(8e^{-0.01t})(-0.01)}{(1 + 8e^{-0.01t})^2}$$

$$= \frac{12.8e^{-0.01t}}{(1 + 8e^{-0.01t})^2}$$

To determine when the population is growing most rapidly, find the absolute maximum of $P'(t)$ by setting its deriative, $P''(t)$, equal to zero and solving for t. By the quotient rule,

$$P''(t) = \frac{(1 + 8e^{-0.01t})^2(12.8e^{-0.01t})(-0.01)}{(1 + 8e^{-0.01t})^4}$$

$$- \frac{(12.8e^{-0.01t})(2)(1 + 8e^{-0.01t})(8e^{-0.01t})(-0.01)}{(1 + 8e^{-0.01t})^4}$$

$$= \frac{(-0.01)(12.8e^{-0.01t})(1 + 8e^{-0.01t} - 16e^{-0.01t})}{(1 + 8e^{-0.01t})^3}$$

$$= -\frac{0.128e^{-0.01t}(1 - 8e^{-0.01t})}{(1 + 8e^{-0.01t})^3}$$

which is zero when $1 - 8e^{-0.01t} = 0$ or $e^{-0.01t} = \frac{1}{8}$. Taking the natural logarithm of each side yields

$$-0.01t = \ln \frac{1}{8} = -\ln 8, \ \text{ or } \ t = \frac{\ln 8}{0.01} \simeq 207.94 \ \text{years}$$

63.   By definition, the percentage rate of change of f with respect to x is $100 \left[\frac{f'(x)}{f(x)} \right]$. Since $\frac{d}{dx}[\ln f(x)] = \frac{f'(x)}{f(x)}$, it follows that the percentage rate of change can be written as $100\frac{d}{dx}[\ln f(x)]$.

65.   The population of the town x years from now will be

$$P(x) = 500\sqrt{x^2 + 4x + 19}$$

From Problem 63, the percentage rate of change x years from now will be

$$100\frac{d}{dx}\left[\ln 500\sqrt{x^2 + 4x + 19} \right]$$
$$= 100\frac{d}{dx}\left[\ln 500 + \ln(x^2 + 4x + 19)^{1/2} \right]$$
$$= 100\frac{d}{dx}\left[\ln 500 + \frac{1}{2}\ln(x^2 + 4x + 19) \right]$$
$$= 100\left[0 + \left(\frac{1}{2}\right)\frac{2x + 4}{x^2 + 4x + 19} \right]$$
$$= \frac{100(x + 2)}{x^2 + 4x + 19}$$

Hence the percentage rate of change 3 years from now will be

$$\frac{100(3 + 2)}{3^2 + 4(3) + 19} = \frac{500}{40} = 12.5 \ \text{percent per year.}$$

Chapter 4, Section 5

1. (a) Simple interest: $B(t) = P(1 + rt)$. If $P = 5,000$, $r = 0.05$, and $t = 20$, then

$$B(20) = 5,000[1 + 0.05(20)] = \$10,000$$

(b) Interest compounded k times per year: $B(t) = P(1 + \frac{r}{k})^{kt}$. If $P = 5,000$, $r = 0.05$, and $t = 20$, and $k = 2$, then

$$B(20) = 5,000 \left[1 + \frac{0.05}{2}\right]^{40} \simeq \$13,425.32$$

(c) Interest compounded continuously: $B(t) = Pe^{rt}$. If $P = 5,000$, $r = 0.05$, and $t = 20$, then

$$B(20) = 5,000e^{0.05(20)} \simeq \$13,591.41$$

3. (a) If the interest rate is 12 percent and the interest is compounded quarterly, the balance after t years is

$$B(t) = P \left(1 + \frac{0.12}{4}\right)^{4t} = P(1.03)^{4t}$$

The doubling time is the value of t for which $B(t) = 2P$, that is,

$$2P = P(1.3)^{4t} \qquad 2 = (1.3)^{4t}$$

$$\ln 2 = 4t \ \ln 1.03 \ \text{ or } \ t = \frac{\ln 2}{4 \ \ln 1.03} \simeq 5.86 \text{ years}$$

(b) If the interest rate is 12 percent and interest is compounded continuously, the balance after t years is

$$B(t) = Pe^{0.12t}$$

The doubling time is the value of t for which $B(t) = 2P$, that is

$$2P = Pe^{0.12t} \qquad 2 = e^{0.12t}$$

$$\ln 2 = 0.12t \ \text{ or } \ t = \frac{\ln 2}{0.12} \simeq 5.78 \text{ years}$$

5. The balance after t years is $B(t) = P(1 + \frac{r}{k})^{kt}$. The doubling time is the value of t for which $B(t) = 2P$, that is,

$$2P = P(1 + \frac{r}{k})^{kt} \qquad 2 = (1 + \frac{r}{k})^{kt}$$

$$\ln 2 = kt \ \ln(1 + \frac{r}{k}) \ \text{ or } \ t = \frac{\ln 2}{k \ \ln (1 + r/k)} \text{ years}$$

7. If P dollars is invested at an annual interest rate of 6 percent compounded semi-annually, the balance after t years will be $B(t) = P(1 + \frac{0.06}{2})^{2t} = P(1.03)^{2t}$. The tripling time is the value of t for which $B(t) = 3P$, that is, for which

$$3P = P(1.03)^{2t} \quad \ln 3 = 2t \ln 1.03 \quad \text{or} \quad t = \frac{\ln 3}{2 \ln 1.03} \simeq 18.58 \text{ years}$$

9. If P dollars is invested at an annual interest rate r and interest is compounded k times per year, the balance after t years will be $B(t) = P(1 + \frac{r}{k})^{kt}$. The tripling time is the value of t for which $B(t) = 3P$, that is, for which

$$3P = P(1 + \frac{r}{k})^{kt} \qquad 3 = (1 + \frac{r}{k})^{kt}$$

$$\ln 3 = kt \ln(1 + \frac{r}{k}) \quad \text{or} \quad t = \frac{\ln 3}{k \ln(1 + r/k)}$$

11. (a) Use $B(t) = P(1 + \frac{r}{k})^{kt}$ with $P = 1,000$, $B = 2,500$, $r = 0.06$, and $k = 4$ to get

$$2,500 = 1,000(1 + \frac{0.06}{4})^{4t}, \ (1,015)^{4t} = 2.5$$

$$4t \ln 1.015 = \ln 2.5 \quad \text{or} \quad t = \frac{\ln 2.5}{4 \ln 1.015} \simeq 15.39 \text{ years}$$

(b) Use $B(t) = Pe^{rt}$ with $P = 1,000$, $B = 2,5000$, and $r = 0.06$ to get

$$2,500 = 1,000e^{0.06t}, \ 0.06t = \ln 2.5, \quad \text{or} \quad t = \frac{\ln 2.5}{0.06} \simeq 15.27 \text{ years}$$

13. (a) With $r = 0.06$ and $k = 4$,

$$\text{Effective rate} = (1 + \frac{r}{k})^k - 1 = (1 + \frac{0.06}{4})^4 \simeq 0.0614$$

or 6.14 percent.

(b) With $r = 0.06$,

$$\text{Effective rate} = e^r - 1 = e^{0.06} - 1 \simeq 0.0618 \text{ or } 6.18 \text{ percent}$$

15. Compare effective interest rates. The effective rate corresponding to 10.25 percent compounded semiannually is

$$(1 + \frac{r}{k})^k - 1 = (1 + \frac{0.1025}{2})^2 - 1 \simeq 0.1051 \text{ or } 10.51 \text{ percent}$$

The effective rate corresponding to 10.20 percent compounded continuously is

$$e^r - 1 = e^{0.1020} - 1 \simeq 0.1074 \text{ or } 10.74 \text{ percent}$$

Hence 10.20 percent compounded continuously is the better investment.

17. Use $B(t) = P(1 + \frac{r}{k})^{kt}$ with $k = 4$, $P = 1,000$, $t = 8$, and $B(8) = 2,203.76$ to get

$$2,203.76 = 1,000(1 + \frac{r}{4})^{32} = 2,203.76$$

$$(1 + \frac{r}{4}) = (2,203.76)^{1/32}, \quad \frac{r}{4} = (2,203.76)^{1/32} - 1$$

$$r = 4(2,203.76)^{1/32} - 4 \simeq 0.10 \text{ or } 10 \text{ percent}$$

19. At 6 percent compounded annually, the effective interest rate is

$$(1 + \frac{r}{k})^k - 1 = (1 + \frac{0.06}{1})^{-1} = 0.06$$

At r percent compounded continuously, the effective interest rate is $e^r - 1$. Setting the two effective rates equal to each other yields

$$e^r - 1 = 0.06, \quad e^r = 1.06, \text{ or } r = \ln 1.06 \simeq 0.0583 \text{ or } 5.83 \text{ percent}$$

21. (a) Using $P(t) = B(1 + \frac{r}{k})^{-kt}$ with $B = 20,000$, $r = 0.08$, $t = 15$, and $k = 4$,

$$P(15) = 20,000(1 + \frac{0.08}{4})^{-4(15)} \simeq \$6,095.65$$

(b) Using $P(t) = Be^{-rt}$ with $B = 20,000$, $r = 0.08$, and $t = 15$,

$$P(15) = 20,000e^{-0.08(15)} \simeq \$6,023.88$$

23. The three deposits of $\$2,000$ and the value of each (at 10 percent compounded continuously) at the end of the term is shown in the accompanying figure.

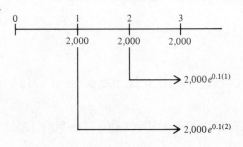

Annuity for Problem 23.

The amount of the annuity at the end of the term is the corresponding sum

$$2,000 + 2,000e^{0.1} + 2,000e^{0.2} \simeq \$6,653.15$$

25. The accompanying figure shows the sequence of $50 deposits and the value of that money (at 6 percent interest compounded semiannually) at the time of the 4 th deposit.

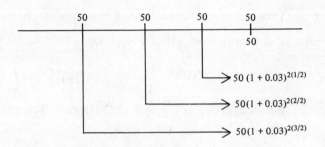

Annuity for Problem 25.

The amount after the 4 th deposit is

$$50(1 + \frac{0.06}{2})^3 + 50(1 + \frac{0.06}{2})^2 + 50(1 + \frac{0.06}{2})^1 + 50 \simeq \$209.18$$

27. The accompanying figure shows the sequence of $500 withdrawals and the present value of each withdrawal at 6 percent interest compounded annually.

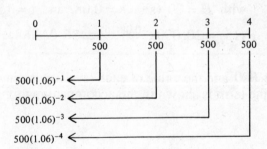

Annuity for Problem 27.

The money is to be invested now (that is, the present value of the sequence of payments) is

$$500(1.06)^{-1} + 500(1.06)^{-2} + 500(1.06)^{-3} + 500(1.06)^{-4} \simeq \$1,732.55$$

29. The percentage rate of chance of the market price $V(t) = 8,000e^{\sqrt{t}}$ of the land (expressed in decimal form) is

$$\frac{V'(t)}{V(t)} = \frac{8,000e^{\sqrt{t}}}{8,000e^{\sqrt{t}}}\left[\frac{1}{2\sqrt{t}}\right] = \frac{1}{2\sqrt{t}}$$

which will be equal to the prevailing interest rate of 6 percent when

$$\frac{1}{2\sqrt{t}} = 0.06 \text{ or } t = (\frac{1}{0.12})^2 \simeq 69.44$$

Moreover, $\frac{1}{2\sqrt{t}} > 0.06$ when $0 < t < 69.44$ and $\frac{1}{2\sqrt{t}} < 0.06$ when $t > 69.44$. Hence the percentage rate of growth of the value of the land is greater than the prevailing interest rate when $0 < t < 69.44$ and less than the prevailing interest rate when $t > 69.44$. Thus the land should be sold in 69.44 years.

31. Since the stamp collection is currently worth $1,200$ and its value increases linearly at the rate of 200 per year, its value t years from now is $V(t) = 1,200 + 200t$. The percentage rate of change of the value (expressed in decimal form) is

$$\frac{V'(t)}{V(t)} = \frac{200}{1,200 + 200t} = \frac{200}{200(6 + t)} = \frac{1}{6 + 2}$$

which will be equal to the prevailing interest rate of 8 percent when

$$\frac{1}{6 + t} = 0.08 \text{ or } t = \frac{1}{0.08} - 6 = 6.5$$

Moreover, $\frac{1}{6+t} > 0.08$ when $0 < t < 6.5$ and $\frac{1}{6+t} < 0.08$ when $t > 6.5$. Hence the percentage rate of growth of the value of the collection is greater than the prevailing interest rate when $0 < t < 6.5$ and less than the prevailing interest rate when $t > 6.5$. Thus the collection should be sold in 6.5 years.

Chapter 4, Review Problems

1.   (a)   If $f(x) = 5e^{-x}$, then $f(x)$ approaches zero as x increases without bound, and $f(x)$ increases without bound as x decreases without bound. The y intercept is $f(0) = 5$.

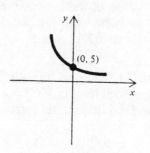

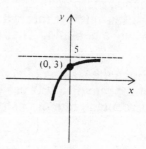

Graph for Problem 1a.   Graph for Problem 1b.

(b)   If $f(x) = 5 - 2e^{-x}$, then $f(x)$ approaches 5 as x increases without bound, and $f(x)$ decreases without bound as x decreases without bound. The y intercept is $f(0) = 3$.

(c)   If $f(x) = 1 - \frac{6}{2 + e^{-3x}}$, the y intercept is $f(0) = 1 - \frac{6}{2 + e^0} = 1 - \frac{6}{2+1} = 1 - 2 = -1$. As x increases without bound, e^{-3x} approaches zero, and so

$$\lim_{x \to \infty} f(x) = 1 - \frac{6}{2 + 0} = 1 - 3 = -2$$

As x decreases without bound, e^{-3x} increases without bound, and so

$$\lim_{x \to -\infty} f(x) = 1 - 0 = 1$$

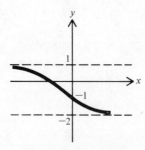

Graph for Problem 1c.

(d)   If $f(x) = \frac{3 + 2e^{-2x}}{1 + e^{-2x}}$, then $f(x)$ approaches 3 as x increases without bound since e^{-2x} approaches zero. Rewriting the function, by multiplying numerator and denominator by e^{2x} yields $f(x) = \frac{3e^{2x}+2}{e^{2x}+1}$. As x decreases without bound e^{2x} approaches zero and thus $f(x)$ approaches 2. The y intercept is $f(0) = \frac{3+2}{1+1} = \frac{5}{2}$. Applying the quotient rule to determine $f'(x)$ yields

$$f'(x) = \frac{2e^{2x}}{(e^{2x}+1)^2}.$$

Since $2e^{2x} > 0$, there are no critical points and $f(x)$ increases for all x. To find $f''(x)$, differentiate $f'(x)$ using the quotient rule obtaining

$$f''(x) = \frac{4e^{2x}(1 - e^{2x})}{(e^{2x}+1)^3}.$$

There is one inflection point $(0, f(0)) = (0, 5/2)$ since $f(0) = 0$. The function $f(x)$ is concave down for $x > 0$ and concave up for $x < 0$.

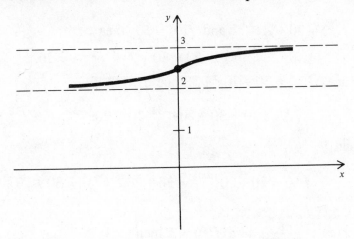

2.   (a)   If $f(x) = Ae^{-kx}$ and $f(x) = 10$ then

$$10 = Ae^0 \text{ or } A = 10.$$

Hence, $f(x) = 10e^{-kx}$. Since $f(1) = 25$

$$25 = 10e^{-k} \text{ or } e^{-k} = \frac{5}{2}.$$

Then

$$f(4) = 10e^{-4k} = 10(e^{-k})^4 = 10(\frac{5}{2})^4 = 390.625$$

   (b)   If $f(x) = Ae^{12x}$ and $f(1) = 3$ and $f(2) = 10$ then

$$3 = Ae^k$$
$$10 = Ae^{2k},$$

two equations and two unknowns. Dividing the first equation by the second yields

$$\frac{3}{10} = \frac{Ae^k}{Ae^{2k}}$$
$$\frac{3}{10} = \frac{1}{e^k}$$

or

$$e^k = \frac{10}{3}.$$

Since $e^k = \frac{10}{3}$, $3 = Ae^k = A(\frac{10}{3})$ and so $A = \frac{9}{10}$. Thus

$$f(3) = \frac{9}{10}e^{3k} = \frac{9}{10}(e^k)^3 = \left(\frac{9}{10}\right)\left(\frac{10}{3}\right)^3 = \frac{100}{3}.$$

(c)   If $f(x) = 30 + Ae^{-kx}$ and $f(0) = 50$, then

$$50 = 30 + Ae^0 \text{ or } A = 20$$

Hence, $f(x) = 30 + 20e^{kx}$. Since $f(3) = 40$,

$$40 = 30 + 20e^{-3k}, 10 = 20e^{-3k}, \text{ or } e^{-3k} = \frac{1}{2}$$

Then

$$f(9) = 30 + 20e^{-9k} = 30 + 20(e^{-3k})^3 = 30 + 20\left(\frac{1}{2}\right)^3 = 32.5$$

(d)   If $f(x) = \frac{6}{1+Ae^{-kx}}$ and $f(0) = 3$ then

$$3 = \frac{6}{1 + Ae^0}$$
$$3 = \frac{6}{1 + A}$$
$$3 + 3A = 6$$
or
$$A = 1.$$

Hence,

$$f(x) = \frac{6}{1 + e^{-kx}}.$$

Since $f(5) = 2$,

$$2 = \frac{6}{1 + e^{-5k}}$$
$$2 + 2e^{5k} = 6$$
or
$$e^{-5k} = 2.$$

Then,

$$f(10) = \frac{6}{1 + e^{-10k}} = \frac{6}{1 + (e^{-5k})^2} = \frac{6}{1 + (2)^2} = \frac{6}{5}.$$

3. Let $V(t)$ denote the value of the machine after t years. Since the value decreases exponentially and was originally $50,000$, it follows that $V(t) = 50,000e^{-kt}$. Since the value after 5 years is $20,000$,

$$20,000 = V(5) = 50,000e^{-5k} \quad e^{-5k} = \frac{2}{5} \quad -5k = \ln\frac{2}{5} \text{ or } k = -\frac{1}{5}\ln\frac{2}{5}$$

Hence $V(t) = 50,000e^{[-(1/5)\ln(2/5)]t}$, and so

$$V(10) = 50,000e^{[-(1/5)\ln(2/5)]10} = 50,000e^{2\ln(2/5)}$$

$$= 50,000e^{\ln(2/5)^2} = 50,000e^{\ln(4/25)} = 50,000\left(\frac{4}{25}\right) = \$8,000$$

4. The sales function is $Q(x) = 50 - 40e^{-0.1x}$ units, where x is the amount (in thousands) spent on advertising.

    (a) As x increases without bound, $Q(x)$ approaches 50. The vertical intercept is $Q(0) = 10$. The graph, shown below, is like that of a leaning curve.

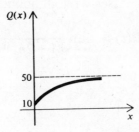

Sales function for Problem 4.

    (b) If no money is spent on advertising, sales will be $Q(0) = 10$ thousand units.

    (c) If $8,000$ is spent on advertising, sales will be

$$Q(8) = 50 - 40e^{-0.8} \simeq 32.027 \text{ thousand or } 32,027 \text{ units}$$

    (d) Sales will be 35 thousand if $Q(x) = 35$, that is, if

$$50 - 40e^{-0.1x} = 35 \quad -40e^{-0.1x} = -15 \quad e^{-0.1x} = \frac{3}{8}$$

$$-0.1x = \ln\frac{3}{8} \text{ or } x = \frac{-\ln(3/8)}{0.1} \simeq 9.808 \text{ thousand or } \$9,808$$

(e)   Since $Q(x)$ approaches 50 as x increases without bound, the most optimistic sales projection is 50,000 units.

5.   The output function is $Q(t) = 120 - Ae^{-kt}$. Since $Q(0) = 30$,

$$30 = 120 - A \text{ or } A = 90$$

Since $Q(8) = 80$,

$$80 = 120 - 90e^{-8k} \quad -40 = -90e^{-8k} \text{ or } e^{-8k} = \frac{4}{9}$$

Hence,

$$Q(4) = 120 - 90e^{-4k} = 120 - 90(e^{-8k})^{1/2} = 120 - 90\left(\frac{4}{9}\right)^{1/2} = 60 \text{ units}$$

6.   The population t years from now will be $P(t) = \frac{30}{1+2e^{-0.05t}}$.

(a)   The vertical intercept is $P(0) = \frac{30}{1+2} = 10$ million. As t increases without bound, $e^{-0.05t}$ approaches zero. Hence,

$$\lim_{t\to\infty} P(t) = \lim_{t\to\infty} \frac{30}{2e^{-0.05t}} = \frac{30}{1} = 30$$

As t decreases without bound, $e^{-0.05t}$ increases without bound. Hence, the denominator $1 + 2e^{-0.05t}$ increases without bound and

$$\lim_{t\to-\infty} P(t) = \lim_{t\to-\infty} \frac{30}{1 + 2e^{-0.05t}} = 0$$

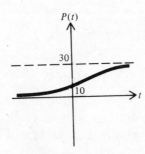

Graph for Problem 6a.

(b)   The current population is $P(0) = 10$ million.

(c)   The population in 20 years will be

$$P(20) = \frac{30}{1 + 2e^{-0.05(20)}} = \frac{30}{1 + 2e^{-1}} \simeq 17.28 \text{ million}$$

(d) In the long run (as t increases without bound), $e^{-0.05t}$ approaches zero and so the population $P(t)$ approaches 30 million.

7.   (a) $\ln e^5 = 5$

       (b) $e^{\ln 2} = 2$

       (c) $e^{3\ln 4 - \ln 2} = e^{\ln 4^3 - \ln 2} = e^{\ln(64/2)} = e^{\ln 32} = 32$

       (d) $\ln(9e^2) + \ln(3e^{-2}) = \ln[(9e^2)(3e^{-2})] = \ln 27 = \ln 3^3 = 3\ln 3$

8.   (a)   If $8 = 2e^{0.04x}$, then

$$e^{0.04x} = 4 \quad 0.04x = \ln 4 \quad \text{or} \quad x = \frac{\ln 4}{0.04} \simeq 34.66$$

       (b)   If $5 = 1 + 4e^{-6x}$, then

$$4e^{-6x} = 4 \quad e^{-6x} = 1 \quad -6x = \ln 1 = 0 \quad \text{or} \quad x = 0$$

       (c)   If $4\ln x = 8$, then

$$\ln x = 2 \quad \text{or} \quad x = e^2 \simeq 7.39$$

       (d)   If $5^x = e^3$, then

$$\ln 5^x = \ln e^3 \quad x\ln 5 = 3 \quad \text{or} \quad x = \frac{3}{\ln 5} \simeq 1.86$$

9.   Let $Q(t)$ denote the number of bacteria after t minutes. Since $Q(t)$ grows exponentially and $5,000$ bacteria were present initially, $Q(t) = 5,000e^{kt}$. Since $8,000$ bacteria were present after 10 minutes,

$$8,000 = Q(10) = 5,000e^{10k} \quad e^{10k} = \frac{8}{5} \quad \text{or} \quad k = \frac{1}{10}\ln\frac{8}{5}$$

The bacteria will double when $Q(t) = 10,000$, that is, when

$$5,000^{kt} = 10,000 \quad e^{kt} = 2 \quad kt = \ln 2$$
$$t = \frac{\ln 2}{k} = \frac{10\ln 2}{\ln 8/5} \simeq 14.75 \text{ minutes}$$

10.   (a)   If $f(x) = 2e^{3x+5}$, then

$$f'(x) = 2e^{3x+5}\frac{d}{dx}(3x + 5) = 6e^{3x+5}$$

(b)   If $f(x) = x^2 e^{-x}$, then

$$f'(x) = x^2 \frac{d}{dx}(e^{-x}) + e^{-x}\frac{d}{dx}(x^2)$$
$$= 2xe^{-x} - x^2 e^{-x} = x(2-x)e^{-x}$$

(c)   If $f(x) = \ln\sqrt{x^2 + 4x + 1} = \ln(x^2 + 4x + 1)^{1/2} = \frac{1}{2}\ln(x^2 + 4x + 1)$, then

$$f'(x) = \frac{1}{2}\left[\frac{1}{x^2 + 4x + 1}\right]\frac{d}{dx}(x^2 + 4x + 1)$$
$$= \frac{1}{2}\left[\frac{2x+4}{x^2 + 4x + 1}\right] = \frac{x+2}{x^2 + 4x + 1}$$

(d)   If $f(x) = x\ln x^2 = 2x\ln x$, then

$$f'(x) = \left[x\frac{d}{dx}(\ln x) + (\ln x)\frac{d}{dx}(x)\right]$$
$$= 2\left[x\left(\frac{1}{x}\right) + \ln x\right] = 2(1 + \ln x)$$

(e)   If $f(x) = \frac{x}{\ln 2x}$, then

$$f'(x) = \frac{(\ln 2x)(1) - x(1/2x)(2)}{(\ln 2x)^2} = \frac{\ln 2x - 1}{(\ln 2x)^2}$$

11.   (a)   Differentiate both sides of the equation $e^{y/x} = 3xy^2$ with respect to x to get

$$\frac{d}{dx}\left(\frac{y}{x}\right)e^{y/x} = 6xy\frac{dy}{dx} + 3y^2 \quad \text{or} \quad \left[\frac{x\frac{dy}{dx} - y}{x^2}\right]e^{y/x} = 6xy\frac{dy}{dx} + 3y^2$$

Solve for $\frac{dy}{dx}$ to get

$$\left(x\frac{dy}{dx} - y\right)e^{y/x} = 6x^3 y\frac{dy}{dx} + 3x^2 y^2$$
$$(xe^{y/x} - 6x^3 y)\frac{dy}{dx} = 3x^2 y^2 + ye^{y/x}$$
$$\frac{dy}{dx} = \frac{3x^2 y^2 + ye^{y/x}}{xe^{y/x} - 6x^3 y} = \frac{y(3x^2 y + e^{y/x})}{x(e^{y/x} - 6x^2 y)}$$

Simplify further by substituting $e^{y/x} = 3xy^2$ (from the original equation) to get

$$\frac{dy}{dx} = \frac{y(3x^2 y + 3xy^2)}{x(3xy^2 - 6x^2 y)} = \frac{3xy^2(x+y)}{3x^2 y(y - 2x)} = \frac{y(x+y)}{x(y - 2x)}$$

(b)   To simplify the calcualtion, rewrite the equation $\ln \frac{x}{y} = x^3 y^2$ as

$$\ln x - \ln y = x^3 y^2$$

Now differentitate both sides with respect to x to get

$$\frac{1}{x} - \frac{1}{y}\frac{dy}{dx} = 2x^3 y \frac{dy}{dx} + 3x^2 y^2$$

and solve for $\frac{dy}{dx}$ to get

$$y - x\frac{dy}{dx} = 2x^4 y^2 \frac{dy}{dx} + 3x^3 y^3 \quad (2x^4 y^2 + x)\frac{dy}{dx} = y - 3x^3 y^3$$

$$\frac{dy}{dx} = \frac{y - 3x^3 y^3}{2x^4 y^2 + x} = \frac{y(1 - 3x^3 y^2)}{x(2x^3 y^2 + 1)}$$

12.   The average level of carbon monoxide in the air t years from now is $Q(t) = 4e^{0.03t}$ parts per million.

(a)   The rate of change of the carbon monoxide level t years from now is $Q'(t) = 0.12e^{0.03t}$, and the rate 2 years from now is $Q'(2) = 0.12e^{0.06} \simeq 0.13$ part per million per year.

(b)   The percentage rate of change of the carbon monoxide level t years from now is

$$100\left[\frac{Q'(t)}{Q(t)}\right] = 100\left[\frac{0.12e^{0.03t}}{4e^{0.03t}}\right] = 3 \text{ percent per year}$$

which is a constant independent of time.

13.   Let $F(p)$ denote the profit, where p is the price per camera. Then

$$F(p) = \text{(number of cameras sold) (profit per camera)}$$
$$= 800e^{-0.01p}(p - 40) = 800(p - 40)e^{-0.01p}$$

The derivative is

$$F'(p) = 800[e^{-0.01p}(1) + (p - 40)e^{-0.01p}(-0.01)]$$
$$= 800e^{-0.01p}(1.4 - 0.01p) = 8e^{-0.01p}(140 - p)$$

which is zero when $p = 140$. Since $F'(p) > 0$ (and F is increasing) for $0 < p < 140$, and $F'(p) < 0$ (and F is decreasing) for $p > 140$, it follows that $F(p)$ has an absolute maximum at $p = 140$. Thus the cameras should be sold for \$140 apiece to maximize profit.

14.   (a)   If $f(x) = xe^{-2x}$, the first derivative is

$$f'(x) = xe^{-2x}(-2) + e^{-2x}(1) = e^{-2x}(1 - 2x)$$

which is zero when $x = \frac{1}{2}$, and the corresponding first-order critical point is $\left(\frac{1}{2}, \frac{1}{2}e^{-1}\right) = \left(\frac{1}{2}, \frac{1}{2e}\right)$. The second derivative is

$$f''(x) = e^{-2x}(-2) + (1 - 2x)e^{-2x}(-2) = 4e^{-2x}(x - 1)$$

which is zero when $x = 1$. The corresponding second-order critical point is $(1, e^{-2}) = \left(1, \frac{1}{e^2}\right)$.

Interval	$f'(x)$	$f''(x)$	Inc/dec	Concavity
$x < \frac{1}{2}$	+	−	increasing	down
$\frac{1}{2} < x < 1$	−	−	decreasing	down
$x > 1$	−	+	decreasing	up

Notice that the first-order critical point $\left(\frac{1}{2}, \frac{1}{2e}\right)$ is as relative maximum and that the second-order critical point $\left(1, \frac{1}{e^2}\right)$ is an inflection point.

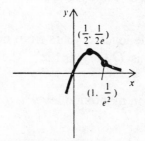

Graph for Problem 14a.

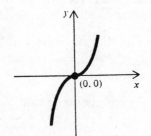

Graph for Problem 14b.

(b) If $f(x) = e^x - e^{-x}$, the first derivative is

$$f'(x) = e^x + e^{-x}$$

which is never zero, and so there are no first-order critical points. The second derivative is

$$f''(x) = e^x - e^{-x}$$

which is zero when $x = 0$. The corresponding second-order critical point is $(0, 0)$.

Interval	$f'(x)$	$f''(x)$	Inc/dec	Concavity
$x < 0$	+	−	increasing	down
$x > 0$	+	+	decreasing	up

Notice that the second-order critical point $(0, 0)$ is an inflection point.

(c) If $f(x) = \frac{4}{1 + e^{-x}}$, the first derivative is

$$f'(x) = \frac{(1 + e^{-x})(0) - 4e^{-x}(-1)}{(1 + e^{-x})^2} = \frac{4e^{-x}}{(1 + e^{-x})^2}$$

which is never zero, and so there are no first-order critical points. The second derivative is

$$f''(x) = \frac{(1 + e^{-x})^2(4e^{-x})(-1) - 4e^{-x}(2)(1 + e^{-x})(e^{-x})(-1)}{(1 + e^{-x})^4}$$

$$= \frac{4e^{-x}(1 + e^{-x})(-1 - e^{-x} + 2e^{-x})}{(1 + e^{-x})^2} = \frac{4e^{-x}(e^{-x} - 1)}{(1 + e^{-x})^3}$$

which is zero when $x = 0$. The corresponding second-order critical point is $(0, 2)$.

Interval	$f'(x)$	$f''(x)$	Inc/dec	Concavity
$x < 0$	+	+	increasing	up
$x > 0$	+	−	increasing	down

Notice that the second-order critical point $(0, 2)$ is an inflection point. Notice also that as x increases without bound, e^{-x} approaches zero and hence $f(x)$ approaches 4. As x decreases without bound, e^{-x} increases without bound and hence $f(x)$ approaches zero.

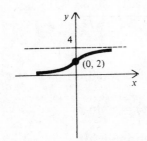

Graph for Problem 14c.

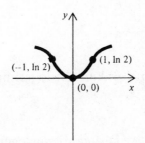

Graph for Problem 14d.

(d)   If $f(x) = \ln(x^2 + 1)$, the first derivative is

$$f'(x) = \frac{2x}{x^2 + 1}$$

which is zero when $x = 0$. The corresponding first-order critical point is $(0, 0)$. The second derivative is

$$f''(x) = \frac{(x^2 + 1)(2) - 2x(2x)}{(x^2 + 1)^2}$$

$$= \frac{2 - 2x^2}{(x^2 + 1)^2} = \frac{2(1 - x)(1 + x)}{(x^2 + 1)^2}$$

Interval	$f'(x)$	$f''(x)$	Inc/dec	Concavity
$x < -1$	−	−	decreasing	down
$-1 < x < 0$	−	+	decreasing	up
$0 < x < 1$	+	+	increasing	up
$x > 1$	+	−	increasing	down

Notice that the first-order critical point $(0,0)$ is a relative minimum and that the second-order critical points are inflection points.

15. (a) Taking the logarithm of each side of the equation

$$f(x) = \sqrt{(x^2+1)(x^2+2)} = [(x^2+1)(x^2+2)]^{1/2}$$

gives

$$\ln f(x) = \frac{1}{2}[\ln(x^2+1) + \ln(x^2+2)]$$

Differentiating each side of this equation yields

$$\frac{f'(x)}{f(x)} = \frac{1}{2}\left[\frac{2x}{x^2+1} + \frac{2x}{x^2+2}\right]$$

$$f'(x) = f(x)\left[\frac{x}{x^2+1} + \frac{x}{x^2+2}\right]$$

$$= \sqrt{(x^2+1)(x^2+2)}\left[\frac{x(2x^2+3)}{(x^2+1)(x^2+2)}\right]$$

$$= \frac{x(2x^2+3)}{(x^2+1)^{1/2}(x^2+2)^{1/2}}$$

(b) Taking the logarithm of each side of the equation $f(x) = x^{x^2}$ gives

$$\ln f(x) = x^2 \ln x$$

Differentiating each side of this equation yields

$$\frac{f'(x)}{f(x)} = x^2\left(\frac{1}{x}\right) + (\ln x)(2x) = x + 2x\ln x = x(1 + 2\ln x)$$

$$f'(x) = f(x)[x(1+2ln\ x)] = x^{x^2}[x(1+2\ln x)]$$

$$= x^{(x^2+1)}(1 + 2\ln x)$$

(c) Taking the natural logarithm of both sides of the equation

$$f(x) = (x^x)(2^x)$$

gives

$$\ln f(x) = \ln x^x + \ln 2^x = x\ln x + x\ln 2$$

Differentiating each side of this equation yields

$$\frac{f'(x)}{f(x)} = x\left(\frac{1}{x}\right) + \ln x + \ln 2 = 1 + \ln x + \ln 2 = 1 + \ln 2x$$

and so

$$f'(x) = f(x)[1 + \ln 2x] = (x^x)(2^x)[1 + \ln 2x]$$

16. (a) Using the formula $B(t) = P(1 + \frac{r}{k})^{kt}$ with $P = 2,000$, $B = 5,000$, $r = 0.08$, and $k = 4$,

$$5,000 = 2,000 \left(1 + \frac{0.08}{4}\right)^{4t} \qquad (1.02)^{4t} = \frac{5}{2}$$

$$4t \ln 1.02 = \ln\frac{5}{2} \quad \text{or} \quad t = \frac{1}{4}\left[\frac{\ln 5/2}{\ln 1.02}\right] \simeq 11.57 \text{ years}$$

(b) Using the formula $B(t) = Pe^{rt}$ with $P = 2,000$, $B = 5,000$, and $r = 0.08$,

$$5,000 = 2,000e^{0.08t} \qquad e^{0.08t} = \frac{5}{2} \quad \text{or} \quad t = \frac{\ln 5/2}{0.08} \simeq 11.45 \text{ years}$$

17. Compare the effective interest rates. The effective interest rate for 8.25 percent compounded quarterly is

$$\left(1 + \frac{r}{k}\right)^k - 1 = \left(1 + \frac{0.0825}{4}\right)^4 - 1 \simeq 0.0805 \quad \text{or} \quad 8.05 \text{ percent}$$

The effective interest rate for 8.20 percent compounded continuously is

$$e^r - 1 = e^{0.082} - 1 \simeq 0.0855 \quad \text{or} \quad 8.55 \text{ percent}$$

Hence the investment at 8.20 percent compounded continuously is the better one.

18. (a) Using the present value formula $P = B\left(1 + \frac{r}{k}\right)^{-kt}$ with $B = 2,000$, $t = 10$, $r = 0.0625$, and $k = 2$

$$P = 2,000 \left(1 + \frac{0.0625}{2}\right)^{-20} \simeq \$1,080.81$$

(b) Using the present value formula $P = Be^{-rt}$ with $B = 2,000$, $t = 10$, and $r = 0.0625$

$$P = 2,000e^{-0.0625(10)} \simeq \$1,070.52$$

19. Since $r = 0.10$ and interest is compounded continuously, the appropriate present value formula is $P(t) = e^{-0.1t}$. The sequence of payments and corresponding present values are shown in the accompanying figure.

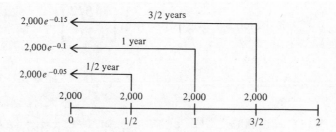

The present value of the annuity is the corresponding sum

$$2,000 + 2,000e^{-0.05} + 2,000e^{-0.1} + 2,000e^{-0.15} \simeq \$7,433.55$$

20. The value of the coin collection in t years is $V(t) = 2,000e^{\sqrt{t}}$. Hence, the percentage rate of change of the value of the collection (expressed in decimal form) is

$$\frac{V'(t)}{V(t)} = \frac{2,000e^{\sqrt{t}}}{2,000e^{\sqrt{t}}}\left[\frac{1}{2\sqrt{t}}\right] = \frac{1}{2\sqrt{t}}$$

which will be equal to the prevailing interest rate of 7 percent when

$$\frac{1}{2\sqrt{t}} = 0.07 \quad\text{or}\quad t = \left[\frac{1}{2(0.07)}\right]^2 \simeq 51.02$$

Moreover, $\frac{1}{2\sqrt{t}} > 0.07$ when $0 < t < 51.02$ and $\frac{1}{2\sqrt{t}} < 0.07$ when $t > 51.02$. Hence the percentage rate of growth of the collection is greater than the prevailing interest rate when $0 < t < 51.02$ and less than the prevailing interest rate when $t > 51.02$. It follows that the coin collection should be sold 51.02 years from now.

CHAPTER 5

ANTIDIFFERENTIATION

---

1 ANTIDERIVATIVES

2 INTEGRATION BY SUBSTITUTION

3 INTEGRATION BY PARTS

4 THE USE OF INTEGRAL TABLES

REVIEW PROBLEMS

Chapter 5, Section 1

1. $\int x^5 dx = \frac{1}{6}x^6 + C$

3. $\int \frac{1}{x^2} dx = \int x^{-2} dx = -x^{-1} + C = -\frac{1}{x} + C$

5. $\int 5 dx = 5x + C$

7. $\int (3x^2 - 5x + 2) dx = 3 \int x^2 dx - 5 \int x \, dx + \int 2 \, dx$

   $= 3 \left(\frac{1}{3}x^3 \right) - 5 \left(\frac{1}{2}x^2 \right) + 2x + C = x^3 - \frac{5}{2}x^2 + 2x + C$

9. $\int \left(3\sqrt{x} - \frac{2}{x^3} + \frac{1}{x} \right) dx = 3 \int x^{\frac{1}{2}} dx - 2 \int x^{-3} dx + \int \frac{1}{x} dx$

   $= 3 \left(\frac{2}{3}x^{3/2} \right) - 2 \left(-\frac{1}{2}x^{-2} \right) + \ln |x| + C = 2x^{3/2} + \frac{1}{x^2} + \ln |x| + C$

11. $\int \left(2e^x + \frac{6}{x} + \ln 2 \right) dx = 2 \int e^x dx + 6 \int \frac{1}{x} dx + \ln 2 \int 1 \, dx$

   $= 2e^x + 6 \ln |x| + x \ln 2 + C = 2e^x + \ln x^6 + x \ln 2 + C$

13.
$$\int \left(\sqrt{x^3} - \frac{1}{2\sqrt{x}} + \sqrt{2}\right) dx = \int \left(x^{3/2} - \frac{1}{2}x^{-1/2} + \sqrt{2}\right) dx$$

$$= \frac{2}{5}x^{5/2} - \frac{1}{2}(2x^{1/2}) + x\sqrt{2} + C = \frac{2}{5}x^{5/2} - \sqrt{x} + x\sqrt{2} + C$$

15.
$$\int \frac{x^2 + 2x + 1}{x^2} dx = \int \left(1 + \frac{2}{x} + \frac{1}{x^2}\right) dx$$

$$= x + 2\ln|x| - x^{-1} + C = x + \ln x^2 - \frac{1}{x} + C$$

17.
$$\int \left[\frac{2x^3 - 5x}{3x}\right]^2 dx = \int \left(\frac{2}{3}x^2 - \frac{5}{3}\right)^2 dx = \int \left(\frac{4}{9}x^4 - \frac{20}{9}x^2 + \frac{25}{9}\right) dx$$

$$= \frac{4}{45}x^5 - \frac{20}{27}x^3 + \frac{25}{9}x + C$$

19.
$$\int \sqrt{x}(x^2 - 1)dx = \int x^{1/2}(x^2 - 1)dx = \int (x^{5/2} - x^{1/2})dx$$

$$= \int x^{5/2}dx - \int x^{1/2}dx = \frac{2}{7}x^{7/2} - \frac{2}{3}x^{3/2} + C$$

21. Let $P(t)$ denote the population of the town t months from now. Since $\frac{dP}{dt} = 4 + 5t^{2/3}$, the function $P(t)$ is an antiderivative of $4 + 5t^{2/3}$. Thus,

$$P(t) = \int (4 + 5t^{2/3})dt = 4t + 5\left(\frac{3}{5}t^{5/3}\right) + C = 4t + 3t^{5/3} + C$$

Since the current population is $10,000$, it follows that

$$10,000 = P(0) = 4(0) + 3(0)^{5/3} + C \text{ or } C = 10,000$$

Thus, the population t months from now will be

$$P(t) = 4t + 3t^{5/3} + 10,000$$

and the population 8 months from now will be $P(8) = 10,128$.

23. Let $V(t)$ denote the value of the machine t years from now. Since $\frac{dV}{dt} = 220(t - 10)$ dollars per year, the function $V(t)$ is an antiderivative of $220(t - 10)$. Thus,

$$V(t) = \int 220(t - 10)dx = 110t^2 - 2,200t + C$$

Since the machine was originally worth $12,000$, it follows that

$$12,000 = V(0) = C$$

Thus, the value of the machine after t years will be

$$V(t) = 110t^2 - 2,200t + 12,000$$

and the value after 10 years will be $V(10) = \$1,000$.

25. Let $C(q)$ denote the total cost of producing q units. Since marginal cost is $\frac{dC}{dq} = 6q+1$ dollars per unit, the function $C(q)$ is an antiderivative of $6q + 1$. Thus,

$$C(q) = \int (6q+1)dq = 3q^2 + q + C$$

Since the cost of producing the first unit is $\$130$, it follows that

$$130 = C(1) = 3 + 1 + C \text{ or } C = 126$$

Thus, the cost of producing q units is

$$C(q) = 3q^2 + q + 126$$

and the cost of producing 10 units is $C(10) = \$436$.

27. Let $P(q)$ denote the profit from the production and sale of q units. Since $\frac{dP}{dq} = 100 - 2q$ dollars per unit, the function $P(q)$ is an antiderivative of $100 - 2q$. Thus,

$$P(q) = \int (100 - 2q)dq = 100q - q^2 + C$$

Since the profit is $\$700$ when 10 units are produced, it follows that

$$700 = P(10) = 100(10) - (10)^2 + C \text{ or } C = -200$$

Thus, the profit function is

$$P(q) = 100q - q^2 - 200$$

which attains its maximum value when its derivative, $\frac{dP}{dq} = 100 - 2q$, is zero, that is, when $x = 50$. Hence the maximum profit is $P(50) = \$2,300$.

29. The slope of the tangent is the derivative. Thus, $f'(x) = 4x + 1$ and so f is an antiderivative of $4x + 1$. That is,

$$f(x) = \int (4x+1)dx = 2x^2 + x + C$$

Since the graph of f passes through the point $(1,2)$, it follows that

$$2 = f(1) = 2(1)^2 + 1 + C \text{ or } C = -1$$

Hence, the function is $f(x) = 2x^2 + x - 1$.

31. The slope of the tangent is the derivative of f. Hence $f'(x) = x^3 - \frac{2}{x^2} + 2$, and so f is an antiderivative of $x^3 - \frac{2}{x^2} + 2$. That is,

$$f(x) = \int (x^3 - \frac{2}{x^2} + 2)dx = \int (x^3 - 2x^{-2} + 2)dx = \frac{1}{4}x^4 + \frac{2}{x} + 2x + C$$

Since the graph of f passes through the point $(1, 3)$ it follows that

$$3 = f(1) = \frac{1}{4} + 2 + 2 + C \text{ or } C = 3 - \frac{17}{4} = -\frac{5}{4}$$

Hence the function is $f(x) = \frac{1}{4}x^4 + \frac{2}{x} + 2x - \frac{5}{4}$.

33. Guess that $\int e^{3x}dx = e^{3x} + C$. If his answer is correct, its derivative should be equal to e^{3x}. However, $\frac{d}{dx}(e^{3x} + C) = 3e^{3x}$, which is too large by a factor of 3. Divide the original guess by 3 to get

$$\int e^{3x}dx = \frac{1}{3}e^{3x} + C$$

which is correct since $\frac{d}{dx}\left(\frac{1}{3}e^{3x} + C\right) = \frac{1}{3}(3e^{3x}) = e^{3x}$.

35. Guess that $\int (2x + 3)^5 dx = \frac{1}{6}(2x + 3)^6 + C$. If this is correct, its derivative should be equal to $(2x + 3)^5$. However,

$$\frac{d}{dx}\left[\frac{1}{6}(2x + 3)^6 + C\right] = 2(2x + 3)^5$$

which is too large by a factor of 2. Divide the original guess by 2 to get

$$\int (2x + 3)^5 dx = \frac{1}{12}(2x + 3)^6 + C$$

which is correct since its derivative is $(2x + 3)^5$.

37. Guess that $\int \frac{1}{2x+1}dx = \ln|2x + 1| + C$. If this answer is correct, its derivative should be equal to $\frac{1}{2x+1}$. However,

$$\frac{d}{dx}(ln|2x + 1| + C) = \frac{2}{2x + 1}$$

which is too large by a factor of 2. Divide the original guess by 2 to get

$$\int \frac{1}{2x + 1}dx = \frac{1}{2}\ln|2x + 1| + C$$

which is correct since its derivative is $\frac{1}{2x+1}$.

Chapter 5, Section 2

1.   Let $u = 2x + 6$. Then $du = 2dx$ or $dx = \frac{1}{2}du$. Hence,

$$\int (2x + 6)^5 dx = \frac{1}{2} \int u^5 du = \frac{1}{12}u^6 + C = \frac{1}{12}(2x + 6)^6 + C$$

3.   Let $u = 4x - 1$. Then $du = 4dx$ or $dx = \frac{1}{4}du$. Hence,

$$\int \sqrt{4x - 1}\,dx = \frac{1}{4} \int u^{1/2} du = \frac{1}{4}\left(\frac{2}{3}u^{3/2}\right) + C = \frac{1}{6}(4x - 1)^{3/2} + C$$

5.   Let $u = 1 - x$. Then $du = -dx$ or $dx = -du$. Hence,

$$\int e^{1-x} dx = -\int e^u du = -e^u + C = -e^{1-x} + C$$

7.   Let $u = x^2$. Then $du = 2x\,dx$ or $x\,dx = \frac{1}{2}du$. Hence

$$\int xe^{x^2} dx = \frac{1}{2} \int e^u du = \frac{1}{2}e^u + C = \frac{1}{2}e^{x^2} + C$$

9.   Let $u = x^2 + 1$. Then $du = 2x\,dx$ or $x\,dx = \frac{1}{2}du$. Hence,

$$\int x(x^2 + 1)^5 dx = \frac{1}{2} \int u^5 du = \frac{1}{12}u^6 + C = \frac{1}{12}(x^2 + 1)^6 + C$$

11. Let $u = x^3 + 1$. Then $du = 3x^2 dx$ or $x^2 dx = \frac{1}{3}du$. Hence,

$$\int x^2(x^3 + 1)^{3/4} dx = \frac{1}{3} \int u^{3/4} du = \frac{1}{3}\left(\frac{4}{7}u^{7/4}\right) + C = \frac{4}{21}(x^3 + 1)^{7/4} + C$$

13. Let $u = x^5 + 1$. Then $du = 5x^4 dx$ or $x^4 dx = \frac{1}{5}du$. Hence,

$$\int \frac{2x^4}{x^5 + 1}dx = \frac{2}{5} \int \frac{1}{u}du = \frac{2}{5}\ln|u| + C = \frac{2}{5}\ln|x^5 + 1| + C$$

15. Let $u = x^2 + 2x + 5$. Then $du = (2x + 2)dx$ or $(x + 1)dx = \frac{1}{2}du$. Hence,

$$\int (x + 1)(x^2 + 2x + 5)^{12} dx = \frac{1}{2} \int u^{12} du$$

$$= \frac{1}{26}u^{13} + C = \frac{1}{26}(x^2 + 2x + 5)^{13} + C$$

17. Let $u = x^5 + 5x^4 + 10x + 12$. Then,

$$du = (5x^4 + 20x^3 + 10)dx \text{ or } (x^4 + 4x^3 + 2)dx = \frac{1}{5}du$$

Hence,

$$\int \frac{3x^4 + 12x^3 + 6}{x^5 + 5x^4 + 10x + 12}dx = 3 \int \frac{x^4 + 4x^3 + 2}{x^5 + 5x^4 + 10x + 12}dx = \frac{3}{5} \int \frac{1}{u}du$$
$$= \frac{3}{5}\ln|u| + C = \frac{3}{5}\ln|x^5 + 5x^4 + 10x + 12| + C$$

19. Let $u = x^2 - 2x + 6$. Then $du = (2x - 2)dx$ or $(3x - 3)dx = \frac{3}{2}du$. Hence,

$$\int \frac{3x - 3}{(x^2 - 2x + 6)^2}dx = \frac{3}{2}\int \frac{1}{u^2}du = \frac{3}{2}\left(-\frac{1}{u}\right) + C$$
$$= \frac{-3}{2(x^2 - 2x + 6)} + C$$

21. Let $u = \ln 5x$. Then $du = \frac{5}{5x}dx = \frac{1}{x}dx$. Hence,

$$\int \frac{\ln 5x}{x}dx = \int u\, du = \frac{1}{2}u^2 + C = \frac{1}{2}(\ln 5x)^2 + C$$

23. Let $u = \ln x$. Then $du = \frac{1}{x}dx$. Hence,

$$\int \frac{1}{x(\ln x)^2}dx = \int \frac{1}{u^2}du = -\frac{1}{u} + C = -\frac{1}{\ln x} + C$$

25. Let $u = \ln(x^2 + 1)$. Then $du = \frac{2x}{x^2+1}dx$. Hence,

$$\int \frac{2x\,\ln(x^2 + 1)}{x^2 + 1}dx = \int u\, du = \frac{1}{2}u^2 + C = \frac{1}{2}[\ln(x^2 + 1)]^2 + C$$

27. Let $u = x - 1$. Then $du = dx$ and $x = u + 1$. Hence,

$$\int \frac{x}{x - 1}dx = \int \frac{u + 1}{u}du = \int (1 + \frac{1}{u})du$$
$$= u + \ln|u| + C = (x - 1) + \ln|x - 1| + C$$

Note that the answer can also be written as $x + \ln|x - 1| + C$, with the term -1 included in the constant C.

29. Let $u = x - 5$. Then $du = dx$ and $x = u + 5$. Hence,

$$\int \frac{x}{(x-5)^6}dx = \int \frac{u+5}{u^6}du = \int (u^{-5} + 5u^{-6})du$$

$$= -\frac{1}{4}u^{-4} - u^{-5} + C = -\frac{1}{4(x-5)^4} - \frac{1}{(x-5)^5} + C$$

31. Let $u = x - 4$. Then $du = dx$ and $x = u + 4$. Hence,

$$\int \frac{x+3}{(x-4)^2}dx = \int \frac{(u+4)+3}{u^2}du = \int \frac{u+7}{u^2}du = \int (\frac{1}{u} + 7u^{-2})du$$

$$= \ln|u| - 7u^{-1} + C = \ln|x-4| - \frac{7}{x-4} + C$$

33. Let $u = 2x - 1$. Then $2x + 3 = u + 4$ and $du = 2x$ or $dx = \frac{1}{2}du$. Hence,

$$\int (2x+3)\sqrt{2x-1}dx = \frac{1}{2}\int (u+4)u^{1/2}du = \frac{1}{2}\int (u^{3/2} + 4u^{1/2})du$$

$$= \frac{1}{2}\left[\frac{2}{5}u^{5/2} + \frac{8}{3}u^{3/2}\right] + C = \frac{1}{5}(2x-1)^{5/2} + \frac{4}{3}(2x-1)^{3/2} + C$$

35. The slope of the tangent is the derivative of f. Thus, $f'(x) = \frac{2x}{1-3x^2}$ and so f is an antiderivative of $\frac{2x}{1-3x^2}$. That is,

$$f(x) = \int \frac{2x}{1-3x^2}dx = -\frac{1}{3}\ln|1-3x^2| + C$$

Since the graph of f passes through the point $(0, 5)$, it follows that

$$5 = f(0) = -\frac{1}{3}\ln(1) + C \text{ or } C = 5$$

Hence, the function is $f(x) = -\frac{1}{3}\ln|1-3x^2| + 5$.

37. Let $V(t)$ denote the value of the machine after t years. Since $\frac{dV}{dt} = -960e^{-t/5}$ dollars per year, $V(t)$ is an antiderivative of $-960e^{-t/5}$. Thus,

$$V(t) = \int -960e^{-t/5}dt = -960(-5)e^{-t/5} + C = 4,800e^{-t/5} + C$$

Since the value of the machine was originally $\$5,000$, it follows that

$$5,000 = V(0) = 4,800e^0 + C = 4,800 + C \text{ or } C = 200$$

Hence, the value of the machine after t years is

$$V(t) = 4,800e^{-t/5} + 200$$

and the value after 10 years is

$$V(10) = 4,800e^{-2} + 200 \simeq \$849.61$$

39.   Let $V(x)$ denote the value of the farm land x years from now. Since

$$\frac{dV}{dx} = \frac{0.4x^3}{\sqrt{0.2x^4 + 8,000}} \quad \text{dollars per year}$$

$V(x)$ is an antiderivative of this expression. Let $u = 0.2x^4 + 8,000$. Then $du = 0.8x^3 dx$ or $0.4^3 \, dx = \frac{1}{2}du$. Hence,

$$V(x) = \int \frac{0.4x^3}{\sqrt{0.2x^4 + 8,000}} dx = \frac{1}{2} \int u^{-1/2} du$$
$$= u^{1/2} + C = (0.2x^4 + 8,000)^{1/2} + C$$

Since the land is currently worth \$500 per acre, it follows that

$$500 = V(0) = (8,000)^{1/2} + C \quad \text{or} \quad C = 500 - (8,000)^{1/2}$$

Hence the value of the land 10 years from now will be

$$V(10) = [0.2(10)^4 + 8,000]^{1/2} + 500 - (8,000)^{1/2}$$
$$= (10,000)^{1/2} + 500 - (8,000)^{1/2} = 600 - (8,000)^{1/2} \simeq \$510.56$$

per acre.

Chapter 5, Section 3

1. Let $g(x) = e^{-x}$ and $f(x) = x$. Then, $G(x) = -e^{-x}$ and $f'(x) = 1$, and so

$$\int xe^{-x}dx = -xe^{-x} - \int 1(-e^{-x})dx = -xe^{-x} + \int e^{-x}dx$$
$$= -xe^{-x} - e^{-x} + C = -(x+1)e^{-x} + C$$

3. Let $g(x) = e^{-x/5}$ and $f(x) = x$. Then, $G(x) = -5e^{-x/5}$ and $f'(x) = 1$, and so

$$\int xe^{-x/5}dx = -5xe^{-x/5} - \int (-5e^{-x/5})(1)dx$$
$$= -5xe^{-x/5} + 5\int e^{-x/5}dx$$
$$= -5xe^{-x/5} - 25e^{-x/5} + C = -5(x+5)e^{-x/5} + C$$

5. Let $g(x) = e^x$ and $f(x) = 1 - x$. Then, $G(x) = e^x$ and $f'(x) = -1$, and so

$$\int (1-x)e^x dx = (1-x)e^x - \int (-1)e^x dx = (1-x)e^x + \int e^x dx$$
$$= (1-x)e^x + e^x + C = (2-x)e^x + C$$

7. Let $g(x) = x$ and $f(x) = \ln 2x$. Then, $G(x) = \frac{1}{2}x^2$ and $f'(x) = \frac{1}{x}$, and so

$$\int x \ln 2x \, dx = \frac{1}{2}x^2 \ln 2x - \int \frac{1}{x}\left(\frac{1}{2}x^2\right)dx = \frac{1}{2}x^2 \ln 2x - \frac{1}{2}\int x\,dx$$
$$= \frac{1}{2}x^2 \ln 2x - \frac{1}{2}\left[\frac{1}{2}x^2\right] + C = \frac{1}{2}x^2\left(\ln 2x - \frac{1}{2}\right) + C$$

9. Let $g(x) = \sqrt{x-6} = (x-6)^{1/2}$ and $f(x) = x$. Then, $G(x) = \frac{2}{3}(x-6)^{3/2}$ and $f'(x) = 1$, and so

$$\int x\sqrt{x-6}\,dx = \frac{2}{3}x(x-6)^{3/2} - \int \frac{2}{3}(x-6)^{3/2}dx$$
$$= \frac{2}{3}x(x-6)^{3/2} - \frac{2}{3}\left[\frac{2}{5}(x-6)^{5/6}\right] + C = \frac{2}{3}x(x-6)^{3/2} - \frac{4}{15}(x-6)^{5/2} + C$$

11. Let $g(x) = (x+1)^8$ and $f(x) = x$. Then, $G(x) = \frac{1}{9}(x+1)^9$ and $f'(x) = 1$, and so

$$\int x(x+1)^8 dx = \frac{1}{9}x(x+1)^9 - \int \frac{1}{9}(x+1)^9 dx$$
$$= \frac{1}{9}x(x+1)^9 - \frac{1}{9}\left[\frac{1}{10}(x+1)^{10}\right] + C = \frac{1}{9}x(x+1)^9 - \frac{1}{90}(x+1)^{10} + C$$

13. Let $g(x) = \frac{1}{\sqrt{x+2}} = (x+2)^{-1/2}$ and $f(x) = x$. Then, $G(x) = 2(x+2)^{1/2}$ and $f'(x) = 1$, and so

$$\int \frac{x}{\sqrt{x+2}} dx = 2x(x+2)^{1/2} - \int 2(x+2)^{1/2} dx$$

$$= 2x(x+2)^{1/2} - 2\left[\frac{2}{3}(x+2)^{3/2}\right] + C = 2x(x+2)^{1/2} - \frac{4}{3}(x+2)^{3/2} + C$$

15. Let $g(x) = e^{-x}$ and $f(x) = x^2$. Then, $G(x) = -e^{-x}$ and $f'(x) = 2x$, and so

$$\int x^2 e^{-x} dx = -x^2 e^{-x} - \int 2x(-e^{-x}) dx = -x^2 e^{-x} + 2\int xe^{-x} dx$$

Now let $g(x) = e^{-x}$ and $f(x) = x$. Then, $G(x) = -e^{-x}$ and $f'(x) = 1$, and so

$$\int xe^{-x} dx = -xe^{-x} - \int -e^{-x} dx = -xe^{-x} - e^{-x}$$

Thus,

$$\int x^2 e^{-x} dx = -x^2 e^{-x} + 2(-xe^{-x} - e^{-x}) + C$$

$$= -(x^2 + 2x + 2)e^{-x} + C$$

17. Let $g(x) = e^x$ and $f(x) = x^3$. Then, $G(x) = e^x$ and $f'(x) = 3x^2$, and so

$$\int x^3 e^x dx = x^3 e^x - \int 3x^2 e^x dx = x^3 e^x - 3\int x^2 e^x dx$$

Now let $g(x) = e^x$ and $f(x) = x^2$. Then $G(x) = e^x$ and $f'(x) = 2x$, and so

$$\int x^2 e^x dx = x^2 e^x - \int 2xe^x dx = x^2 e^x - 2\int xe^x dx$$

$$= x^2 e^x - 2(xe^x - e^x) = x^2 e^x - 2xe^x + 2e^x$$

Hence,

$$\int x^3 e^x dx = x^3 e^x - 3(x^2 e^x - 2xe^x + 2e^x) + C$$

$$= (x^3 - 3x^2 + 6x - 6)e^x + C$$

19. Let $g(x) = x^2$ and $f(x) = \ln x$. Then, $G(x) = \frac{1}{3}x^3$ and $f'(x) = \frac{1}{x}$, and so

$$\int x^2 \ln x\, dx = \frac{1}{3}x^3 \ln x - \int \frac{1}{x}\left(\frac{1}{3}x^3\right) dx$$

$$= \frac{1}{3}x^3 \ln x - \frac{1}{3}\int x^2 dx = \frac{1}{3}x^3 \ln x - \frac{1}{9}x^3 + C$$

21. Rewrite the integral as

$$\int \frac{\ln x}{x^2}\,dx = \int \left(\frac{1}{x^2}\right) \ln x\, dx$$

so that the integrand is a product. Let $g(x) = \frac{1}{x^2}$ and $f(x) = \ln x$. Then, $G(x) = -\frac{1}{x}$ and $f'(x) = \frac{1}{x}$, and so

$$\int \frac{\ln x}{x^2}\,dx = -\frac{1}{x}\ln x - \int \left(-\frac{1}{x}\right)\left(\frac{1}{x}\right)dx = -\frac{1}{x}\ln x + \int \frac{1}{x^2}\,dx$$

$$= -\frac{1}{x}\ln x - \frac{1}{x} + C = -\frac{1}{x}(\ln x + 1) + C$$

23. Rewrite the integral as

$$\int x^3 e^{x^2}\,dx = \int x^2 (xe^{x^2})\,dx$$

Let $g(x) = xe^{x^2}$ and $f(x) = x^2$. Then, $G(x) = \frac{1}{2}e^{x^2}$ and $f'(x) = 2x$, and so

$$\int x^3 e^{x^2}\,dx = \frac{1}{2}x^2 e^{x^2} - \int 2x\left(\frac{1}{2}e^{x^2}\right)dx$$

$$= \frac{1}{2}x^2 e^{x^2} - \int xe^{x^2}\,dx = \frac{1}{2}e^{x^2} - \frac{1}{2}e^{x^2} + C$$

25. Rewrite the integral as

$$\int x^7(x^4 + 5)^8\,dx = \int x^4[x^3(x^4 + 5)^8]\,dx$$

Let $g(x) = x^3(x^4 + 5)^8$ and $f(x) = x^4$. Then $G(x) = \frac{1}{36}(x^4 + 5)^9$ and $f'(x) = 4x^3$, and so

$$\int x^7(x^4 + 5)^8\,dx = \frac{1}{36}x^4(x^4 + 5)^9 - \int 4x^3\left(\frac{1}{36}\right)(x^4 + 5)^9\,dx$$

$$= \frac{1}{36}x^4(x^4 + 5)^9 - \frac{4}{36}\int (x^4 + 5)^9 x^3\,dx$$

$$= \frac{1}{36}x^4(x^4 + 5)^9 - \frac{1}{360}(x^4 + 5)^{10} + C$$

27. Let $f(x)$ be the function whose tangent has slope $x\ln\sqrt{x}$. Then

$$f'(x) = x\ln\sqrt{x}, \quad x > 0$$

and

$$f(x) = \int x \ln \sqrt{x}\, dx$$

$$= \int x \ln x^{1/2}\, dx$$

$$= \int x \left(\frac{1}{2}\right) \ln x\, dx$$

$$= \frac{1}{2} \int x \ln x\, dx$$

To integrate by parts let $g(x) = x$ and $f(x) = \ln x$. Then, $G(x) = \frac{x^2}{2}$ and $f'(x) = \frac{1}{x}$, and so

$$f(x) = \frac{1}{2}\int x \ln x\, dx = \frac{1}{2}\left[\left(\frac{x^2}{2}\right)\ln x - \int \left(\frac{x^2}{2}\right)\left(\frac{1}{x}\right) dx\right]$$

$$= \frac{1}{2}\left[\frac{x^2 \ln x}{2} - \frac{1}{2}\int x\, dx\right]$$

$$= \frac{1}{2}\left[\frac{x^2 \ln x}{2} - \frac{1}{4}x^2\right]$$

$$= \frac{x^2 \ln x}{4} - \frac{1}{8}x^2 + C$$

Since $(2, f(2)) = (2, -3)$, that is, when $x = 2$, $f = -3$,

$$-3 = \frac{(2)^2 \ln 2}{4} - \frac{1}{8}(2)^2 + C$$

$$-3 = \ln 2 - \frac{1}{2} + C$$
or
$$C = -\frac{5}{2} - \ln 2.$$

Thus, the desired function is

$$f(x) = \frac{x^2 \ln x}{4} - \frac{1}{8}x^2 - \frac{5}{2} - \ln 2.$$

29. Let t denote time and $Q(t)$ the number of units produced. Then, $\frac{dQ}{dt} = 100te^{-0.5t}$, and

$$Q(t) = 100\int te^{-0.5t}\, dt$$

Let $g(t) = e^{-0.5t}$ and $f(t) = t$. Then, $G(t) = -\frac{1}{0.5}e^{-0.5t} = -2e^{-0.5t}$ and $f'(t) = 1$, and so

$$Q(t) = 100(-2te^{-0.5t} - \int -2e^{-0.5t}\, dt)$$

$$= -200te^{-0.5t} - 400e^{-0.5t} + C = -200(t + 2)e^{-0.5t} + C$$

Since no units have been produced when $t = 0$,

$$0 = Q(0) = -200(2) + C \text{ or } C = 400$$

Hence, $Q(t) = -200(t+2)e^{-0.5t} + 400$, and the number of units produced during the first 3 hours is $Q(3) \simeq 176.87$.

31. Let q denote the number of units produced and $C(q)$ the cost of producing the first q units. Then $\frac{dC}{dq} = 0.5(q+1)e^{0.3q}$, and

$$C(q) = 0.5 \int (q+1)e^{0.3q} dq = \frac{1}{2} \int (q+1)e^{3/10q} dq$$

To integrate by parts let $g(q) = e^{\frac{3}{10}q}$ and $f(q) = q + 1$. Then, $G(q) = \frac{10}{3}e^{\frac{3}{10}q}$ and $f'(q) = 1$ and so

$$C(q) = \frac{1}{2}\left[\frac{10}{3}(q+1)e^{\frac{3}{10}q} - \frac{10}{3}\int e^{\frac{3}{10}q} dq \right]$$

$$= \frac{1}{2}\left[\frac{10}{3}(q+1)e^{\frac{3}{10}q} - \frac{100}{9}e^{\frac{3}{10}q} \right]$$

$$= \frac{5}{3}(q+1)e^{\frac{3}{10}q} - \frac{50}{9}e^{\frac{3}{10}q} + C$$

When $q = 10$ then $C = 200$ yields

$$200 = \frac{5}{3}(11)e^{(\frac{3}{10})(10)} - \frac{50}{9}e^{(\frac{3}{10})(10)} + C$$

$$= \frac{55}{3}e^3 - \frac{50}{9}e^3 + C$$

$$= \frac{65}{9}e^3 + C$$

or

$$C = 200 - \frac{65}{9}e^3.$$

Thus, $C(q) = \frac{5}{3}(q+1)e^{\frac{3}{10}q} - \frac{50}{9}e^{\frac{3}{10}q} - \frac{65}{9}e^3 + 200$ and

$$C(20) = \frac{5}{3}(21)e^6 - \frac{50}{9}e^6 - \frac{65}{9}e^3 + 200$$

$$= \frac{13}{9}e^6 - \frac{65}{9}e^3 + 200 \approx \$637.67$$

is the total cost of producing the first 20 units.

33. (a) Let $g(x) = e^{ax}$ and $f(x) = x^n$. Then, $G(x) = \frac{1}{a}e^{ax}$ and $f'(x) = nx^{n-1}$, and so

$$\int x^n e^{ax} dx = \frac{1}{a}x^n e^{ax} - \int (nx^{n-1})\left(\frac{1}{a}e^{ax}\right) dx$$

$$= \frac{1}{a}x^n e^{ax} - \frac{n}{a}\int x^{n-1}e^{ax} dx$$

(b)   Let $n = 3$ and $a = 5$. Then, by the formula from part (a),

$$\int x^3 e^{5x} dx = \frac{1}{5}x^3 e^{5x} - \frac{3}{5}\int x^2 e^{5x} dx$$

Now let $n = 2$ and $a = 5$ to get

$$\int x^2 e^{5x} dx = \frac{1}{5}x^2 e^{5x} - \frac{2}{5}\int x e^{5x} dx$$

Finally, let $n = 1$ and $a = 5$ to get

$$\int x e^{5x} dx = \frac{1}{5}x e^{5x} - \frac{1}{5}\int e^{5x} dx = \frac{1}{5}x e^{5x} - \frac{1}{25}e^{5x}$$

Putting it all together,

$$\begin{aligned}
\int x^3 e^{5x} dx &= \frac{1}{5}x^3 e^{5x} - \frac{3}{5}\left[\frac{1}{5}x^2 e^{5x} - \frac{2}{5}\int x e^{5x} dx\right]\\
&= \frac{1}{5}x^3 e^{5x} - \frac{3}{25}x^2 e^{5x} + \frac{6}{25}\int x e^{5x} dx\\
&= \frac{1}{5}x^3 e^{5x} - \frac{3}{25}x^2 e^{5x} + \frac{6}{25}\left[\frac{1}{5}x e^{5x} - \frac{1}{25}e^{5x}\right]\\
&= \left(\frac{1}{5}x^3 - \frac{3}{25}x^2 + \frac{6}{125}x - \frac{6}{625}\right)e^{5x} + C
\end{aligned}$$

Chapter 5, Section 4

A small table of integrals:

(1) $\int \frac{dx}{p^2-x^2} = \frac{1}{2p} \ln \left| \frac{p+x}{p-x} \right|$

(2) $\int \frac{dx}{x(ax+b)} = \frac{1}{b} \ln \left| \frac{x}{ax+b} \right|$

(3) $\int \frac{dx}{\sqrt{x^2 \pm p^2}} = \ln \left| x + \sqrt{x^2 \pm p^2} \right|$

(4) $\int x^n e^{ax} dx = \frac{1}{a} x^n e^{ax} - \frac{n}{a} \int x^{n-1} e^{ax} dx$

1. Apply formula 2 with $a = 2$ and $b = -3$ to get

$$\int \frac{1}{x(2x-3)} dx = -\frac{1}{3} \ln \left| \frac{x}{2x-3} \right| + C$$

3. Apply formula 3 with $p = 5$ to get

$$\int \frac{1}{\sqrt{x^2+25}} dx = \ln \left| x + \sqrt{x^2+25} \right| + C$$

5. Apply formula 1 with $p = 2$ to get

$$\int \frac{1}{4-x^2} dx = \frac{1}{4} \ln \left| \frac{2+x}{2-x} \right| + C$$

7. Apply formula 2 with $a = 3$ and $b = 2$ to get

$$\int \frac{1}{3x^2+2x} dx = \int \frac{1}{x(3x+2)} dx = \frac{1}{2} \ln \left| \frac{x}{3x+2} \right| + C$$

9. Apply formula 4 with $n = 2$ and $a = 3$ to get

$$\int x^2 e^{3x} dx = \frac{1}{3} x^2 e^{3x} - \frac{2}{3} \int x e^{3x} dx$$

Now apply formula 4 with $n = 1$ and $a = 3$ to get

$$\int x e^{3x} dx = \frac{1}{3} x e^{3x} - \frac{1}{3} \int e^{3x} dx = \frac{1}{3} x e^{3x} - \frac{1}{9} e^{3x}$$

Hence,

$$\int x^2 e^{3x} dx = \frac{1}{3} x^2 e^{3x} - \frac{2}{3} \left[\frac{1}{3} x e^{3x} - \frac{1}{9} e^{3x} \right] + C$$

$$= (\frac{1}{3} x^2 - \frac{2}{9} x + \frac{2}{27}) e^{3x} + C$$

11.   Apply the formula $\int \frac{x}{ax^2 + c} dx = \frac{1}{2a} \ln \left| ax^2 + c \right|$ with $a = -1$ and $c = 2$ to get

$$\int \frac{x}{2 - x^2} dx = -\frac{1}{2} \ln \left| 2 - x^2 \right| + C$$

13.   Apply the formula

$$\int (\ln ax)^2 dx = x (\ln ax)^2 - 2x (\ln ax) + 2x$$

with $a = 2$ to get

$$\int (\ln 2x)^2 dx = x (\ln 2x)^2 - 2x (\ln 2x) + 2x + C$$

15.   Apply the formula

$$\int \frac{dx}{x \sqrt{ax + b}} = \frac{1}{\sqrt{b}} \ln \left| \frac{\sqrt{ax + b} - \sqrt{b}}{\sqrt{ax + b} + \sqrt{b}} \right|$$

with $a = 2$ and $b = 5$ to get

$$\int \frac{1}{3x \sqrt{2x + 5}} dx = \frac{1}{3} \int \frac{1}{x \sqrt{2x + 5}} dx$$

$$= \frac{1}{3\sqrt{5}} \ln \left| \frac{\sqrt{2x + 5} - \sqrt{5}}{\sqrt{2x + 5} + \sqrt{5}} \right| + C$$

17.   Apply the formula

$$\int \frac{dx}{b + ce^{ax}} = \frac{1}{ab} \left[ax - \ln \left| b + ce^{ax} \right| \right]$$

with $a = -1$, $b = 2$, and $c = -3$ to get

$$\int \frac{1}{2 - 3e^{-x}} dx = -\frac{1}{2} \left[-x - \ln \left| 2 - 3e^{-x} \right| \right] + C$$

$$= \frac{1}{2} x + \frac{1}{2} \ln \left| 2 - 3e^{-x} \right| + C$$

19. Rewrite the two integral formulas with the constants of integration:

(1) $\int \frac{dx}{\sqrt{x^2 \pm p^2}} = \ln \left| \frac{x + \sqrt{x^2 \pm p^2}}{p} \right| + C_1$ and

(2) $\int \frac{dx}{\sqrt{x^2 \pm p^2}} = \ln \left| x + \sqrt{x^2 \pm p^2} \right| + C_2$

Now rewrite the right-hand side of (1) using properties of logarithms to get

$$\ln \left| \frac{x + \sqrt{x^2 \pm p^2}}{p} \right| + C_1 = \ln \left| x + \sqrt{x^2 \pm p^2} \right| - \ln \left| p \right| + C_1$$

$$= \ln \left| x + \sqrt{x^2 \pm p^2} \right| + C_2$$

where C_2 replaces the constant $- \ln \left| p \right| + C_1$. In other words, the two given antiderivatives of $\frac{1}{\sqrt{x^2 \pm p^2}}$ differ by only a constant.

ANTIDIFFERENTIATION

CHAPTER 5

Chapter 5, Review Problems

1. $\int (x^5 - 3x^2 + \frac{1}{x^2})dx = \int (x^5 - 3x^2 + x^{-2})dx = \frac{1}{6}x^6 - x^3 - \frac{1}{x} + C$

2. $\int (x^{2/3} - \frac{1}{x} + 5 + \sqrt{x})dx = \frac{3}{5}x^{5/3} - \ln|x| + 5x + \frac{2}{3}x^{3/2} + C$

3. Let $u = 3x + 1$. Then $du = 3dx$ or $dx = \frac{1}{3}du$. Hence

$$\int \sqrt{3x + 1}dx = \frac{1}{3}\int u^{1/2}du$$

$$= \frac{1}{3}\left[\frac{2}{3}u^{3/2}\right] + C = \frac{2}{9}(3x + 1)^{3/2} + C$$

4. Let $u = 3x^2 + 2x + 5$. Then $du = (6x + 2)dx$ or $(3x + 1)dx = \frac{1}{2}du$. Hence,

$$\int (3x + 1)\sqrt{3x^3 + 2x + 5}dx = \frac{1}{2}\int u^{1/2}du$$

$$= \frac{1}{2}\left[\frac{2}{3}u^{3/2}\right] + C = \frac{1}{3}(3x^2 + 2x + 5)^{3/2} + C$$

5. Let $u = x^2 + 4x + 2$. Then $du = (2x + 4)dx$ or $(x + 2)dx = \frac{1}{2}du$. Hence,

$$\int (x + 2)(x^2 + 4x + 2)^5 dx = \frac{1}{2}\int u^5 du$$

$$= \frac{1}{2}\left[\frac{1}{6}u^6\right] + C = \frac{1}{12}(x^2 + 4x + 2)^6 + C$$

6. Let $u = x^2 + 4x + 2$. Then $du = (2x + 4)dx$ or $(x + 2)dx = \frac{1}{2}du$. Hence,

$$\int \frac{x + 2}{x^2 + 4x + 2}dx = \frac{1}{2}\int \frac{1}{u}du$$

$$= \frac{1}{2}\ln|u| + C = \frac{1}{2}\ln|x^2 + 4x + 2| + C$$

7. Let $u = 2x^2 + 8x + 3$. Then $du = (4x + 8)dx$ or $(3x + 6)dx = \frac{3}{4}du$. Hence,

$$\int \frac{3x + 6}{(2x^2 + 8x + 3)^2}dx = \frac{3}{4}\int \frac{1}{u^2}du = \frac{3}{4}\int u^{-2}du$$

$$= -\frac{3}{4}u^{-1} + C = -\frac{3}{4(2x^2 + 8x + 3)} + C$$

8.   Let $u = x - 5$. Then $du = dx$ and

$$\int (x - 5)^{12} dx = \int u^{12} du = \frac{1}{13} u^{13} + C = \frac{1}{13}(x - 5)^{13} + C$$

9.   *Method 1* : Integration by substitution. Let $u = x - 5$. Then $du = dx$ and $x = u + 5$.
Hence,

$$\int x(x - 5)^{12} dx = \int (u + 5)u^{12} du = \int (u^{13} + 5u^{12}) du$$

$$= \frac{1}{14} u^{14} + 5\left[\frac{1}{13} u^{13}\right] + C = \frac{1}{14}(x - 5)^{14} + \frac{5}{13}(x - 5)^{13} + C$$

Method 2 : Integration by parts. Let $g(x) = (x - 5)^{12}$ and $f(x) = x$. Then, $G(x) = \frac{1}{13}(x - 5)^{13}$ and $f'(x) = 1$. Hence,

$$\int x(x - 5)^{12} dx = \frac{1}{13} x(x - 5)^{13} - \int \frac{1}{13}(x - 5)^{13} dx$$

$$= \frac{1}{13} x(x - 5)^{13} - \frac{1}{13}\left[\frac{1}{14}(x - 5)^{14}\right] + C$$

$$= \frac{1}{13} x(x - 5)^{13} - \frac{1}{182}(x - 5)^{14} + C$$

10.   Let $u = 3x$. Then $du = 3dx$ or $dx = \frac{1}{3} du$. Hence,

$$\int 5e^{3x} dx = \frac{5}{3} \int e^u du = \frac{5}{3} e^u + C = \frac{5}{3} e^{3x} + C$$

11.   Let $g(x) = e^{3x}$ and $f(x) = x$. Then, $G(x) = \frac{1}{3} e^{3x}$ and $f'(x) = 1$. Hence,

$$\int 5xe^{3x} dx = 5\left(\frac{1}{3} xe^{3x} - \int \frac{1}{3} e^{3x} dx\right)$$

$$= 5\left[\frac{1}{3} xe^{3x} - \frac{1}{9} e^{3x}\right] + C = \left(\frac{5}{3} x - \frac{5}{9}\right) e^{3x} + C$$

12.   Let $g(x) = e^{-x/2}$ and $f(x) = x$. Then, $G(x) = -2e^{-x/2}$ and $f'(x) = 1$. Hence,

$$\int xe^{-x/2} dx = -2xe^{-x/2} - \int (-2e^{-x/2})(1) dx = -2xe^{-x/2} + 2\int e^{-x/2} dx$$

$$= -2xe^{-x/2} - 4e^{-x/2} + C = -2(x + 2)e^{-x/2} + C$$

13. Rewrite the integral as

$$\int x^5 e^{x^3} dx = \int x^3 [x^2 e^{x^3}] dx$$

and let $g(x) = x^2 e^{x^3}$ and $f(x) = x^3$. Then $G(x) = \frac{1}{3} e^{x^3}$ and $f'(x) = 3x^2$. Hence,

$$\int x^3 [x^2 e^{x^3}] dx = \frac{1}{3} x^3 e^{x^3} - \frac{1}{3} \int 3x^2 e^{x^3} dx = \frac{1}{3} x^3 e^{x^3} - \int x^2 e^{x^3} dx$$

$$= \frac{1}{3} x^3 e^{x^3} - \frac{1}{3} e^{x^3} + C = \frac{1}{3}(x^3 - 1) e^{x^3} + C$$

14. Let $g(x) = e^{0.1x}$ and $f(x) = 2x + 1$. Then $G(x) = 10 e^{0.1x}$ and $f'(x) = 2$. Hence,

$$\int (2x + 1) e^{0.1x} dx = 10 e^{0.1x}(2x + 1) - 20 \int e^{0.1x} dx$$

$$= 10(2x + 1) e^{0.1x} - 200 e^{0.1x} + C = 10(2x - 19) e^{0.1x} + C$$

15. Let $g(x) = x$ and $f(x) = \ln 3x$. Then, $G(x) = \frac{1}{2} x^2$ and $f'(x) = \frac{1}{x}$. Hence,

$$\int x \ln 3x \, dx = \frac{1}{2} x^2 \ln 3x - \int \frac{1}{x} \left[\frac{1}{2} x^2 \right] dx$$

$$= \frac{1}{2} x^2 \ln 3x - \frac{1}{2} \int x \, dx = \frac{1}{2} x^2 \ln 3x - \frac{1}{4} x^2 + C$$

16. Let $g(x) = 1$ and $f(x) = \ln 3x$. Then, $G(x) = x$ and $f'(x) = \frac{1}{x}$. Hence,

$$\int \ln 3x \, dx = \int 1 (\ln 3x) dx = x \ln 3x - \int \frac{1}{x}(x) dx$$

$$= x \ln 3x - \int 1 \, dx = x \ln 3x - x + C$$

17. Let $u = \ln 3x$. Then $du = \frac{1}{x} dx$ and so

$$\int \frac{\ln 3x}{x} dx = \int u \, du = \frac{1}{2} u^2 + C = \frac{1}{2}(\ln 3x)^2 + C$$

18. Let $g(x) = \frac{1}{x^2}$ and $f(x) = \ln 3x$. Then $G(x) = -\frac{1}{x}$ and $f'(x) = \frac{1}{x}$. Hence,

$$\int \frac{\ln 3x}{x^2} dx = -\frac{1}{x} \ln 3x - \int (-\frac{1}{x})(\frac{1}{x}) dx = -\frac{1}{x} \ln 3x + \int \frac{1}{x^2} dx$$

$$= -\frac{1}{x} \ln 3x - \frac{1}{x} + C = -\frac{1}{x}(\ln 3x + 1) + C$$

19.   Rewrite the integral as

$$\int x^3(x^2+1)^8 dx = \int x^2 [x(x^2+1)^8] dx$$

and let $g(x) = x(x^2+1)^8$ and $f(x) = x^2$. Then, $G(x) = \frac{1}{18}(x^2+1)^9$ and $f'(x) = 2x$. Hence,

$$\int x^3(x^2+1)^8 dx = \frac{1}{18}x^2(x^2+1)^9 - \int 2x \left[\frac{1}{18}(x^2+1)^9 \right] dx$$

$$= \frac{1}{18}x^2(x^2+1)^9 - \frac{1}{9}\int x(x^2+1)^9 dx$$

$$= \frac{1}{18}x^2(x^2+1)^9 - \frac{1}{180}(x^2+1)^{10} + C$$

20.   Let $u = x^2 + 1$. Then $du = 2x dx$. Hence,

$$\int 2x \ln(x^2+1) dx = \int \ln u\ du = \int 1(\ln u) du$$

Now apply integration by parts letting $g(u) = 1$ and $f(u) = \frac{1}{u}$. Then $G(u) = u$ and $f'(u) = \frac{1}{u}$. Hence,

$$\int 2x \ln(x^2+1) dx = \int 1(\ln u) du = u \ln u - \int u(\frac{1}{u}) du$$

$$= u \ln u - \int 1 du = u \ln u - u + C = u(\ln u - 1) + C$$

$$= (x^2+1)[\ln(x^2+1) - 1] + C$$

21.   The slope of the tangent is the derivative. Hence, $f'(x) = x(x^2+1)^3$ and so f is an antiderivative of $x(x^2+1)^3$. That is,

$$f(x) = \int x(x^2+1)^3 dx = \frac{1}{8}(x^2+1)^4 + C$$

Since the graph of f passes through the point $(1,5)$,

$$5 = f(1) = \frac{1}{8}(2)^4 + C = 2 + C \text{ or } C = 3$$

Hence, $f(x) = \frac{1}{8}(x^2+1)^4 + 3$.

22.   Let $Q(x)$ denote the number of commuters using the new subway line x weeks from now. It is given that $\frac{dQ}{dx} = 18x^2 + 500$ commuters per week. Hence $Q(x)$ is an antideriative of $18x^2 + 500$. That is,

$$Q(x) = \int (18x^2 + 500) dx = 6x^3 + 500x + C$$

Since $8,000$ commuters currently use the subway,

$$8,000 = Q(0) = C$$

Hence, $Q(x) = 6x^3 + 500x + 8,000$, and the number of commuters who will be using the subway in 5 years is $Q(5) = 11,250$.

23. Let $Q(x)$ denote the number of inmates in county prisons x years from now. It is given that $\frac{dQ}{dx} = 280e^{0.2x}$ inmates per year. Hence $Q(x)$ is an antiderivative of $280e^{0.2x}$. That is,

$$Q(x) = \int 280e^{0.2x}dx = \frac{280}{0.2}e^{0.2x} + C = 1,400e^{0.2x} + C$$

Since the prisons currently houses $2,000$ inmates,

$$2,000 = Q(0) = 1,400 + C \ \text{ or } \ C = 600$$

Hence, $Q(x) = 1,400e^{0.2x} + 600$, and the number of inmates 10 years from now will be

$$Q(10) = 1,400e^2 + 600 \simeq 10,945$$

24. Let $P(q)$ denote the profit, $R(q)$ the revenue, and $C(q)$ the cost when the level of production is q units. Since marginal revenue $= 200q^{-1/2} = \frac{dR}{dq}$, marginal cost $= 0.4q = \frac{dC}{dq}$, and profit equals revenue minus cost, it follows that the marginal profit is

$$\frac{dP}{dq} = \frac{dR}{dq} - \frac{dC}{dq} = 200q^{-1/2} - 0.4q \ \text{ dollars per unit}$$

The profit function $P(q)$ is an antiderivative of the marginal profit. That is,

$$P(q) = \int (200q^{-1/2} - 0.4q)dq = 400q^{1/2} - 0.2q^2 + C$$

Since profit is $\$2,000$ when the level of production is 25 units,

$$2,000 = P(25) = 400(5) - 0.2(25)^2 + C \ \text{ or } \ C = 125$$

Hence, $P(q) = 400q^{1/2} - 0.2q^2 + 125$, and the profit when 36 units are produced is $P(36) = \$2,265.80$.

25. Apply the formula

$$\int \frac{dx}{p^2 - x^2} = \frac{1}{2p}\ln\left|\frac{p+x}{p-x}\right|$$

with $p = 2$ to get

$$\int \frac{5}{8 - 2x^2}dx = \frac{5}{2}\int \frac{1}{4 - x^2}dx = \frac{5}{2}\int \frac{dx}{2^2 - x^2}$$

$$= \frac{5}{2}(\frac{1}{4})\ln\left|\frac{2+x}{2-x}\right| + C = \frac{5}{8}\ln\left|\frac{2+x}{2-x}\right| + C$$

26.   Apply the formula

$$\int \frac{dx}{\sqrt{x^2 \pm p^2}} = \ln|x + \sqrt{x^2 \pm p^2}|$$

with $p = \frac{4}{3}$ to get

$$\int \frac{2}{\sqrt{9x^2 + 16}} dx = \frac{2}{3} \int \frac{1}{\sqrt{x^2 + 16/9}} dx = \frac{2}{3} \ln \left| x + \sqrt{x^2 + \frac{16}{9}} \right| + C$$

27.   Apply the formula

$$\int x^n e^{ax} dx = \frac{1}{a} x^n e^{ax} - \frac{n}{a} \int x^{n-1} e^{ax} dx$$

with $n = 2$ and $a = -\frac{1}{2}$ to get

$$\int x^2 e^{-x/2} dx = -2x^2 e^{-x/2} + 4 \int x e^{-x/2} dx$$

Now apply the formula with $n = 1$ and $a = -\frac{1}{2}$ to get

$$\int x e^{-x/2} dx = -2x e^{-x/2} + 2 \int e^{-x/2} dx = -2x e^{-x/2} - 4 e^{-x/2}$$

Putting it all together,

$$\int x^2 e^{-x/2} dx = -2x^2 e^{-x/2} + 4[-2x e^{-x/2} - 4 e^{-x/2}] + C$$
$$= (-2x^2 - 8x - 16) e^{-x/2} + C$$

CHAPTER 6

FURTHER TOPICS IN INTEGRATION

---

1 THE DEFINITE INTEGRAL

2 AREA AND INTEGRATION

3 APPLICATIONS TO BUSINESS AND ECONOMICS

4 THE INTEGRAL AS THE LIMIT OF A SUM

5 FURTHER APPLICATIONS OF THE DEFINITE INTEGRAL

6 NUMERICAL INTEGRATION

REVIEW PROBLEMS

Chapter 6, Section 1

1. $\int_0^1 (x^4 - 3x^3 + 1)dx = (\frac{1}{5}x^5 - \frac{3}{4}x^4 + x)\Big|_0^1 = (\frac{1}{5} - \frac{3}{4} + 1) - 0 = \frac{9}{20}$

3. $\int_2^5 (2 + 2t + 3t^2)dt = (2t + t^2 + t^3)\Big|_2^5 = (10 + 25 + 125) - (4 + 4 + 8) = 144$

5. $\int_1^3 (1 + \frac{1}{x} + \frac{1}{x^2})dx = (x + ln|x| - \frac{1}{x})\Big|_1^3$

     $= (3 + ln|3| - \frac{1}{3}) - (1 + ln|1| - 1) = \frac{8}{3} + ln3$

7. $\int_{-3}^{-1} \frac{t+1}{t^3}dt = \int_{-3}^{-1} (t^{-2} + t^{-3})dt$

     $= (-\frac{1}{t} - \frac{1}{2t^2})\Big|_{-3}^{-1} = (1 - \frac{1}{2}) - (\frac{1}{3} - \frac{1}{18}) = \frac{2}{9}$

9.   Let $u = 2x - 4$. Then $du = 2dx$ or $dx = \frac{1}{2}du$. When $x = 1, u = -2$, and when $x = 2, u = 0$. Hence,

$$\int_1^2 (2x - 4)^5 dx = \frac{1}{2}\int_{-2}^0 u^5 du = \frac{1}{12}u^6\Big|_{-2}^0 = -\frac{64}{12} = -\frac{16}{3}$$

11.   Let $u = 6t + 1$. Then $du = 6dt$ or $dt = \frac{1}{6}du$. When $t = 0, u = 1$, and when $t = 4, u = 25$. Hence,

$$\int_0^4 \frac{1}{\sqrt{6t+1}}dt = \frac{1}{6}\int_1^{25} u^{-1/2}du = \frac{1}{3}u^{1/2}\Big|_1^{25} = \frac{5}{3} - \frac{1}{3} = \frac{4}{3}$$

13.   Let $u = t^4 + 2t^2 + 1$. Then $du = (4t^3 + 4t)dt$ or $(t^3 + t)dt = \frac{1}{4}du$. When $t = 0, u = 1$, and when $t = 1, u = 4$. Hence,

$$\int_0^1 (t^3 + t)\sqrt{t^4 + 2t^2 + 1}\,dt = \frac{1}{4}\int_1^4 u^{1/2}du = \frac{1}{6}u^{3/2}\Big|_1^4$$

$$= \frac{1}{6}(4^{3/2} - 1^{3/2}) = \frac{1}{6}(8 - 1) = \frac{7}{6}$$

15.   Let $u = x - 1$. Then $du = dx$ and $x = u + 1$. When $x = 2, u = 1$, and when $x = e + 1, u = e$. Hence,

$$\int_2^{e+1} \frac{x}{x-1}dx = \int_1^e \frac{u+1}{u}du = \int_1^e (1 + \frac{1}{u})du$$

$$= (u + \ln|u|)\Big|_1^3 = (e + \ln e) - (1 + \ln 1) = e$$

17.   Let $g(t) = 1$ and $f(t) = \ln t$. Then $G(t) = t$ and $f'(t) = \frac{1}{t}$, and so

$$\int_1^{e^2} \ln t\,dt = (t\ln t)\Big|_1^{e^2} - \int_1^{e^2} \frac{1}{t}(t)dt$$

$$= (t\ln t)\Big|_1^{e^2} - \int_1^{e^2} 1\,dt = (t\ln t - t)\Big|_1^{e^2}$$

$$= (e^2\ln e^2 - e^2) - (\ln 1 - 1) = e^2 + 1$$

19.   Let $g(x) = e^{-x}$ and $f(x) = x$. The $G(x) = -e^{-x}$ and $f'(x) = 1$. Hence

$$\int_{-2}^2 xe^{-x}dx = -xe^{-x}\Big|_{-2}^2 - \int_{-2}^2 -e^{-x}dx$$

$$= (-xe^{-x} - e^{-x})\Big|_{-2}^2 = (-2e^{-2} - e^{-2}) = (2e^2 - e^2) = -3e^{-2} - e^2$$

21.   Let $u = \ln x$. Then $du = \frac{1}{x}dx$. When $x = 1, u = 0$, and when $x = e^2, u = 2$. Hence,

$$\int_1^{e^2} \frac{(\ln x)^2}{x}dx = \int_0^2 u^2 du = \frac{1}{3}u^3\Big|_0^2 = \frac{8}{3}$$

23.   Let $g(t) = e^{-0.1t}$ and $f(t) = (20+t)$. Then $G(t) = -10e^{-0.1t}$ and $f'(x) = 1$. Hence,

$$\int_0^{10} (20+t)e^{-0.1t}dt = -10(20_t)e^{0.1t}\Big|_0^{10} + 10\int_0^{10} e^{-0.1t}dt$$

$$= [-10(20+t)e^{-0.1t} - 100e^{-0.1t}]\Big|_0^{10} = -(300+10t)e^{-0.1t}\Big|_0^{10}$$

$$= [-300+100)e^{-1}] - [-(300+0)e^0] = -400e^{-1}+300 \simeq 152.85$$

25.   Let $P(x)$ denote the population of the town x months from now. Then $\frac{dP}{dx} = 5 + 3x^{2/3}$, and the amount by which the population will increase during the next 8 months is

$$P(8) - P(0) = \int_0^8 (5+3x^{2/3})dx = (5x + \frac{9}{5}x^{5/3})\Big|_0^8$$

$$= \left[5(8) + \frac{9}{5}(8)^{5/3}\right] - 0 = 97.6 \simeq 98 \text{ people}$$

27.   Let $V(x)$ denote the value of the machine after x years. Then $\frac{dV}{dx} = 220(x-10)$, and the amount by which the machine depreciates during the 2nd year is

$$V(2) - V(1) = \int_1^2 220(x-10)dx = 220(\frac{1}{2}x^2 - 10x)\Big|_1^2$$

$$= 220\left[(2-20) - (\frac{1}{2} - 10)\right] = -\$1,870$$

where the minus sign indicates that the value of the machine has decreased.

29.   Let $C(q)$ denote the total cost of producing q units. Then marginal cost is $\frac{dC}{dq} = 6(q-5)^2$, and the increase in cost is

$$C(13) - C(10) = \int_{10}^{13} 6(q-5)^2 dq = 2(q-5)^3\Big|_{10}^{13}$$

$$= 2[(8)^3 - (5)^3] = \$774$$

31.   Let $N(t)$ denote the number of bushels that are produced over the next t days. Then $\frac{dN}{dt} = 0.3t^2 + 0.6t + 1$, and the increase in the crop over the next 5 days is

$$N(5) - N(0) = \int_0^5 (0.3t^2 + 0.6t + 1)dt = (0.1t^3 + 0.3t^2 + t)\Big|_0^5$$

$$= [0.1(125) + 0.3(25) + 5] - 0 = 25 \text{ bushels}$$

If the price remains fixed at \$3 per bushel, the corresponding increase in the value of the crop is \$75.

33. Let $D(t)$ denote the demand for the product. Since the current demand is $5,000$ and the demand increases exponentially, $D(t) = 5,000e^{0.02t}$ units per year. Let $R(t)$ denote the total revenue t years from now. Then the rate of change of revenue is

$$\frac{dR}{dt} = \frac{\text{dollars}}{\text{year}} = \frac{\text{dollars}}{\text{unit}}\frac{\text{unit}}{\text{year}} = 400D(t)$$

$$= 400(5,000e^{0.02t}) = 2,000,000e^{0.02t}$$

The increase in revenue over the next 2 years is

$$R(2) - R(0) = \int_0^2 2,000,000e^{0.02t}\,dt = 100,000,000e^{0.02t}\,\Big|_0^2$$

$$= 100,000,000(e^{0.04} - 1) \simeq \$4,081,077.40$$

35. (a) If $F(x)$ is an antiderivative of $f(x)$, then

$$\int_a^b f(x)dx + \int_b^c f(x)dx = [F(b) - F(a)] + [F(c) - F(b)]$$

$$= F(c) - F(a) = \int_a^c f(x)dx$$

(b) Since

$$|x| = \begin{cases} -x & \text{if } x < 0 \\ x & \text{if } x > 0 \end{cases}$$

write the integral as

$$\int_{-1}^1 |x|dx = \int_{-1}^0 -x\,dx + \int_0^1 x\,dx$$

$$= (-\frac{1}{2}x^2)\,\Big|_{-1}^0 + (\frac{1}{2}x^2)\,\Big|_0^1 = -(0 - \frac{1}{2}) + (\frac{1}{2} - 0) = 1$$

(c) Since

$$|x - 3| = \begin{cases} -(x - 3) & \text{if } x < 3 \\ x - 3 & \text{if } x > 3 \end{cases}$$

write the integral as

$$\int_0^4 (1 + |x - 3|)^2 dx = \int_0^3 [1 - (x - 3)]^2 dx + \int_3^4 [1 + (x - 3)]^2 dx$$

$$= \int_0^3 (4 - x)^2 dx + \int_3^4 (x - 2)^2 dx$$

$$= \left[-\frac{1}{3}(4 - x)^3\right]\Big|_0^3 + \left[\frac{1}{3}(x - 2)^3\right]\Big|_3^4$$

$$= -\frac{1}{3}[(1)^3 - (4)^3] + \frac{1}{3}[(2)^3 - (1)^3] = \frac{70}{3}$$

Chapter 6, Section 2

1.   Area $= \int_0^{4/3}(4 - 3x)dx = (4x - \frac{3}{2}x^2)\Big|_0^{4/3} = 4(\frac{4}{3}) - \frac{3}{2}(\frac{4}{3})^2 = \frac{8}{3}$

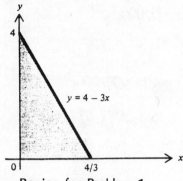

Region for Problem 1.

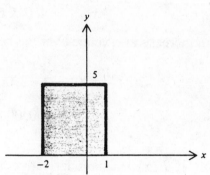

Region for Problem 3.

3.   Area $= \int_{-2}^1 5dx = 5x\Big|_2^1 = 5 - (-10) = 15$

5.   Area $= \int_4^9 \sqrt{x}dx = \frac{2}{3}x^{3/2}\Big|_4^9 = \frac{2}{3}(27 - 8) = \frac{38}{3}$

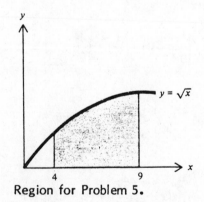

Region for Problem 5.

Region for Problem 7.

7.   Area $= \int_{-1}^1(1 - x^2)dx = (x - \frac{1}{3}x^3)\Big|_{-1}^1 = (1 - \frac{1}{3}) - (-1 - \frac{-1}{3}) = \frac{4}{3}$

9.   Area $= \int_{\ln 1/2}^0 e^x dx = e^x\Big|_{\ln 1/2}^0 = 1 - e^{\ln 1/2} = \frac{1}{2}$

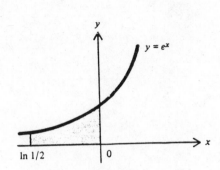

Region for Problem 9.

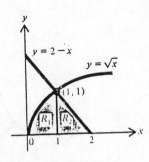

Region for Problem 11.

11. Break R into two subregions R_1 and R_2 as shown in the accompanying figure. Then,

$$\text{Area of } R = \text{area of } R_1 + \text{area of } R_2 = \int_0^1 \sqrt{x}\,dx + \int_1^2 (2-x)\,dx$$

$$= \frac{2}{3}x^{3/2} \Big|_0^1 + (2x - \frac{1}{2}x^2) \Big|_1^2 = \frac{2}{3} + \left[(4-2) - (2 - \frac{1}{2})\right] = \frac{7}{6}$$

13. Since $y = x^2 + 4$ and $y = -x + 10$, the curves intersect when

$$x^2 + 4 = -x + 10 \qquad x^2 + x - 6 = 0 \qquad (x+3)(x-2) = 0 \text{ or } x = -3 \text{ and } x = 2$$

Since the upper boundary of the region changes when $x = 2$, break the region into two pieces to get

$$\text{Area} = \int_0^2 (x^2 + 4)\,dx + \int_2^{10} (-x + 10)\,dx$$

$$= (\frac{1}{3}x^3 + 4x) \Big|_0^2 + (-\frac{1}{2}x^2 + 10x) \Big|_2^{10}$$

$$= \left[(\frac{8}{3} + 8) - 0\right] + \left[(-50 + 100) - (-2 + 20)\right] = \frac{128}{3}$$

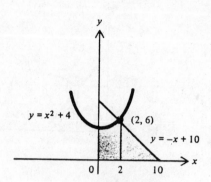

$y = x^2 + 4$

(2, 6)

$y = -x + 10$

Region for Problem 13.

15. The probability density function is $f(x) = 0.01e^{0.01x}$.

(a) $P(50 \leq x \leq 60) = \int_{50}^{60} 0.01e^{0.01x}\,dx$

$$= -e^{-0.01x}\Big|_{50}^{60} = -e^{-0.6} + e^{-0.5} \simeq 0.0577$$

(b) $P(0 \leq x \leq 60) = \int_{0}^{60} 0.01e^{-0.01x}\,dx$

$$= -e^{-0.01x}\Big|_{0}^{60} = -e^{-0.6} + 1 \simeq 0.4512$$

(c) Using the result from part (b),

$$P(x \geq 60) = 1 - P(0 \leq x \leq 60) \simeq 1 - 0.4512 = 0.5488$$

17. The probability density function is $f(x) = 0.2e^{-0.2x}$.

(a) $P(0 \leq x \leq 5) = \int_{0}^{5} 0.2e^{-0.2x}\,dx$

$$= -e^{-0.2x}\Big|_{0}^{5} = -e^{-1} + 1 \simeq 0.6321$$

(b) $P(x \geq 6) = 1 - P(0 \leq x \leq 6) = 1 - \int_{0}^{6} 0.2e^{-0.2x}\,dx$

$$= 1 - (-e^{-0.2x})\Big|_{0}^{6} = 1 - (-e^{-1.2} + 1) \simeq 0.3012$$

19.   Area $= \int_0^3 [(x^2 + 5) - (-x^2)]dx = \int_0^3 (2x^2 + 5)dx$

            $= (\frac{2}{3}x^3 + 5x) \Big|_0^3 = \frac{2}{3}(27) + 5(3) = 33$

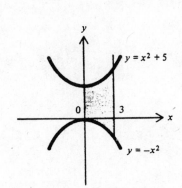

Region for Problem 19.

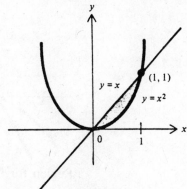

Region for Problem 21.

21.   Since $y = x$ and $y = x^2$, the curves intersect when

         $x = x^2 \quad x^2 - x = 0 \quad x(x-1) = 0$ or $x = 0$ and $x = 1.$

Hence,

$$\text{Area} = \int_0^1 (x - x^2)dx = (\frac{1}{2}x^2 - \frac{1}{3}x^3) \Big|_0^1 = \frac{1}{2} - \frac{1}{3} = \frac{1}{6}$$

23.   Since $y = x^2$ and $y = \sqrt{x}$, the curves intersect when

         $x^2 = \sqrt{x} \quad x^4 = x \quad x^4 - x = 0 \quad x(x^3 - 1)$ or $x = 0$ and $x = 1$

Hence,

$$\text{Area} = \int_0^1 (\sqrt{x} + x^2)dx = (\frac{2}{3}x^{3/2} - \frac{1}{3}x^3) \Big|_0^1 = \frac{2}{3} - \frac{1}{3} = \frac{1}{3}$$

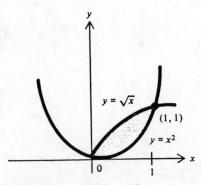

Region for Problem 23.

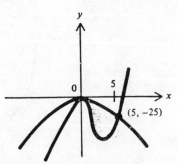

Region for Problem 25.

25. Since $y = -x^2$ and $y = x^3 - 6x^2$, the curves intersect when

$$-x^2 = x^3 - 6x^2 \qquad x^3 - 5x^2 = 0 \qquad x^2(x - 5) = 0 \text{ or } x = 0 \text{ and } x = 5$$

Hence,

$$\text{Area} = \int_0^5 [-x^2 - (x^3 - 6x^2)]dx = \int_0^5 (5x^2 - x^3)dx$$

$$= (\frac{5}{3}x^3 - \frac{1}{4}x^4) \Big|_0^5 = \frac{5}{3}(5)^3 - \frac{1}{4}(5)^4 = \frac{625}{12}$$

27. Break R into two subregions R_1 and R_2 as shown in the accompanying figure. Then,

$$\text{Area of } R = \text{area of } R_1 + \text{area of } R_2$$

$$= \int_0^1 (x - \frac{x}{8})dx + \int_1^2 (\frac{1}{x^2} - \frac{x}{8})dx$$

$$= \frac{7}{8} \int_0^1 xdx + \int_1^2 (\frac{1}{x^2} - \frac{x}{8})dx = \frac{7}{16}x^2 \Big|_0^1 + (-\frac{1}{x} - \frac{1}{16}x^2) \Big|_1^2$$

$$= \left[\frac{7}{16} - 0\right] + \left[(-\frac{1}{2} - \frac{1}{4}) - (-1 - \frac{1}{16})\right] = \frac{3}{4}$$

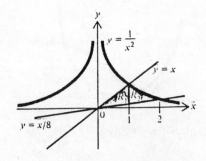

Chapter 6, Section 3

1. The first plan generates profit at the rate of $R_1(x) = 100 + x^2$ dollars per year and the second plan generates profit at the rate of $R_2(x) = 220 + 2x$ dollars per year.

    (a) The second plan will be the more profitable one until $R_1(x) = R_2(x)$, that is, until

    $$100 + x^2 = 220 + 2x \qquad x^2 - 2x - 120 = 0 \qquad (x - 12)(x + 10) = 0$$

    or $x = 12$ years.

    (b) For $0 \le x \le 12$, the rate at which the profit generated by the second plans exceeds that of the first plan is $R_2(x) - R_1(x)$. Hence, the net excess profit generated by the second plan over the 12-year period is the definite integral

    $$\int_0^{12} [R_2(x) - R_1(x)]dx = \int_0^{12} [(220 + 2x) - (100 + x^2)]dx$$

    $$= \int_0^{12} (120 + 2x - x^2)dx = (120x + x^2 - \frac{1}{3}x^3) \Big|_0^{12} = \$1,008.00.$$

    (c) In geometric terms, the net excess profit generated by the second plan is the area of the region between the curves $y = R_2(x)$ and $y = R_1(x)$ from $x = 0$ to $x = 12$.

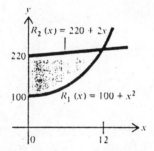

Graph for Problem 1c.

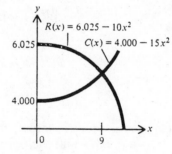

Graph for Problem 3c.

3. The machine generates revenue at the rate of $R(x) = 6,025 - 10x^2$ dollars per year and results in costs that accumulate at the rate of $C(x) = 4,000 + 15x^2$ dollars per year.

    (a) The use of the machine will be profitable as long as the rate at which revenue is generated is greater than the rate at which costs accumulate, that is, until

    $$R(x) = C(x) \qquad 6,025 - 10x^2 = 4,000 + 15x^2$$
    $$25x^2 = 2,025 \qquad x^2 = 81 \text{ or } x = 9 \text{ years}$$

    (b) The difference $R(x) = C(x)$ represents the rate of change of the net earnings generated by the machine. Hence, the net earnings over the next 9 years is the definite integral

$$\int_0^9 [R(x) - C(x)]dx = \int_0^9 [(6,025 - 10x^2) - (4,000 + 15x^2)]dx$$

$$= \int_0^9 (2,025 - 25x^2)dx = \left(2,025x - \frac{25}{3}x^3\right)\Big|_0^9 = \$12,150$$

(c)   In geometric terms, the net earnings in part (b) is the are of the region between the curves $y = R(x)$ and $y = C(x)$ from $x = 0$ to $x = 9$.

5.   (a)   The campaign generates revenue at the rate of $R(t) = 5,000e^{-0.2t}$ dollars per week and accumulates expenses at the rate of \$676 per week. The campaign will be profitable as long as $R(t)$ is greater than 676, that is, until

$$5,000e^{-0.2t} = 676 \qquad e^{-0.2t} = \frac{676}{5,000}$$

$$-0.2t = \ln\frac{676}{5,000} \qquad t = -\frac{\ln(676/5,000)}{0.2} \simeq 10 \text{ weeks}$$

(b)   For $0 \le t \le 10$, the difference $R(t) - 676$ is the rate of change with respect to time of the net earnings generated by the campaign. Hence, the net earnings during the 10 week period is the definite integral

$$\int_0^{10} [R(t) - 676]dt = \int_0^{10} (5,000e^{0.2t} - 676)dt$$

$$= -25,000e^{-0.2t}\Big|_0^{10} - 676t\Big|_0^{10}$$

$$= -25,000(e^{-2} - 1) - 6,760 \simeq \$14,857$$

(c)   In geometric terms, the net earnings in part (b) is the area between the curve $y = R(t)$ and the horizontal line $y = 676$ from $x = 0$ to $x = 10$.

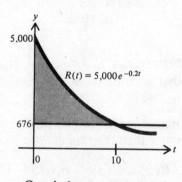

Graph for Problem 5c.

7.   (a)   If the consumers' demand function is $D(q) = \frac{300}{(0.1q+1)^2}$ dollars per unit, the total

amount that consumers are willing to spend to get 5 units is the definite integral

$$\int_0^5 D(q)dq = 300 \int_0^5 (0.1q + 1)^{-2} dq$$

$$= -3,000(0.1q + 1)^{-1} \Big|_0^5 = -3,000(\frac{1}{1.5} - 1) \simeq \$1,000$$

(b) The total willingness to spend in part (a) is the area of the region under the demand curve from $q = 0$ to $q = 5$.

9. (a) If the consumers' demand function is $D(q) = 40e^{-0.05q}$ dollars per unit, the total amount that consumers are willing to spend to get 10 units is the definite integral

$$\int_0^{10} D(q)dq = 40 \int_0^{10} e^{-0.05q} dq$$

$$= -800e^{-0.05q} \Big|_0^{10} = -800(e^{-0.5} - 1) \simeq \$314.78$$

(b) The total willingness to spend in part (a) is the area under the demand curve from $q = 0$ to $q = 10$.

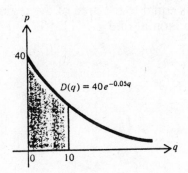

Graph for Problem 9b.

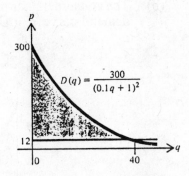

Graph for Problem 11b.

11. (a) The consumers' demand function is $D(q) = \frac{300}{(0.1q+1)^2}$ and the market price is \$12 per unit. To find the number of units that will be bought at this price, solve the equation $p = D(q)$ for q when $p = 12$ to get

$$12 = \frac{300}{(0.1q + 1)^2} \qquad (0.1q + 1)^2 = 25$$

$$0.1q + 1 = 5 \text{ or } q = 40 \text{ units}$$

The corresponding consumers' surplus is

$$\int_0^{40} D(q)dq - (12)(40) = 300 \int_0^{40} \frac{1}{(0.1q + 1)^2} dq - 480$$

$$= \frac{-3,000}{0.1q + 1} \Big|_0^{40} - 480 = -3,000(\frac{1}{5} - 1) - 480 = \$1,920$$

(b)   the consumers' surplus in part (a) is equal to the area of the region between the demand cruve $p = D(q)$ and the horizontal line $p = 12$.

13.   (a)   The consumers' demand function is $D(q) = 40e^{-0.05q}$ and the market price is $11.46 per unit. To find the number of units that will be bought at this price, solve the equation $p = D(q)$ for q when $p = 11.46$ to get

$$11.46 = 40e^{-0.05q} \qquad \frac{11.46}{40} = e^{-0.05q}$$

$$\ln\frac{11.46}{40} = 0.05q \qquad q = -\frac{\ln(11.46/40)}{0.05} \simeq 25 \text{ units}$$

The corresponding consumers' surplus is

$$\int_0^{25} D(q)dq - (11.46)(25) = 40\int_0^{25} e^{-0.05q}dq - 286.50$$

$$= -800e^{-0.05q} \Big|_0^{25} - 286.50$$

$$= -800(e^{-1.25} - 1) - 286.50 \simeq \$284.30$$

(b)   The consumers' surplus in part (a) is equal to the area of the region between the demand curve $p = D(q)$ and the horizontal line $p = 11.46$.

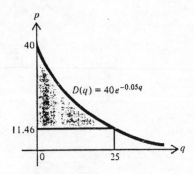

Graph for Problem 13b.

Chapter 6, Section 4

1. Divide the interval $2 \leq t \leq 3$ into n equal subintervals of length Δt minutes, and let t_j denote the beginning of the j th subinterval. During the j th subinterval,

$$\text{Distance traveled} = (\text{speed at time } t_j)(\text{number of minutes})$$

$$= S(t_j)\Delta t$$

Hence,

$$\text{Total distance} = \sum_{j=1}^{n} S(t_j)\Delta t$$

Now, as n increases without bound, this approximation improves while the sum approaches the corresponding integral. Hence,

$$\text{Total distance} = \int_2^3 S(t)dt = \int_2^3 (1 + 4t + 3t^2)dt$$

$$= (t + 2t^2 + t^3)\Big|_2^3 = (3 + 18 + 27) - (2 + 8 + 8) = 30 \text{ meters}$$

3. Divide the interval $0 \leq t \leq 5$ into n equal subintervals of length Δt years, and let t_j denote the beginning of the j th subinterval. During the j th subinterval,

$$\text{Increase in value} = (\text{rate of increase at time } t_j)(\text{number of years})$$

$$= r(t_j)\Delta t$$

Hence,

$$\text{Total increase in value} = \sum_{j=1}^{n} r(t_j)\Delta t$$

Now, as n increases without bound, this approximation improves while the sum approaches the corresponding integral. Hence,

$$\text{Total increase in value} = \int_0^5 r(t)dt$$

5. Divide the interval $0 \leq x \leq 16$ into n equal subintervals of length Δx months, and let x_j denote the beginning of the j th subinterval. During the j th subinterval,

$$\text{Revenue} = (\text{monthly demand})(\text{price per bike at time } x_j)(\text{number of months})$$

$$= 5,000P(x_j)\Delta x$$

Hence,

$$\text{Total revenue} = \sum_{j=1}^{n} 5,000 P(x_j)\Delta x$$

Now, as n increases without bound, this approximation improves while the sum approaches the corresponding integral. Hence,

$$\text{Total revenue} = \int_0^{16} 5,000 P(x)dx = 5,000 \int_0^{16} (80 + 3\sqrt{x})dx$$

$$= 5,000(80x + 2x^{3/2})\Big|_0^{16}$$

$$= 5,000[80(16) + 2(16)^{3/2}] = \$7,040,000$$

7. Divide the interval $0 \le x \le 12$ into n equal subintervals of length Δx months, and let x_j denote the beginning of the j th subinterval. During the j th subinterval,

Revenue = (monthly demand at time x_j)(price at time x_j)(number of months)

$$= n(x_j)p(x_j)\Delta x$$

Hence,

$$\text{Total revenue} = \sum_{j=1}^{n} n(x_j)p(x_j)\Delta x$$

Now, as n increases without bound, this approximation improves while the sum approaches the corresponding integral. Hence,

$$\text{Total revenue} = \int_0^{12} n(x)p(x)dx$$

9. Let $R(t) = 0.3t^2 + 0.6t + 1$ denote the rate of growth of the crop (in bushels per day) t days from now. Divide the interval $0 \le t \le 5$ into n equal subintervals of length Δt days, and let t_j denote the beginning of the j th subinterval. During the j th subinterval,

Number of additional bushels = (rate of growth at time t_j)(number of days)

$$= R(t_j)\Delta t$$

Hence,

$$\text{Total additional bushels} = \sum_{j=1}^{n} R(t_j)\Delta t$$

Now, as n increases without bound, this approximation improves while the sum approaches the corresponding integral. Hence,

$$\text{Total additional bushels} = \int_0^5 R(t)dt$$

and at \$3 per bushel,

$$\text{Total increase in value} = 3\int_0^5 R(t)dt = 3\int_0^5 (0.3t^2 + 0.6t + 1)dt$$

$$= 3(0.1t^3 + 0.3t^2 + t)\Big|_0^5$$

$$= 3[0.1(125) + 0.3(25) + 5] = \$75$$

11. Let $Q(t)$ denote the number of pounds of soybeans in storage after t weeks. Then, $Q(t) = 12,000 - 300t$, (a linear function that decreases from $12,000$ to 0 in 40 weeks at a constant rate of 300 pounds per week). Divide the interval $0 \leq t \leq 40$ into n equal subintervals of length Δt, and let t_j denote the beginning of the j th subinterval. During the j th interval,

$$\text{Storage cost} = (\text{cost/pound} - \text{week})(\text{pounds at time } t_j)(\text{number of weeks})$$

$$= 0.2Q(t_j)\Delta t$$

Hence,

$$\text{Total storage cost} = \sum_{j=1}^{n} 0.2Q(t_j)\Delta t$$

Now, as n increases without bound, this approximation improves while the sum approaches the corresponding integral. Hence,

$$\text{Total storage cost} = 0.2\int_0^{40} Q(t)dt = 0.2\int_0^{40}(12,000 - 300t)dt$$

$$= 60\int_0^{40}(40 - t)dt = 60(40t - \frac{1}{2}t^2)\Big|_0^{40}$$

$$= 60\left[40(40) - \frac{1}{2}(40)^2\right] = 48,000 \text{ cents} = \$480$$

13. If the region under the line $y = 3x + 1$ from $x = 0$ to $x = 1$ is revolved about the x axis, the volume of the resulting solid is

$$\text{Volume} = \pi \int_0^1 (3x+1)^2 dx = \frac{\pi}{9}(3x+1)^3 \Big|_0^1 = \frac{\pi}{9}[(4)^3 - 1] = 7\pi$$

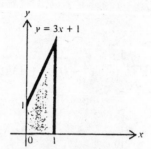

Figure for Problem 13.

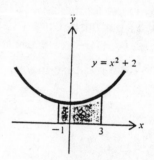

Figure for Problem 15.

15. If the region under the curve $y = x^2 + 2$ from $x = -1$ to $x = 3$ is revolved about the x axis, the volume of the resulting solid is

$$\text{Volume} = \pi \int_{-1}^3 (x^2+2)^2 dx = \pi \int_{-1}^3 (x^4 + 4x^2 + 4) dx$$

$$= \pi \left(\frac{1}{5}x^5 + \frac{4}{3}x^3 + 4x \right) \Big|_{-1}^3$$

$$= \pi \left[\left(\frac{243}{5} + 36 + 12 \right) - \left(-\frac{1}{5} - \frac{4}{3} - 4 \right) \right] = \frac{1,532\pi}{15}$$

17. If the region under the curve $y = \sqrt{4 - x^2}$ from $x = -2$ to $x = 2$ is revolved about the x axis, the volume of the resulting solid (a sphere of radius 2) is

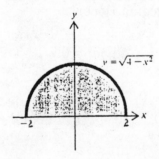

Figure for Problem 17.

$$\text{Volume} = \pi \int_{-2}^{2} (\sqrt{4 - x^2})^2 dx = \pi \int_{-2}^{2} (4 - x^2) dx$$

$$= \pi (4x - \frac{1}{3}x^3)\Big|_{-2}^{2} = \pi \left[(8 - \frac{8}{3}) - (-8 + \frac{8}{3}) \right] = \frac{32\pi}{3}$$

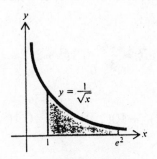

Figure for Problem 19.

19.   If the region under the curve $y = \frac{1}{\sqrt{x}}$ from $x = 1$ to $x = e^2$ is revolved about the x axis, the volume of the resulting solid is

$$\text{Volume} = \int_{1}^{e^2} \left[\frac{1}{\sqrt{x}} \right]^2 dx = \pi \int_{1}^{e^2} \frac{1}{x} dx$$

$$= \pi \ln |x| \Big|_{1}^{e^2} = \pi (\ln e^2 - \ln 1) = 2\pi$$

21.   To form a sphere of radius r, revolve the semicircular region under the curve $y = \sqrt{r^2 - x^2}$ from $x = -r$ to $x = r$ about the x axis. The resulting volume is

$$\text{Volume} = \pi \int_{-r}^{r} (\sqrt{r^2 - x^2})^2 dx = \pi \int_{-r}^{r} (r^2 - x^2) dx$$

$$= \pi (r^2 x - \frac{1}{3}x^3)\Big|_{-r}^{r} = \pi [(r^3 - \frac{1}{3}r^3) - (-r^3 + \frac{1}{3}r^3)] = \frac{4}{3}\pi r^3$$

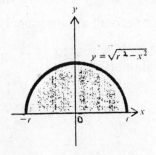

Figure for Problem 21.

Chapter 6, Section 5

1.   Average value $= \frac{1}{4-0}\int_0^4 x\,dx = \frac{1}{4}\left(\frac{1}{2}x^2\right)\big|_0^4 = \frac{1}{4}(8) = 2$

3.   Average value $= \frac{1}{0-(-4)}\int_{-4}^0 (x+2)^2\,dx = \frac{1}{12}(x+2)^3\big|_{-4}^0$

$= \frac{1}{12}[(2)^3 - (-2)^3] = \frac{4}{3}$

5.   The average temperature between 9:00 A.M. and noon is

$$\frac{1}{12-9}\int_9^{12}(-0.3t^2 + 4t + 10)\,dt = \frac{1}{3}(-0.1t^3 + 2t^2 + 10t)\Big|_9^{12}$$

$$= \frac{1}{3}[-0.1(12)^3 + 2(12)^2 + 10(12)] - \frac{1}{3}[-0.1(9)^3 + 2(9)^2 + 10(90)] = 18.7$$

degrees Celsius.

7.   The average rate during the first 3 months is

$$\frac{1}{3-0}\int_0^3 (700 - 400e^{-0.5t})\,dt = \frac{1}{3}(700t + 800e^{-0.5t})\Big|_0^3$$

$$\frac{1}{3}[700(3) + 800e^{-1.5}] - \frac{1}{3}[0 + 800] = 492.83 \text{ letters per hour}$$

9.   (a)   Average speed $= \frac{1}{N}\int_0^N S(t)\,dt$

(b)   Total distance $= \int_0^N S(t)\,dt$

(c)   Notice from parts (a) and (b) that

$$\text{Average speed} = \frac{\text{total distance}}{\text{number of hours}}$$

11. Recall that P dollars invested at an annual interest rate of 6 percent compounded continuously will be worth $Pe^{0.06t}$ dollars t years later. To approximate the future value of the income stream, divide the 5-year time interval $0 \le t \le 5$ into n equal subintervals of length Δt years and let t_j denote the beginning of the j th subinterval. Then,

$$\text{Money deposited during } j\text{th subinterval} = 2{,}400\Delta t$$

This money will remain in the account approximately $5 - t_j$ years hence,

$$\begin{array}{c}\text{Future value of Money deposited} \\ \text{during } j\text{th subinterval}\end{array} \simeq (2{,}400)e^{0.06(5-t_j)}\Delta t$$

and

$$\text{Future value of income stream} = \lim_{n \to \infty} \sum_{j=1}^{n} 2{,}400e^{0.06(5-t_j)} \Delta t$$

$$= \int_0^5 2{,}400e^{0.06(5-t)} \, dt = 2{,}400e^{0.3} \int_0^5 e^{-0.06t} \, dt$$

$$= \frac{2{,}400}{-0.06} e^{0.3} (e^{-0.06t}) \Big|_0^5 = -\frac{2{,}400}{0.06}(e^{0.3} - 1) = \$13{,}994.35$$

13. Recall that P dollars invested at an annual interest rate of 8 percent compounded continuously will be worth $Pe^{0.08t}$ dollars t years later. To approximate the future value of the income stream, divide the 3-year time interval $0 \le t \le 3$ into n equal subintervals of length Δt years and let t_j denote the beginning of the jth interval. Then,

$$\text{Money deposited during } j\text{th subinterval} = 8{,}000 \, \Delta t$$

This money will remain in the account for approximately $3 - t_j$ years. Hence,

$$\begin{matrix}\text{Future value of money deposited} \\ \text{during } j \text{th subinterval}\end{matrix} \simeq 8{,}000e^{0.08(3-t_j)} \Delta t$$

and

$$\text{Future value of income stream} = \lim_{n \to \infty} \sum_{j=1}^{n} 8{,}000e^{0.08t(3-t_j)} \Delta t$$

$$= \int_0^3 8{,}000e^{0.08(3-t)} dt = 8{,}000e^{0.24} \int_0^3 e^{-0.08t} dt$$

$$= \frac{8{,}000}{-0.08} e^{0.24} (e^{-0.08t}) \Big|_0^3 = -\frac{8{,}000}{0.08} e^{0.24} (e^{-0.24} - 1) \simeq \$27{,}124.92$$

15. Recall that the present value of B dollar payable t years from now with an annual interest rate of 6 percent compounded continuously is $Be^{-0.06t}$. Divide the interval $0 \le t \le 5$ into n equal subintervals of length Δt years. Then,

$$\text{Income from } j\text{th subinterval} = 2{,}400 \, \Delta t$$

and

$$\begin{matrix}\text{Present value of income} \\ \text{from } j \text{th subinterval}\end{matrix} \simeq 2{,}400e^{-0.06t_j} \Delta t$$

Hence,

$$\text{Present value of investment} = \lim_{n \to \infty} \sum_{j=1}^{n} 2{,}400e^{-0.06t_j}\,\Delta t$$

$$= 2{,}400 \int_0^5 e^{-0.6t}dt = \frac{2{,}400}{-0.06}(e^{-0.06t})\Big|_0^5$$

$$= -\frac{2{,}400}{0.06}(e^{-0.3} - 1) \simeq \$10{,}367.27$$

17. Recall that the present value of B dollars payable t years from now with an annual interest rate of 12 percent compounded continuously is $Be^{-0.12t}$. Divide the interval $0 \le t \le 5$ into n equal subintervals of length Δt years. Then,

$$\text{Income from } j \text{ th subinterval} = 1{,}200\,\Delta t$$

and

$$\begin{matrix}\text{Present value of income}\\ \text{from } j \text{ th subinterval}\end{matrix} \simeq 1{,}200e^{-0.12t_j}\,\Delta t$$

Hence,

$$\text{Present value of investment} = \lim_{n \to \infty} \sum_{j=1}^{n} 1{,}200e^{-0.12t_j}\,\Delta t$$

$$= \int_0^5 1{,}200e^{-0.12t}dt = \frac{1{,}200}{0.12}(e^{-0.12t})\Big|_0^5$$

$$= -10{,}000(e^{-0.6} - 1) \simeq \$4{,}511.88$$

19. Recall that the present value of B dollars payable t years from now with an annual interest rate of 10 percent compounded continuously is $Be^{-0.1t}$. To find the present value of the 10-year pension, divide the interval $0 \le t \le 10$ into n equal subintervals of length Δt years and let t_j denote the beginning of the j th subinterval. Then,

$$\text{Income from } j \text{ th subinterval} = 5{,}000\,\Delta t$$

and

$$\text{Present value of income from } j \text{ th subinterval} \simeq 5{,}000e^{-0.1t_j}\Delta t$$

Hence,

$$\text{Present value of pension} = \lim_{n \to \infty} \sum_{j=1}^{n} 5,000 e^{-0.1 t_j} \, \Delta t$$

$$= \int_{0}^{10} 5,000 e^{-0.1t} dt = -\frac{5,000}{0.1} (e^{-0.1t}) \Big|_{0}^{10}$$

$$= -\frac{5,000}{0.1}(e^{-1} - 1) \simeq 31,606 \text{ pounds}$$

Hence, the spy should take the flat sum of $35,000$ pounds since the present value of the pension is only $31,606$ pounds.

21. Since $f(t) = e^{-t/10}$ is the fraction of members active after t months, and since there were $8,000$ charter members, the number of charter members still active at the end of 10 months is

$$8,000 f(10) = 8,000 e^{-1}$$

Now, divide the interval $0 \le t \le 10$ into n equal subintervals of length Δt months and let t_j denote the beginning of the jth subinterval. During the jth subinterval, $200 \, \Delta t$ new members join, and at the end of the 10 months ($10 - t_j$ months later), the number of these retaining membership is

$$200(\Delta t) f(10 - t_j) = 200 e^{-(10 - t_j)/10} \Delta t$$

Hence the number of new members still active 10 months from now is approximately

$$\lim_{n \to \infty} \sum_{j=1}^{n} 200 e^{-(10 - t_j)/10} \, \Delta t = \int_{0}^{10} 200 e^{-(10 - t)/10} dt$$

Hence, the total number N of active members 10 months from now is

$$N = 8,000 e^{-1} + \int_{0}^{10} 200 e^{-(10 - t)/10} dt$$

$$= 8,000 e^{-1} + 200 e^{-1} \int_{0}^{10} e^{t/10} dt$$

$$= 8,000 e^{-1} + 200 e^{-1} (10 e^{t/10}) \Big|_{0}^{10} = 8,000 + 2,000 e^{-1}(e^{1} - 1)$$

$$= 8,000 e^{-1} + 2,000 - 2,000 e^{-1} = 6,000 e^{-1} + 2,000 \simeq 4,207$$

23. Let $f(t)$ denote the fraction of the membership of the group that will remain active for at least t years, P_0 the initial membership, and $r(t)$ the rate per year at which additional members are added to the group. Then, the size of the group N years from now is the number of initial members still active plus the number of new members still active. Of the P_0 initial members, $f(N)$ is the fraction remaining active for N years. Hence, the number of initial members still active after N years is

$$P_0 f(N)$$

To find the number of new members still active after N years, divide the interval $0 \leq t \leq N$ into n equal subintervals of length Δt years and let t_j denote the beginning of the jth subinterval. During the jth subinterval, approximately $r(t_j)\Delta t$ new members joined the group. Of these, the fraction still active at time $t = N$ (i.e., $N - t_j$ years later) is $f(N - t_j)$, and so the number of these still active at time $t = N$ is $r(t_j)(\Delta t)f(N - t_j)$. Hence, the total number of new members still active after N years is

$$\lim_{n \to \infty} \sum_{j=1}^{n} r(t_j) f(N - t_j)\,\Delta t = \int_0^N r(t) f(N - t)\,dt$$

Putting it all together, the total number of active members N years from now is

$$P_0 f(N) = \int_0^N r(t) f(N - t)\,dt$$

25. Let $S(r)$ denote the speed of the fluid r centimeters from the central axis. Divide the interval $0 \leq r \leq R$ into n equal subintervals of length Δr, and let r_j denote the beginning of the jth subinterval. This divides the cross section into n concentric rings as shown in the figure. If Δr is small,

$$\text{Area of } j\text{th ring} = 2\pi r_j\,\Delta r$$

where $2\pi r_j$ is the circumference of the circle of radius r_j that forms the inner boundary and Δr is the width of the ring. Now,

$$\text{Rate of flow through } j\text{th ring} \simeq (\text{area of ring})(\text{speed of fluid})$$

$$\simeq 2\pi r_j S(r_j)\,\Delta r$$

and so

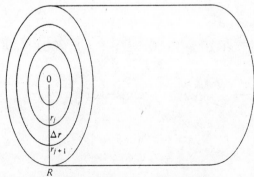

Cross section of pipe for Problem 25.

$$\text{Rate of flow through pipe} = \lim_{n \to \infty} \sum_{j=1}^{n} 2\pi r_j S(r_j) \Delta r$$

$$= \int_0^R 2\pi r S(r)\, dr$$

27. Divide the interval $0 \le r \le 3$ into n equal subintervals of length Δr, and let r_j denote the beginning of the j th subinterval. This divides the circular disk of radius 3 into n concentric circles, as shown in the figure. If Δr is small,

$$\text{Area of } j \text{ th ring} \simeq 2\pi r_j \Delta r$$

where $2\pi r_j$ is the circumference of the circle of radius r_j that forms the inner boundary of the ring and Δr is the width of the ring. Then, since $D(r) = 5,000e^{-0.1r}$ is the population density (people per square mile) r miles from the center, it follows that

$$\text{Number of people in } j \text{ th ring} = D(r_j)(\text{area of } j \text{ th ring})$$

$$= 5,000e^{-0.1r_j}(2\pi r_j \Delta r)$$

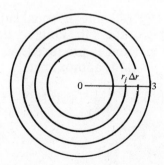

Concentric rings for Problem 27.

Hence, if N is the total number of people within 3 miles of the center of the city,

$$N = \lim_{n \to \infty} \sum_{j=1}^{n} 5,000e^{-0.1r_j}(2\pi r_j \, \Delta r)$$

$$= \int_0^3 5,000(2\pi)re^{-0.1r} \, dr = 10,000\pi \int_0^3 re^{-0.1r} \, dr$$

Applying integration by parts,

$$\int_0^3 re^{-0.1r} \, dr = -10re^{-0.1r}\Big|_0^3 - \int_0^3 -10e^{-0.1r} \, dr$$

$$= (-10re^{-0.1r} - 100e^{-0.1r})\Big|_0^3$$

$$= (-30e^{-0.3} - 100e^{-0.3}) - (-100)$$

$$= -130e^{-0.3} + 100 \simeq 3.6936$$

Hence the total number of people within 3 miles of the center of the city is

$$N = 10,000\pi \int_0^3 re^{-0.1r} \, dr = 10,000\pi(3.6936) \simeq 116,039$$

29. Radioactive material decays exponentially so that if $A(t)$ denotes the amount of radioactive material present after t years, $A(t) = A_0 e^{-kt}$, where A_0 is the amount present initially and k is a positive constant. Since the half life is 28 years,

$$\frac{1}{2}A_0 = A(28) = A_0 e^{-28k} \qquad e^{-28k} = \frac{1}{2}$$

$$-28k = \ln\frac{1}{2} = -\ln 2 \qquad \text{or} \qquad k = \frac{\ln 2}{28}$$

Now, divide the interval $0 \le t \le 140$ into n equal subintervals of length Δt years and let t_j denote the begbinning of the j th subinterval. During the j th subinterval, $500 \, \Delta t$ pounds of radioactive material is produced, and the amount left 140 years from now (i.e., after $140 - t_j$ years) is approximately

$$(500 \, \Delta t)A(140 - t_j) = 500A(140 - t_j)\Delta t$$

Hence,

Total waste after 140 years $= N = \displaystyle\lim_{n \to \infty} \sum_{j=1}^{n} 500 A(140 - t_j)\, \Delta t$

$$= \int_0^{140} 500 A(140 - t)\, dt = 500 \int_0^{140} e^{-(140-t)k}\, dt$$

$$= 500 e^{-140k} \int_0^{140} e^{kt}\, dt = \frac{500}{k} e^{-140k + kt} \Big|_0^{140}$$

$$= \frac{500}{k} e^{-140k}(e^{140k} - 1) = \frac{500}{k}(1 - e^{-140k})$$

Since $k = \frac{\ln 2}{28}$, it follows that

$$N = \frac{28(500)}{\ln 2}[1 - e^{-(140 \ln 2)/28}]$$

$$= \frac{28(500)}{\ln 2}[1 - e^{-5 \ln 2}] = \frac{28(500)}{\ln 2}[1 - e^{\ln(2^{-5})}]$$

$$= \frac{28(500)}{\ln 2}(1 - 2^{-5}) = \frac{28(500)}{\ln 2}\left(\frac{31}{32}\right) \simeq 19{,}567 \text{ pounds}$$

Chapter 6, Section 6

1. For $\int_1^2 x^2\,dx$ with $n = 4$, $\Delta x = \frac{2-1}{4} = 0.25$, and

$$x_1 = 1, \quad x_2 = 1.25, \quad x_3 = 1.5, \quad x_4 = 1.75, \quad \text{and} \quad x_5 = 2$$

(a) Trapezoidal rule:

$$\int_1^2 x^2\,dx \simeq \frac{\Delta x}{2}[f(x_1) + 2f(x_2) + 2f(x_3) + 2f(x_4) + f(x_5)]$$

$$= \frac{0.25}{2}[1^2 + 2(1.25)^2 + 2(1.5)^2 + 2(1.75)^2 + 2^2] \simeq 2.3438$$

(b) Simpson's rule:

$$\int_1^2 x^2\,dx \simeq \frac{\Delta x}{3}[f(x_1) + 4f(x_2) + 2f(x_3) + 4f(x_4) + f(x_5)]$$

$$= \frac{0.25}{3}[1^2 + 4(1.25)^2 + 2(1.5)^2 + 4(1.75)^2 + 2^2] \simeq 2.3333$$

3. For $\int_0^1 \frac{1}{1+x^2}\,dx$ with $n = 4$, $\Delta x = \frac{1-0}{4} = 0.25$, and

$$x_1 = 0, \quad x_2 = 0.25, \quad x_3 = 0.5, \quad x_4 = 0.75, \quad \text{and} \quad x_5 = 1$$

(a) Trapezoidal rule:

$$\int_0^1 \frac{1}{1+x^2}\,dx \simeq \frac{\Delta x}{2}[f(x_1) + 2f(x_2) + 2f(x_3) + 2f(x_4) + f(x_5)]$$

$$= \frac{0.25}{2}\left[1 + \frac{2}{1+(0.25)^2} + \frac{2}{1+(0.5)^2} + \frac{2}{1+(0.75)^2} + \frac{1}{2}\right] \simeq 0.7828$$

(b) Simpson's rule:

$$\int_0^1 \frac{1}{1+x^2}\,dx \simeq \frac{\Delta x}{3}[f(x_1) + 4f(x_2) + 2f(x_3) + 4f(x_4) + f(x_5)]$$

$$= \frac{0.25}{3}\left[1 + \frac{4}{1+(0.25)^2} + \frac{2}{1+(0.5)^2} + \frac{4}{1+(0.75)^2} + \frac{1}{2}\right] \simeq 0.7854$$

5. For $\int_{-1}^0 \sqrt{1+x^2}\,dx$ with $n = 4$, $\Delta t = \frac{0-(-1)}{4} = 0.25$, and

$$x_1 = -1, \quad x_2 = -0.75, \quad x_3 = -0.5, \quad x_4 = -0.25, \quad \text{and} \quad x_5 = 0$$

(a) Trapezoidal rule:

$$\int_{-1}^{0} \sqrt{1+x^2}\, dx \simeq \frac{\Delta x}{2}[f(x_1) + 2f(x_2) + 2f(x_3) + 2f(x_4) + f(x_5)]$$

$$= \frac{0.25}{2}(\sqrt{1+(-1)^2} + 2\sqrt{1+(-0.75)^2} + 2\sqrt{1+(-0.5)^2}$$

$$+ 2\sqrt{1+(-0.25)^2} + \sqrt{1+0^2}) \simeq 1.1515$$

(b)   Simpson's rule:

$$\int_{-1}^{0} \sqrt{1+x^2}\, dx \simeq \frac{\Delta x}{3}[f(x_1) + 4f(x_2) + 2f(x_3) + 4f(x_4) + f(x_5)]$$

$$= \frac{0.25}{3}(\sqrt{1+(-1)^2} + 4\sqrt{1+(-0.75)^2} + 2\sqrt{1+(-0.5)^2}$$

$$+ 4\sqrt{1+(-0.25)^2} + \sqrt{1+0^2}) \simeq 1.1478$$

7.   For $\int_{0}^{1} e^{-x^2}\, dx$ with $n = 4$, $\Delta x = \frac{1-0}{4} = 0.25$, and

$$x_1 = 0, \quad x_2 = 0.25, \quad x_3 = 0.5, \quad x_4 = 0.75, \quad \text{and} \quad x_5 = 1$$

(a)   Trapezoidal rule:

$$\int_{0}^{1} e^{-x^2}\, dx \simeq \frac{\Delta x}{2}[f(x_1) + 2f(x_2) + 2f(x_3) + 2f(x_4) + f(x_5)]$$

$$= \frac{0.25}{2}[1 + 2e^{-(0.25)^2} + 2e^{-(0.5)^2} + 2e^{-(0.75)^2} + e^{-1}] \simeq 0.7430$$

(b)   Simpson's rule:

$$\int_{0}^{1} e^{-x^2}\, dx \simeq \frac{\Delta x}{3}[f(x_1) + 4f(x_2) + 2f(x_3) + 4f(x_4) + f(x_5)]$$

$$= \frac{0.25}{3}[1 + 4e^{-(0.25)^2} + 2e^{-(0.5)^2} + 4e^{-(0.75)^2} + e^{-1}] \simeq 0.7469$$

9.   For $\int_{1}^{2} \frac{1}{x^2}\, dx$ with $n = 4$, $\Delta x = \frac{2-1}{4} = 0.25$, and

$$x_1 = 1, \quad x_2 = 1.25, \quad x_3 = 1.5, \quad x_4 = 1.75, \quad \text{and} \quad x_5 = 2$$

(a)   Trapezoidal rule:

$$\int_1^2 \frac{1}{x^2}\,dx \simeq \frac{\Delta x}{2}[f(x_1) + 2f(x_2) + 2f(x_3) + 2f(x_4) + f(x_5)]$$

$$= \frac{0.25}{2}\left[1 + \frac{2}{(1.25)^2} + \frac{2}{(1.5)^2} + \frac{2}{(1.75)^2} + \frac{1}{2^2}\right] \simeq 0.5090$$

The error estimate is $|E_n| \le \frac{M(b-a)^3}{12n^2}$. For $n = 4$, $a = 1$, and $b = 2$,

$$|E_n| \le \frac{M(2 - 1)^3}{12(4)^2} = \frac{M}{192}$$

where M is the maximum value of $|f''(x)|$ on $1 \le x \le 2$. Now

$$f(x) = \frac{1}{x^2}, \quad f'(x) = -\frac{2}{x^3}, \quad \text{and} \quad f''(x) = \frac{6}{x^4}$$

and, for $1 \le x \le 2$,

$$|f''(x)| = \frac{6}{x^4} \le \frac{1}{1^4} = 6$$

Hence,

$$|E_4| \le \frac{6}{192} = 0.0313$$

(b)   Simpson's rule:

$$\int_1^2 \frac{1}{x^2}\,dx \simeq \frac{\Delta x}{3}[f(x_1) + 4f(x_2) + 2f(x_3) + 4f(x_4) + f(x_5)]$$

$$= \frac{0.25}{3}\left[1 + \frac{4}{(1.25)^2} + \frac{2}{(1.5)^2} + \frac{4}{(1.75)^2} + \frac{1}{2^2}\right] \simeq 0.5004$$

The error estimate is $|E_n| \le \frac{M(b-a)^5}{180n^4}$. For $n = 4$, $a = 1$, and $b = 2$,

$$|E_4| \le \frac{M(1)^5}{180(4)^4} = \frac{M}{46,080}$$

where M is the maximum value of $|f^{(4)}(x)|$ on $1 \le x \le 2$. Now,

$$f''(x) = \frac{6}{x^4}, \quad f^{(3)}(x) = -\frac{24}{x^5}, \quad \text{and} \quad f^{(4)}(x) = \frac{120}{x^6}$$

and, for $1 \le x \le 2$,

$$|f^{(4)}(x)| = \frac{120}{x^6} \le \frac{120}{1^6} = 120$$

Hence,

$$|E_4| \leq \frac{120}{46,080} \simeq 0.0026$$

11. For $\int_1^3 \sqrt{x}\, dx$ with $n = 10$, $\Delta x = \frac{3-1}{10} = 0.2$, and

$$x_1 = 1, \quad x_2 = 1.2, \quad x_3 = 1.4, \ldots, \quad x_{10} = 2.8, \quad \text{and} \quad x_{11} = 3$$

(a) Trapezoidal rule:

$$\int_1^3 \sqrt{x}\, dx \simeq \frac{\Delta x}{2}[f(x_1) + 2f(x_2) + 2f(x_3) + \cdots + 2f(x_{10}) + f(x_{11})]$$

$$= \frac{0.2}{2}(1 + 2\sqrt{1.2} + 2\sqrt{1.4} + 2\sqrt{1.6} + 2\sqrt{1.8} + 2\sqrt{2}$$

$$+ 2\sqrt{2.2} + 2\sqrt{2.4} + 2\sqrt{2.6} + 2\sqrt{2.8} + \sqrt{3}) \simeq 2.7967$$

The error estimate is $|E_n| \leq \frac{M(b-a)^3}{12n^2}$. For $n = 10$, $a = 1$, and $b = 3$,

$$|E_{10}| \leq \frac{M(3-1)^3}{12(10)^2} = \frac{8M}{1,200} = \frac{M}{150}$$

where M is the maximum value of $|f''(x)|$ on $1 \leq x \leq 3$. Now,

$$f(x) = x^{1/2}, \quad f'(x) = \frac{1}{2}x^{-1/2}, \quad \text{and} \quad f''(x) = -\frac{1}{4}x^{-3/2}$$

and for $1 \leq x \leq 3$,

$$|f''(x)| = |-\frac{1}{4}x^{-3/2}| \leq \frac{1}{4}(1)^{-3/2} = \frac{1}{4}$$

Hence,

$$|E_{10}| \leq \frac{1}{150}\left[\frac{1}{4}\right] = 0.0017$$

(b) Simpson's rule:

$$\int_1^3 \sqrt{x}\, dx \simeq \frac{\Delta x}{3}[f(x_1) + 4f(x_2) + 2f(x_3) + \cdots + 4f(x_{10}) + f(x_{11})]$$

$$= \frac{0.2}{3}(1 + 4\sqrt{1.2} + 2\sqrt{1.4} + 4\sqrt{1.6} + 2\sqrt{1.8} + 4\sqrt{2}$$

$$+ 2\sqrt{2.2} + 4\sqrt{2.4} + 2\sqrt{2.6} + 4\sqrt{2.8} + \sqrt{3}) \simeq 2.7974$$

The error estimate is $|E_n| \leq \frac{M(b-a)^5}{180n^4}$. For $n = 4$, $a = 1$, and $b = 3$,

$$|E_{10}| \leq \frac{M(2)^5}{180(10)^4} = \frac{32M}{180(10)^4}$$

where M is the maximum value of $|f^{(4)}(x)|$ on $1 \leq x \leq 3$. Now,

$$f''(x) = -\frac{1}{4}x^{-3/2}, \quad f^{(3)}(x) = \frac{3}{8}x^{-5/2}, \quad \text{and} \quad f^{(4)}(x) = -\frac{15}{16}x^{-7/2}$$

and for $1 \leq x \leq 3$,

$$|f^{(4)}(x)| = \left|-\frac{15}{16}x^{-7/2}\right| \leq \frac{15}{16}(1)^{-7/2} = \frac{15}{16}$$

Hence,

$$|E_{10}| \leq \frac{32}{180(10)^4}\left[\frac{15}{16}\right] = 0.00002$$

13. For $\int_0^1 e^{x^2}\, dx$ with $n = 4$, $\Delta x = \frac{1-0}{4} = 0.25$ and

$$x_1 = 0, \quad x_2 = 0.25, \quad x_3 = 0.5, \quad x_4 = 0.75, \quad \text{and} \quad x_5 = 1$$

(a) Trapezoidal rule:

$$\int_0^1 e^{x^2}\, dx \simeq \frac{\Delta x}{2}[f(x_1) + 2f(x_2) + 2f(x_3) + 2f(x_4) + f(x_5)]$$

$$= \frac{0.25}{2}[1 + 2e^{(0.25)^2} + 2e^{(0.5)^2} + 2e^{(0.75)^2} + e] \simeq 1.4907$$

The error estimate is $|E_n| \leq \frac{M(b-a)^3}{12n^2}$. For $n = 4$, $a = 0$, and $b = 1$,

$$|E_4| \leq \frac{M(1-0)^3}{12(4)^2} = \frac{M}{192}$$

where M is the maximum value of $|f''(x)|$ on $0 \leq x \leq 1$. Now,

$$f(x) = e^{x^2}, \quad f'(x) = 2xe^{x^2}, \quad \text{and} \quad f''(x) = (4x^2 + 2)e^{x^2}$$

and for $0 \leq x \leq 1$,

$$|f''(x)| \leq [4(1)^2 + 2]e^{(1)^2} = 6e$$

Hence,

$$|E_4| \leq \frac{6e}{192} \simeq 0.0849$$

(b) Simpson's rule:

$$\int_0^1 e^{x^2}\, dx \simeq \frac{\Delta x}{3}[f(x_1) + 4f(x_2) + 2f(x_3) + 4f(x_4) + f(x_5)]$$

$$= \frac{0.25}{3}[1 + 4e^{(0.25)^2} + 2e^{(0.5)^2} + 4e^{(0.75)^2} + e] \simeq 1.4637$$

The error estimates is $|E_n| \le \frac{M(b-a)^5}{180n^4}$. For $n = 4$, $a = 0$, and $b = 1$,

$$|E_4| \le \frac{M}{180(4)^4} = \frac{M}{46,080}$$

where M is the maximum value of $|f^{(4)}(x)|$ on $0 \le x \le 1$. Now,

$$f''(x) = (4x^2 + 2)e^{x^2}, \quad f^{(3)}(x) = (8x^3 + 12x)e^{x^2},$$

and $\quad f^{(4)}(x) = (16x^4 + 48x^2 + 12)e^{x^2}$

and for $0 \le x \le 1$,

$$|f^{(4)}(x)| \le [16(1)^4 + 48(1)^2 + 12]e^{(1)^2} = 76e$$

Hence,

$$|E_4| \le \frac{76e}{46,080} \simeq 0.0045$$

15. The integral to be approximated is $\int_1^3 \frac{1}{x}\, dx$. The derivatives of $f(x) = \frac{1}{x}$ are

$$f'(x) = -\frac{1}{x^2}, \quad f''(x) = \frac{2}{x^3}, \quad f^{(3)}(x) = -\frac{6}{x^4}, \quad \text{and} \quad f^{(4)}(x) = \frac{24}{x^5}$$

(a) For the trapezoidal rule, $|E_n| \le \frac{M(b-a)^3}{12n^2}$, where M is the maximum value of $|f''(x)|$ on $1 \le x \le 3$. Since

$$|f''(x)| = \frac{2}{x^3} \le \frac{2}{1^3} = 2$$

on $1 \le x \le 3$,

$$|E_n| \le \frac{2(3-1)^3}{12n^2} = \frac{4}{3n^2}$$

which will be less than 0.00005 if

$$4 < 3(0.00005)n^2 \quad \text{or} \quad n > \sqrt{\frac{4}{3(0.00005)}} \simeq 163.3$$

Hence 164 intervals should be used.

(b) For Simpson's rule, $|E_n| \le \frac{M(b-a)^5}{180n^4}$, where M is the maximum value of $|f^{(4)}(x)|$ on $1 \le x \le 3$. Since

$$|f^{(4)}(x)| = \left|\frac{24}{x^5}\right| \le \frac{24}{(1)^5} = 24$$

on $1 \le x \le 3$,

$$|E_n| \le \frac{24(2)^5}{180n^4} = \frac{768}{180n^4}$$

which will be less than 0.00005 if

$$768 < 180(0.00005)n^4 \quad \text{or} \quad n > \left[\frac{768}{180(0.00005)}\right]^{1/4} \simeq 17.1$$

Hence, 18 subintervals should be used.

17. The integral to be approximated is $\int_1^2 \frac{1}{\sqrt{x}}\, dx$. The derivatives of $f(x) = \frac{1}{\sqrt{x}}$ are

$$f'(x) = -\frac{1}{2}x^{-3/2}, \qquad\qquad f''(x) = \frac{3}{4}x^{-5/2},$$

$$f^{(3)}(x) = -\frac{15}{8}x^{-7/2}, \qquad \text{and} \quad f^{(4)}(x) = \frac{105}{16}x^{-9/2}$$

(a) For the trapezoidal rule, $|E_n| \le \frac{M(b-a)^3}{12n^2}$, where M is the maximum value of $|f''(x)|$ on $1 \le x \le 2$. Since

$$|f''(x)| = |\frac{3}{4}x^{-5/2}| \le \frac{3}{4}$$

on $1 \le x \le 2$,

$$|E_n| \le \frac{3}{4}\left[\frac{(2-1)^3}{12n^2}\right] = \frac{1}{16n^2}$$

which is less than 0.00005 if

$$1 < 16(0.00005)n^2 \quad \text{or} \quad n > \sqrt{\frac{1}{16(0.00005)}} \simeq 35.4$$

Hence 36 subintervals should be used.

(b) For Simpson's rule, $|E_n| \le \frac{M(b-a)^5}{180n^4}$, where M is the maximum value of $|f^{(4)}(x)|$ on $1 \le x \le 2$. Since

$$|f^{(4)}(x)| = |\frac{105}{16}x^{9/2}| \le \frac{105}{16}$$

on $1 \le x \le 2$,

$$|E_n| \le \frac{105}{16}\left[\frac{(2-1)^5}{180n^4}\right] = \frac{7}{192n^4}$$

which is less than 0.00005 if

$$7 < 192(0.00005)n^4 \quad \text{or} \quad n > \left[\frac{7}{192(0.00005)}\right]^{1/4} \simeq 5.2$$

Hence, 6 subintervals should be used.

19.   The integral to be approximated is $\int_{1.2}^{2.4} e^x \, dx$.

(a)   For the trapezoidal rule, $|E_n| \leq \frac{M(b-a)^3}{12n^2}$, where M is the maximum value of $|f''(x)|$ on $1.2 \leq x \leq 2.4$. Since

$$|f''(x)| = |e^x| \leq e^{2.4}$$

on $1.2 \leq x \leq 2.4$,

$$|E_n| \leq \frac{e^{2.4}(2.4 - 1.2)^3}{12n^2} = \frac{1.728e^{2.4}}{12n^2}$$

which is less than 0.00005 if

$$1.728e^{2.4} < 12(0.00005)n^2 \quad \text{or} \quad n > \sqrt{\frac{1.728(e^{2.4})}{12(0.00005)}} \simeq 178.2$$

Hence, 179 subintervals should be used.

(b)   For Simpson's rule, $|E_n| \leq \frac{M(b-a)^5}{180n^4}$, where M is the maximum value of $|f^{(4)}(x)|$ on $1.2 \leq x \leq 2.4$. Since

$$|f^{(4)}(x)| = |e^x| \leq e^{2.4}$$

on $1.2 \leq x \leq 2.4$,

$$|E_n| \leq \frac{e^{2.4}(1.2)^5}{180n^4}$$

which is less than 0.00005 if

$$e^{2.4}(1.2)^5 < 180(0.00005)n^4 \quad \text{or} \quad n > \left[\frac{e^{2.4}(1.2)^5}{180(0.00005)}\right]^{1/4} \simeq 7.4$$

Hence, 8 subintervals should be used.

Chapter 6, Review Problems

1. $\int_0^1 (5x^4 - 8x^3 + 1)dx = (x^5 - 2x^4 + x) \Big|_0^1 = (1 - 2 + 1) - 0 = 0$

2. $\int_1^4 (\sqrt{x} + x^{-3/2})dx = (\frac{2}{3}x^{3/2} - 2x^{-1/2}) \Big|_1^4$

        $= \left[\frac{2}{3}(8) - \frac{2}{2}\right] - \left[\frac{2}{3}(1) - \frac{2}{1}\right] = \frac{17}{3}$

3.   Let $u = 5x - 2$. Then $du = 5dx$ or $dx = \frac{1}{5}du$. When $x = -1, u = -7$, and when $x = 2, u = 8$. Hence,

$$\int_{-1}^2 30(5x - 2)^2 dx = 6\int_{-7}^8 u^2 du$$

$$= 2u^3 \Big|_{-7}^8 = 2[(8)^3 - (-7)^3] = 1,710$$

4.   Let $u = x^2 - 1$. Then $du = 2xdx$. When $x = 0$, $u = -1$, and when $x = 1$, $u = 0$. Hence,

$$\int_0^1 2xe^{x^2-1}dx = \int_{-1}^0 e^u du = e^u \Big|_{-1}^0 = e^0 - e^{-1} \simeq 0.63$$

5.   Let $u = x^2 - 6x + 2$. Then $du = (2x - 6)dx$ or $(x - 3)dx = \frac{1}{2}du$. When $x = 0, u = 2$, and when $x = 1$, $u = -3$. Hence,

$$\int_0^1 (x - 3)(x^2 - 6x + 2)^3 dx = \frac{1}{2}\int_2^{-3} u^3 du$$

$$= \frac{1}{8}u^4 \Big|_2^{-3} = \frac{1}{8}[(-3)^4 - (2)^4] = \frac{65}{8}$$

6.   Let $u = x^2 + 4x + 5$. Then $du = (2x + 4)dx$ or $(3x + 6)dx = \frac{3}{2}du$. When $x = -1, u = 2$, and when $x = 1$, $u = 10$. Hence,

$$\int_{-1}^1 \frac{3x + 6}{x^2 + 4x + 1}dx = \frac{3}{2}\int_2^{10} u^{-2}du$$

$$= -\frac{3}{2u} \Big|_2^{10} = -\frac{3}{2}(\frac{1}{10} - \frac{1}{2}) = \frac{3}{5}$$

7.   Let $g(x) = e^x$ and $f(x) = x$. Then, $G(x) = e^x$ and $f'(x) = 1$ and so

$$\int_{-1}^1 xe^x dx = xe^x \Big|_{-1}^1 - \int_{-1}^1 e^x dx$$

$$= (xe^x - e^x) \Big|_{-1}^1 = (e - e) - (-e^{-1} - e^{-1}) = 2e^{-1} \simeq 0.74$$

8. Let $u = \ln x$. Then $du = \frac{1}{x}dx$. When $x = e, u = 1$, and when $x = e^2$, $u = 2$. Hence,

$$\int_e^{e^2} \frac{1}{x(\ln x)^2}dx = \int_1^2 \frac{1}{u^2}du = -\frac{1}{u}\Big|_1^2 = -\frac{1}{2} - (-1) = \frac{1}{2}$$

9. Let $g(x) = x^2$ and $f(x) = \ln x$. Then $G(x) = \frac{1}{3}x^3$ and $f'(x) = \frac{1}{x}$. Hence,

$$\int_1^e x^2 \ln x \, dx = \frac{1}{3}x^3 \ln x \Big|_1^e - \frac{1}{3}\int_1^e x^3(\frac{1}{x})dx$$

$$= \frac{1}{3}x^3 \ln x \Big|_1^e - \frac{1}{3}\int_1^e x^2 dx$$

$$= (\frac{1}{3}x^3 \ln x - \frac{1}{9}x^3)\Big|_1^e = \frac{1}{3}x^3(\ln x - \frac{1}{3})\Big|_1^e$$

$$= \left[\frac{1}{3}e^3(1 - \frac{1}{3})\right] - \left[\frac{1}{3}(0 - \frac{1}{3})\right] = \frac{2}{9}e^3 + \frac{1}{9} = \frac{1}{9}(2e^3 + 1)$$

10. Let $g(x) = e^{0.2x}$ and $f(x) = 2x + 1$. Then $G(x) = 5e^{0.2x}$ and $f'(x) = 2$. Hence,

$$\int_0^{10} (2x + 1)e^{0.2x}dx = 5(2x + 1)e^{0.2x}\Big|_0^{10} - 10\int_0^{10} e^{0.2x}dx$$

$$= [5(2x + 1)e^{0.2x} - 50e^{0.2x}]\Big|_0^{10}$$

$$= 5(2x - 9)e^{0.2x}\Big|_0^{10} = [5(11)e^2] - [5(-9)] = 55e^2 + 45$$

11. Let $P(x)$ denote the population x months from now. Then $\frac{dP}{dx} = 10 + 2\sqrt{x}$, and the amount by which the population will increase during the next 9 months is

$$P(9) - P(0) = \int_0^9 (10 + 2x^{1/2})dx$$

$$= (10x + \frac{4}{3}x^{3/2})\Big|_0^9 = 90 + \frac{4}{3}(27) = 126 \text{ people}$$

12. Let $N(t)$ denote the size of the crop (in bushels) t days from now. Then, $\frac{dN}{dt} = 0.3t^2 + 0.6t + 1$ bushels per day, and the increase in the size of the crop over the next 6 days is

$$N(6) - N(0) = \int_0^6 (0.3t^2 + 0.6t + 1)dt$$

$$= (0.1t^3 + 0.3t^2 + t)\Big|_0^6 = 0.1(6)^3 + 0.3(6)^2 + 6 = 38.4 \text{ bushels}$$

17. Break R into two subregions R_1 and R_2 as shown in the accompanying figure. Then,

$$\text{Area of } R = \text{ area of} R_1 + \text{ area of } R_2 = \int_0^4 \sqrt{x}\,dx = \int_4^8 \frac{8}{x}\,dx$$

$$= \frac{2}{3}x^{3/2}\Big|_0^4 + 8\ln|x|\Big|_4^8 = \frac{16}{3} + 8\ln 8 - 8\ln 4 = \frac{16}{3} + 8\ln 2$$

18. Area $= \int_0^1 (x - x^4)\,dx = (\frac{1}{2}x^2 - \frac{1}{5}x^5)\Big|_0^1 = \frac{1}{2} - \frac{1}{5} = \frac{3}{10}$

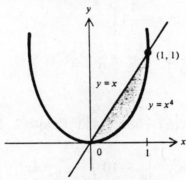

Region for Problem 18.

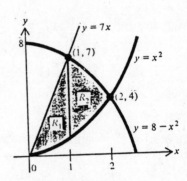

Graph for Problem 19.

19. Break R into two subregions R_1 and R_2 as shown in the accompanying figure. Then,

$$\text{Area of } R = \text{area of } R_1 + \text{area of} R_2$$

$$= \int_0^1 (7x - x^2)\,dx + \int_1^2 [(8 - x^2) - x^2]\,dx$$

$$= \int_0^1 (7x - x^2)\,dx + \int_1^2 (8 - 2x^2)\,dx$$

$$= (\frac{7}{2}x^2 - \frac{1}{3}x^3)\Big|_0^1 + (8x - \frac{2}{3}x^3)\Big|_1^2$$

$$= \left[(\frac{7}{2} - \frac{1}{3}) - 0\right] + \left[(16 - \frac{16}{3}) - (8 - \frac{2}{3})\right] = \frac{13}{2}$$

20. The probability density function is $f(x) = 0.4e^{-0.4x}$.

(a) $P(1 \leq x \leq 2) = \int_1^2 0.4e^{-0.4x}\,dx = -e^{-0.4x}\Big|_1^2 = -e^{-0.8} + e^{-0.4} \simeq 0.2210$
which corresponds to 22.10 percent.

(b) $P(x \leq 2) = \int_0^2 0.4e^{-0.4x}\,dx = -e^{-0.4x}\Big|_0^2 = -e^{-0.8} + 1 \simeq 0.5507$
which corresponds to 55.07 percent.

(c)　$P(x \geq 2) = 1 - P(x \leq 2) \simeq 1 - 0.5507 = 0.4493$
which corresponds to 44.93 percent.

21.　(a)　The machine is profitable as long as $R(x) \geq C(x)$, where $R(x) = 4,575 - 5x^2$
dollars per year is the rate of revenue and $C(x) = 1,200 + 10x^2$ dollars per year
is the rate of cost. $R(x)$ will be equal to $C(x)$ when

$$4,575 - 5x^2 = 1,200 + 10x^2 \qquad 15x^2 = 3,375 \ \text{ or } \ x = 15$$

That is, the machine will be profitable for 15 years.

(b)　Let $P(x)$ denote the profit from the use of the machine. Then,

$$\frac{dP}{dx} = R(x) - C(x) = (4,575 - 5x^2) - (1,200 + 10x^2) = 3,375 - 15x^2$$

The total net profit generated over the next 5 years is

$$P(15) - P(0) = \int_0^{15} (3,375 - 15x^2)dx$$

$$= (3,375x - 5x^3)\Big|_0^{15} = 3,375(15) - 5(15)^3 = \$33,750$$

22.　The demand function is $D(q) = 30 - 2q - q^2$ dollars per unit.

(a)　To find the number of units bought when the price is $p = 15$, solve the equation
$15 = D(q)$ for q to get

$$15 = 30 - 2q - q^2 \qquad q^2 + 2q - 15 = 0$$
$$(q + 5)(q - 3) = 0 \quad \text{or} \quad q = 3 \ \text{ units}$$

(b)　The amount that consumers are willing to spend to get 3 units of the commodity
is

$$\int_0^3 D(q)dq = \int_0^3 (30 - 2q - q^2)dq$$

$$= (30q - q^2 - \frac{1}{3}q^3)\Big|_0^3 = \$72$$

(c)　When the market price is \$15 per unit, 3 units will be bought and the con-
sumers' surplus will be

$$\int_0^3 D(q)dq - (15)(3) = \int_0^3 (30 - 2q - q^2)dq - 45$$

$$= 72 - 45 = \$27.$$

(d)　The consumers' willingness to spend in part (b) is equal to the area under the
demand curve $p = D(q)$ from $q = 0$ to $q = 3$. The consumers' surplus in part
(c) is equal to the area of the region between the demand curve and the horizontal
line $p = 15$.

CHAPTER 7

DIFFERENTIAL EQUATIONS

---

1 ELEMENTARY DIFFERENTIAL EQUATIONS

2 SEPARABLE DIFFERENTIAL EQUATIONS

REVIEW PROBLEMS

Chapter 7, Section 1

1. Let Q denote the number of bacteria. Then, $\frac{dQ}{dt}$ is the rate of change of Q, and since this rate of change is proportional to Q, it follows that $\frac{dQ}{dt} = kQ$, where k is a positive constant of proportionality.

3. Let Q denote the value of the investment. Then $\frac{dQ}{dt}$ is the rate of change of Q, and since this rate is equal to 7 percent of the size of Q, it follows that $\frac{dQ}{dt} = 0.07Q$.

5. Let P denote the population. Then, $\frac{dP}{dt}$ is the rate of change of P, and since this rate of change is the constant 500, it follows that $\frac{dP}{dt} = 500$.

7. Let Q denote the temperature of the object and M the temperature of the surrounding medium. Then, $\frac{dQ}{dt}$ is the rate of change of Q. Since this rate of change is proportional to the difference between the temperature of the object and that of the surrounding medium, it follows that $\frac{dQ}{dt} = k(M - Q)$, where k is a positive constant. (Notice that if $M > Q$, $\frac{dQ}{dt}$ is positive and the temperature Q is increasing, and if $M < Q$, $\frac{dQ}{dt}$ is negative and the temperature Q is decreasing.)

9. Let Q denote the number of facts recalled and N the total number of relevant facts in the person's memory. Then, $\frac{dQ}{dt}$ is the rate of change of Q, and $N - Q$ is the number of relevant facts not recalled. Since the rate of change is proportional to $N - Q$, it follows that $\frac{dQ}{dt} = k(N - Q)$, where k is a positive constant of proportionality.

11. Let Q denote the number of people implicated and N the total number of people involved. Then, $\frac{dQ}{dt}$ is the rate of change of Q, and $N - Q$ is the number involved but not implicated. Since the rate of change is jointly proportional to Q and $N - Q$, it follows that $\frac{dQ}{dt} = kQ(N - Q)$, where k is a positive constant.

13. If $y = Ce^{kx}$, then $\frac{dy}{dx} = kCe^{kx} = ky$.

15. If $y = C_1e^x + C_2xe^x$, then

$$\frac{dy}{dx} = C_1 e^x + C_2(xe^x + e^x) = (C_1 + C_2)e^x + C_2 xe^x$$

and

$$\frac{d^2y}{dx^2} = (C_1 + C_2)e^x + C_2(xe^x + e^x) = (C_1 + 2C_2)e^x + C_2 xe^x$$

Thus,

$$\frac{d^2y}{dx^2} - 2\frac{dy}{dx} + y$$

$$= (C_1 + 2C_2)e^x + C_2 xe^x - 2[(C_1 + C_2)e^x + C_2 xe^x] + (C_1 e^x + C_2 xe^x)$$

$$= (C_1 + 2C_2 - 2C_1 - 2C_2 + C_1)e^x + (C_2 - 2C_2 + C_2)xe^x = 0$$

17. If $\frac{dy}{dx} = 3x^2 + 5x - 6$, then

$$y = \int (3x^2 + 5x - 6)\, dx = x^3 + \frac{5}{2}x^2 - 6x + C$$

19. If $\frac{dV}{dx} = \frac{2}{x+1}$, then

$$V = \int \frac{2}{x+1}\, dx = 2\ln|x+1| + C = \ln(x+1)^2 + C$$

21. If $\frac{d^2P}{dt^2} = 50$, then

$$\frac{dP}{dt} = \int 50\, dt = 50t + C_1$$

and

$$P = \int (50t + C_1)\, dt = 25t^2 + C_1 t + C_2$$

23. If $\frac{dy}{dx} = e^{5x}$, then

$$y = \int e^{5x}\, dx = \frac{1}{5}e^{5x} + C$$

Since $y = 1$ when $x = 0$, $1 = \frac{1}{5}e^0 + C$ or $C = \frac{4}{5}$. Hence, $y = \frac{1}{5}e^{5x} + \frac{4}{5}$.

25. If $\frac{dV}{dt} = 16t(t^2 + 1)^3$, then

$$V = \int 16t(t^2 + 1)^3 = 2(t^2 + 1)^4 + C$$

Since $V = 1$ when $t = 0$, $1 = 2 + C$ or $C = -1$. Hence, $V = 2(t^2 + 1)^4 - 1$.

27. If $\frac{d^2A}{dt^2} = e^{-t/2}$, then

$$\frac{dA}{dt} = \int e^{-t/2}\, dt = -2e^{-t/2} + C$$

Since $\frac{dA}{dt} = 1$ when $t = 0$,

$$1 = -2e^0 + C = -2 + C \qquad \text{or} \qquad C = 3$$

Hence, $\frac{dA}{dt} = -2e^{-t/2} + 3$, and

$$A = \int (-2e^{-t/2} + 3)\, dt = 4e^{-t/2} + 3t + C$$

Since $A = 2$ when $t = 0$,

$$2 = 4e^0 + 3(0) + C \qquad \text{or} \qquad C = -2.$$

Hence, $A = 4e^{-t/2} + 3t - 2$.

29.   (a)   Let $V(t)$ denote the value of the machine after t years. Since $\frac{dV}{dt} = -960e^{-t/5}$ dollars per year,

$$V(t) = \int -960e^{-t/5}\, dt = 4,800e^{-t/5} + C$$

If V_0 denotes the initial value of the machine, then

$$V_0 = V(0) = 4,800e^0 + C \qquad \text{or} \qquad C = V_0 - 4,800$$

Thus,

$$V(t) = 4,800e^{-t/5} + V_0 - 4,800$$

(b)   If $V_0 = \$5,200$, then

$$V(10) = 4,800e^{-2} + 5,200 - 4,800 \simeq \$1,049.61$$

31.   Let $P(x)$ denote the population x years after 1970. Since $\frac{dP}{dx} = 1,500x^{-1/2}$ people per year,

$$P(x) = \int 1,500x^{-1/2}\, dx = 3,000x^{1/2} + C$$

If P_0 denotes the population in 1970, then $P_0 = P(0) = C$, and so
$P(x) = 3,000x^{1/2} + P_0$. Since the population was $39,000$ in 1979 (when $x = 9$),

$$39,000 = P(9) = 3,000(3) + P_0 \qquad \text{or} \qquad P_0 = 30,000$$

Hence,

$$P(x) = 3,000x^{1/2} + 30,000$$

(a)   The population in 1970 was $P(0) = 30,000$.
(b)   The population in 1995 will be $P(25) = 45,000$.

33.   (a)   Let $Q(t)$ denote the ozone level t hours after $7:00$ A.M. Since

$$\frac{dQ}{dt} = \frac{0.24 - 0.03t}{\sqrt{36 + 16t - t^2}} = (0.24 - 0.03t)(36 + 16t - t^2)^{-1/2}$$

parts per million per hour, then (by substitution with $u = 36 + 16t - t^2$),

$$Q(t) = \int (0.24 - 0.03t)(36 + 16t - t^2)^{-1/2} \, dt$$

$$= 0.03 \int (8 - t)(36 + 16t - t^2)^{-1/2} \, dt$$

$$= 0.03(36 + 16t - t^2)^{1/2} + C$$

Since the ozone level was 0.25 at $7:00$ A.M.,

$$0.25 = Q(0) = 0.03(36)^{1/2} + C = 0.18 + C \qquad \text{or} \qquad C = 0.07$$

Hence,

$$Q(t) = 0.03(36 + 16t - t^2)^{1/2} + 0.07$$

(b)   The peak ozone level occurs when $\frac{dQ}{dt} = 0$, that is, when

$$0.24 - 0.035t = 0 \qquad \text{or} \qquad t = 8$$

Note that $Q(t)$ has an absolute maximum at this critical point since $\frac{dQ}{dt}$ is positive (and Q is increasing) for $0 < t < 8$, and $\frac{dQ}{dt}$ is negative (and Q is decreasing) for $t > 8$.) Thus, the peak ozone level is

$$Q(8) = 0.03[36 + 16(8) - (8)^2]^{1/2} + 0.07 = 0.03(10) + 0.07 = 0.37$$

parts per million, which occur at $3:00$ P.M. (8 hours after $7:00$ A.M.).

35.   Let $N(t)$ denote the number of bushels t days from now. Since $\frac{dN}{dt} = 0.3t^2 + 0.6t + 1$ bushels per day,

$$N(t) = \int (0.3t^2 + 0.6t + 1) \, dt = 0.1t^3 + 0.3t^2 + t + C$$

The increase in the number of bushels during the next 5 days is

$$N(5) - N(0) = [(0.1)(t)^3 + (0.3)(5)^2 + 5 + C] - [C] = 25 \text{ bushels}$$

and at \$3 per bushel, the value of the crop will increase by \$75.

37.   (a)   Let $D(t)$ denote the distance the car has traveled and $S(t)$ its speed t seconds after the brakes were applied. Since the acceleration is the second derivative of distance, $\frac{d^2D}{dt^2} = -28$, and so

$$S(t) = \frac{dD}{dt} = \int -28 \, dt = -28t + C_1$$

If S_0 denotes the speed at the time the brakes were applied,

$$S_0 = S(0) = C_1$$

Hence, $\frac{dD}{dt} = -28t + S_0$, and, integrating again,

$$D(t) = \int (-28t + S_0)\, dt = -14t^2 + S_0 t + C_2$$

At the moment the brakes are applied, the additional distance traveled is zero. Hence, $0 = D(0) = C_2$, and so

$$D(t) = -14t^2 + S_0 t$$

(b)   Since $S_0 = 60$ miles per hour $= 88$ feet per second, $D(t) = -14t^2 + 88t$. The car stops when the speed, $S(t) = \frac{dD}{dt}$, is zero, that is, when

$$0 = -28t + 88 \qquad \text{or} \qquad t = \frac{22}{7} \text{ seconds}$$

The stopping distance is the distance traveled during these $\frac{22}{7}$ seconds, that is, $D(\frac{22}{7}) \simeq 138.29$ feet.

Chapter 7, Section 2

1. Separate the variables of $\frac{dy}{dx} = 3y$ and integrate to get

$$\int \frac{1}{y}\, dy = \int 3\, dx, \qquad \ln|y| = 3x + C_1,$$

$$|y| = e^{3x+C_1} = e^{C_1} e^{3x} \quad \text{or} \quad y = \pm e^{C_1} e^{3x} = Ce^{3x}$$

where C is the constant $\pm e^{C_1}$.

3. Separate the variables of $\frac{dy}{dx} = e^y$ and integrate to get

$$\int e^{-y}\, dy = \int 1\, dx, \qquad -e^{-y} = x + C_1, \quad \text{or} \quad e^{-y} = C - x$$

where C is the constant $-C_1$. Hence,

$$\ln e^{-y} = \ln(C - x), \quad -y = \ln(C - x), \quad \text{or} \quad y = -\ln(C - x)$$

5. Separate the variables of $\frac{dy}{dx} = \frac{x}{y}$ and integrate to get

$$\int y\, dy = \int x\, dx, \quad \frac{1}{2}y^2 = \frac{1}{2}x^2 + C_1, \quad y^2 = x^2 + C, \quad y = \pm\sqrt{x^2 + C}$$

where C is the constant $2C_1$.

7. Separate the variables of $\frac{dy}{dx} = y + 10$ and integrate to get

$$\int \frac{1}{y+10}\, dy = \int 1\, dx, \quad \ln|y+10| = x + C_1, \quad |y+10| = e^{C_1+x}$$

$$y + 10 = \pm e^{C_1+x} = \pm e^{C_1} e^x, \quad y = \pm e^{C_1} e^x - 10 = Ce^x - 10$$

where C is the constant $\pm e^{C_1}$.

9. Separate the variables of $\frac{dy}{dx} = y(1 - 2y)$ and integrate to get

$$\int \frac{1}{y(1-2y)}\,dy = \int 1\,dx$$

$$\int \frac{1}{y}\,dy + ②\int \frac{1}{1-2y}\,dy = \int 1\,dx$$

$$\ln|y| - \ln|1-2y| = x + C, \qquad \ln\left|\frac{y}{1-2y}\right| = x + C$$

$$\left|\frac{y}{1-2y}\right| = e^{x+C} = e^x e^C$$

$$\frac{y}{1-2y} = Ae^x \qquad \text{(where } A = \pm e^C)$$

$$y = Ae^x - 2yAe^x$$

$$y = \frac{Ae^x}{1+2Ae^x} = \frac{1}{(1/A)e^{-x}+2} = \frac{1}{2+Be^{-x}}$$

11. Separate the variables of $\frac{dy}{dx} = 4x^3y^2$ and integrate to get

$$\int \frac{1}{y^2}\,dy = \int 4x^3\,dx, \qquad -\frac{1}{y} = x^4 + C \qquad \text{or} \qquad y = -\frac{1}{x^4+C}$$

Since $y = 2$ when $x = 1$,

$$2 = -\frac{1}{1+C} \qquad 2 + 2C = -1 \qquad \text{or} \qquad C = -\frac{3}{2}$$

Hence,

$$y = -\frac{1}{x^4 - 3/2} = \frac{2}{3 - 2x^4}.$$

13. Separate the variables of $\frac{dy}{dx} = 5(8-y)$ and integrate to get

$$\int \frac{1}{8-y}\,dy = \int 5\,dx, \qquad -\ln|8-y| = 5x + C$$

$$|8-y| = e^{-5x-C} = e^{-5x}e^{-C}$$

$$8 - y = Ae^{-5x} \qquad \text{(where } A = \pm e^{-C})$$

$$y = 8 - Ae^{-5x}$$

Since $y = 6$ when $x = 0$, it follows that $6 = 8 - A$ or $A = 2$. Hence,

$$y = 8 - 2e^{-5x}$$

15. Let t denote time and Q the size of the investment. Then $\frac{dQ}{dt}$ is the rate at which the investment changes. Since the investment changes at the rate of 7 percent per year, $\frac{dQ}{dt} = 0.07Q$. Separate the variables and integrate to get

$$\int \frac{1}{Q}\,dQ = \int 0.07\,dt \qquad \ln|Q| = 0.07t + C_1$$

$$|Q| = e^{0.07t + C_1} = e^{0.07t}e^{C_1} \qquad \text{or} \qquad Q = Ce^{0.07t}$$

where $C = e^{C_1}$, and the absolute values can be dropped since $Q > 0$ in the context of this problem. Since the initial investment was \$1,000,

$$1,000 = Q(0) = Ce^0 \qquad \text{or} \qquad C = 1,000$$

Hence, $Q(t) = 1,000e^{0.07t}$.

17. Let t denote time and Q the size of the quantity. Then $\frac{dQ}{dt}$ is the rate of change of Q. Since the quantity decays at a rate proportional to its size, $\frac{dQ}{dt} = -kt$, there k is a positive constant of proportionality. Separate the variables and integrate to get

$$\int \frac{1}{Q}\,dQ = \int -k\,dt \qquad \ln|Q| = -kt + C_1$$

$$|Q| = e^{-kt + C_1} = e^{-kt}e^{C_1} \qquad \text{or} \qquad Q = Ce^{-kt}$$

(where C is the constant $\pm e^{C_1}$), which shows that Q decays exponentially.

19. Let t denote time, Q the number of facts that have been recalled, and B the total number of relevant facts in the subject's memory. then the number of relevant facts that have not been recalled is $B - Q$, and the differential equation describing the recall of the facts is $\frac{dQ}{dt} = k(B - Q)$, where k is a positive constant of proportionality. Separate the variables and integrate to get

$$\int \frac{1}{B - Q}\,dQ = \int k\,dt \qquad -\ln|B - Q| = kt + C_1$$

$$\ln|B - Q| = -kt - C_1 \qquad B - Q = e^{-kt - C_1} = e^{-kt}e^{-C_1} = Ce^{-kt}$$

where C is the constant e^{-C_1}, and the absolute values can be dropped since $B - Q > 0$. Hence,

$$Q = B - Ce^{-kt}$$

Since no facts have been recalled when $t = 0$,

$$0 = Q(0) = B - C \qquad \text{or} \qquad C = B$$

Hence,

$$Q(t) = B - Be^{-kt} = B(1 - e^{-kt})$$

As t increases without bound, e^{-kt} approaches zero and Q approaches B. The vertical intercept is $Q(0) = 0$. The graph is sketched below.

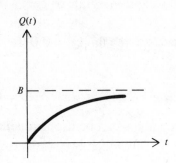

Graph for Problem 19.

21. Let t denote time (in minutes), Q the temperature of the object, and M the temperature of the surrounding medium. The differential equation describing the rate of change of Q is $\frac{dQ}{dt} = k(M - Q)$, where k is a positive constant of proportionality. Separate the variables and integrate to get

$$\int \frac{1}{M - Q}\, dQ = \int k\, dt \quad -\ln|M - Q| = kt + C_1 \quad \ln|M - Q| = -kt - C_1$$

$$M - Q = e^{C_1}e^{-kt} \qquad \text{or} \qquad Q(t) = M - Ce^{-kt}$$

where the absolute values can be dropped since $M - Q > 0$, and C is the constant e^{-C_1}. Since the temperature of the room is 80 degrees, $M = 80$, and so

$$Q(t) = 80 - Ce^{-kt}$$

Since the temperature of the drink was 40 degrees when it left the refrigerator,

$$40 = Q(0) - 80 - C \qquad \text{or} \qquad C = 40$$

Hence,

$$Q(t) = 80 - 40e^{-kt}$$

Finally, since the temperature of the drink was 50 degrees after 20 minutes,

$$50 = Q(20) = 80 - 40e^{-20k} \qquad -30 = -40e^{-20k} \qquad e^{-20k} = \frac{3}{4}$$

$$-20k = \ln\frac{3}{4} \qquad \text{or} \qquad k = -\frac{\ln(3/4)}{20} \simeq 0.014$$

Hence, $Q(t) = 80 - 40e^{-0.014t}$.

23. Let t denote time, A the area of the cell wall, S the concentration of the solute outside the cell, and Q the concentration of the solute inside the cell. It is given that A and S are constants and that $S > Q$. The differential equation describing the rate of change of the concentration is $\frac{dQ}{dt} = kA(S - Q)$, where k is the positive constant of proportionality. Separate the variables and integrate to get

$$\int \frac{1}{S - Q} dQ = \int kA \, dt \qquad -\ln|S - Q| = kAt + C_1$$

$$\ln|S - Q| = -kAt - C_1 \qquad S - Q = e^{-kAt}e^{-C_1} \qquad \text{or} \qquad Q = S - Ce^{-kAt}$$

where the absolute values can be dropped since $S > Q$, and C is the constant e^{-C_1}. If Q_0 denotes the initial concentration inside the cell,

$$Q_0 = Q(0) = S - C \qquad \text{or} \qquad C = S - Q_0$$

Hence, $Q(t) = S - (S - Q_0)e^{-kAt}$.

25. Let Q denote the amount of salt in the tank and t the number of minutes that have elapsed. Then,

$$\frac{dQ}{dt} = \text{rate of change of salt with respect to time(pounds per minute)}$$

$$= \text{(rate at which salt enters tank)} - \text{(rate at which salt leaves tank)}$$

$$= \frac{\text{pounds entering}}{\text{gallon}} \frac{\text{gallons entering}}{\text{minute}} - \frac{\text{pounds leaving}}{\text{gallon}} \frac{\text{gallons leaving}}{\text{minute}}$$

Since

$$\frac{\text{pounds leaving}}{\text{gallon}} = \frac{\text{pounds of salt in tank}}{\text{gallons of mixture in tank}} = \frac{Q}{200}$$

it follows that

$$\frac{dQ}{dt} = 2(t) - (\frac{Q}{200})(5) = 10 - \frac{Q}{40} = \frac{400 - Q}{40}$$

Separate the variables and integrate to get

$$\int \frac{1}{400 - Q} dQ = \int \frac{1}{40} dt \qquad -\ln|400 - Q| = \frac{1}{40}t + C_1$$

$$\ln|400 - Q| = -\frac{1}{40}t - C_1 \qquad 400 - Q = e^{-C_1}e^{-t/40} \qquad Q(t) = 400 - Ce^{-t/40}$$

23. Divide the interval $0 \le x \le 9$ into n equal subintervals of length Δx months, and let x_j denote the beginning of the jth subinterval. During the jth subinterval,

$$\text{Revenue} \simeq 50(\Delta x)P(x_j)$$

Hence,

$$\text{Total revenue} \simeq \sum_{j=1}^{n} 50P(x_j)\Delta x$$

Now, as n increases without bound, this approximation improves while the sum approaches the corresponding integral. Hence,

$$\text{Total revenue} = \int_0^9 50P(x)dx = \int_0^9 50(40 + 3\sqrt{x})dx$$

$$= 50\left[40x + 2x^{3/2}\right]_0^9 = 50[360 + 2(27)] = \$20,700$$

24. Divide the interval $0 \le x \le 12$ into n equal subintervals of length Δx months, and let x_j denote the beginning of the jth subinterval. During the jth subinterval,

$$\text{Amount spent} = \text{(demand per month) (number of months) (price)}$$
$$= D(x_j)(\Delta x)P(x_j) = D(x_j)P(x_j)\Delta x$$

Hence,

$$\text{Total amount spent} = \sum_{j=1}^{n} D(x_j)P(x_j)\Delta x$$

Now, as n increases without bound this approximation improves while the sum approaches the corresponding integral. Hence,

$$\text{Total amount spent} = \int_0^{12} D(x)P(x)dx$$

25. Since $y = x^2 + 1$ is nonnegative over the interval $x = -1$ to $x = 2$ then volume of the solid of revolution is given by

$$\pi \int_a^b [f(x)]^2 dx = \pi \int_{-1}^2 (x^2 + 1)^2 dx$$

$$= \pi \int_{-1}^2 (x^4 + 2x^2 + 1)dx$$

$$= \pi \left(\frac{1}{5}x^5 + \frac{2}{3}x^3 + x\right)\Big|_{-1}^2$$

$$= \pi\left[\left(\frac{32}{5} + \frac{16}{3} + 2\right) - \left(-\frac{1}{5} - \frac{2}{3} - 1\right)\right]$$

$$= \frac{234}{15}\pi \approx 49.01$$

26. Since $y = e^{-x/20}$ is nonnegative over the interval $x = 0$ to $x = 10$ then the volume of the solid of revolution is given by

$$\pi \int_a^b [f(x)]^2 dx = \pi \int_0^{10} (e^{-x/20})^2 dx$$

$$= \pi \int_0 e^{-x/10} dx$$

$$= \pi(-10)e^{-x/10}\Big|_0^{10}$$

$$= -10\pi(e^{-1} - e^0)$$

$$= 10\pi(1 - \frac{1}{e}) \approx 19.86$$

27. Since the price of chicken t months after the beginning of the year is $P(t) = 0.06t^2 - 0.2t + 1.2$ dollars per pound, the average price during the first 6 months is

$$\frac{1}{6-0} \int_0^6 (0.06t^2 - 0.2t + 1.2)dt = \frac{1}{6}\left[0.02t^3 - 0.1t^2 + 1.2t\right]_0^6$$

$$= \frac{1}{6}[0.02(6)^3 - 0.1(6)^2 + 1.2(6)] = \$1.32 \text{ per pound}$$

28. Recall that P dollars invested at an annual interest rate of 8 percent compounded continously will be worth $Pe^{0.08t}$ dollars after t years. Let t_j denote the time (in years) of the j th deposit of 100. This deposit will remain in the account for $5 - t_j$ years and hence will grow to $100e^{0.08(5-t_j)}$ dollars. At the end of 5 years (after 60 deposits),

$$\text{Amount in account} = \sum_{j=1}^{60} 100e^{0.08(5-t_j)}$$

Now observe that $\Delta t = \frac{1}{12}$ (since the deposits are monthly), and rewrite the sum as

$$\text{Amount in account} = 12\sum_{j=1}^{60} 100e^{0.08(5-t_j)}\Delta t$$

This sum can be approximated by the definite integral:

$$\text{Amount in account} \simeq 12 \int_0^5 100e^{0.08(5-t)} dt$$

$$= 1,200e^{0.4} \int_0^5 e^{-0.08t} dt$$

$$= \frac{1,200}{-0.08}e^{0.4}\left[e^{-0.08t}\right]_0^5 = -\frac{1.200}{0.08}e^{0.4}\left[e^{-0.4} - 1\right]$$

$$= -\frac{1,200}{0.08}\left[1 - e^{0.4}\right] \simeq \$7,377.37$$

PAGE MIXUP

29. Recall that the present value of B dollars payable t years from now with an annual interest rate of 7 percent compounded continuously is $Be^{-0.07t}$. Divide the interval $0 \leq t \leq 10$ into n equal subintervals of length Δt years, and let t_j denote the beginning of the jth subinterval. During the jth subinterval,

$$\text{Income} = (\text{dollars per year})(\text{number of years}) \simeq 1{,}000\Delta t$$

and

$$\text{Present value of income} \simeq 1{,}000(\Delta t)e^{-0.07t_j} = 1{,}000e^{-0.07t_j}\Delta t$$

Hence, over the entire 10 years,

$$\text{Present value of investment} \simeq \sum_{j=1}^{n} 1{,}000e^{-0.07t_j}\Delta t$$

Now as n increases without bound, this approximation improves and the sum approaches the corresponding integral. Hence,

$$\text{Present value} = \int_{0}^{10} 1{,}000e^{-0.07t}\,dt = \frac{1{,}000}{-0.07}\left[e^{-0.07t}\right]_{0}^{10}$$

$$= -\frac{1{,}000}{0.07}\left(e^{-0.7} - 1\right) \simeq \$7{,}191.64$$

30. The fraction of homes that remain unsold for t weeks is $f(t) = e^{-0.2t}$. Of the 200 homes currently on the market, the number that will still be on the market 10 weeks from now is

$$200f(10) = 200e^{-2}$$

To find the number of additional homes on the market 10 weeks from now, divide the interval $0 \leq t \leq 10$ into n equal subintervals of length Δt weeks, and let t_j denote the beginning of the jth subinterval. During the jth subinterval, $8\Delta t$ additional homes are placed on the market, and 10 weeks from now (that is, $10 - t_j$ weeks later) the number of these still on the market will be $8(\Delta t)f(10 - t_j)$. Hence, the total number N of homes on the market 10 weeks from now will be approximately

$$N \simeq 200f(10) + \sum_{j=1}^{n} 8f(10 - t_j)\Delta t$$

Now, as n increases without bound, this approximation improves while the sum approaches the corresponding integral. Hence,

$$N = 200f(10) + \int_{0}^{10} 8f(10 - t)\,dt = 200e^{-2} + 8\int_{0}^{10} e^{-0.2(10-t)}\,dt$$

$$= 200e^{-2} + 8e^{-2}\int_{0}^{10} e^{0.2t}\,dt = 200e^{-2} + \frac{8}{0.2}e^{-2}\left[e^{0.2t}\right]_{0}^{10}$$

$$= 200e^{-2} + 40e^{-2}(e^{2} - 1) = 200e^{-2} + 40(1 - e^{-2})$$

$$= 160e^{-2} + 40 \simeq 61.65 \simeq 62 \text{ homes}$$

PAGE MIXUP

31. For $\int_1^3 \frac{1}{x} dx$ with $n = 10, \Delta x = \frac{3-1}{10} = 0.2$ and

$$x_1 = 1, \quad x_2 = 1.2, \quad x_3 = 1.3, \cdots, x_{10} = 2.8, \quad \text{and} \quad x_{11} = 3$$

(a) Trapezoidal rule:

$$\int_1^3 \frac{1}{x} dx \simeq \frac{0.2}{2}(1 + \frac{2}{1.2} + \frac{2}{1.4} + \frac{2}{1.6} + \frac{2}{1.8} + \frac{2}{2.0}$$
$$+ \frac{2}{2.2} + \frac{2}{2.4} + \frac{2}{2.6} + \frac{2}{2.8} + \frac{1}{3}) \simeq 1.1016$$

The error estimate is $|E_n| \le \frac{M(b-a)^3}{12n^2}$. For $n = 10$, $a = 1$, and $b = 3$,

$$|E_{10}| \le \frac{M(3-1)^3}{12(10)^2} = \frac{8M}{1,200} = \frac{M}{150}$$

where M is the maximum value of $|f''(x)|$ on $1 \le x \le 3$. Now,

$$f(x) = \frac{1}{x}, f'(x) = -\frac{1}{x^2}, \quad \text{and} f''(x) = \frac{2}{x^3}$$

and for $1 \le x \le 3$,

$$|f''(x)| \le \frac{2}{(1)^3} = 2$$

Hence,

$$|E_n| \le \frac{2}{150} = 0.0133$$

(b) Simpson's rule:

$$\int_1^3 \frac{1}{x} dx \simeq \frac{0.2}{3}(1 + \frac{4}{1.2} + \frac{2}{1.4} + \frac{4}{1.6} + \frac{2}{1.8} + \frac{4}{2.0}$$
$$+ \frac{2}{2.2} + \frac{4}{2.4} + \frac{2}{2.6} + \frac{4}{2.8} + \frac{1}{3}) \simeq 1.0987$$

The error estimate is $|E_n| \le \frac{M(b-a)^5}{180n^4}$. For $n = 10$, $a = 1$, and $b = 3$,

$$|E_{10}| \le \frac{M(2)^5}{180(10)^4} = \frac{32M}{1,800,000}$$

where M is the absolute value of $|f^{(4)}(x)|$ on $1 \le x \le 3$. Now,

$$f''(x) = \frac{2}{x^3}, \quad f^{(3)}(x) = -\frac{6}{x^4}, \quad \text{and} \quad f^{(4)}(x) = \frac{24}{x^5}$$

and for $1 \le x \le 3$,

$$|f^{(4)}(x)| \le \frac{24}{(1)^5} = 24$$

Hence,

$$|E_{10}| \leq \frac{(32)(24)}{1,800,000} = 0.0004$$

32. For $\int_0^2 e^{x^2}\,dx$ with $n = 8$, $\Delta x = \frac{2-0}{8} = 0.25$ and

$$x_1 = 0, \quad x_2 = 0.25, \quad x_3 = 0.5, \cdots, x_8 = 1.75 \text{ and } x_9 = 2$$

(a) Trapezoidal rule:

$$\int_0^2 e^{x^2}\,dx \simeq \frac{0.25}{2}(e^0 + 2e^{(0.25)^2} + 2e^{(0.5)^2} + 2e^{(0.75)^2}$$
$$+ 2e^{(1)^2} + 2e^{(1.25)^2} + 2e^{(1.5)^2} + 2e^{(1.75)^2} + e^{(2)^2}) \simeq 17.5651$$

The error estimate is $|E_n| \leq \frac{M(b-a)^3}{12n^2}$. For $n = 8$, $a = 0$, and $b = 2$,

$$|E_8| \leq \frac{M(2-0)^3}{12(8)^2} = \frac{M}{96}$$

where M is the maximum value of $|f''(x)|$ on $0 \leq x \leq 2$. Now,

$$f(x) = e^{x^2}, \quad f'(x) = 2xe^{x^2}, \quad \text{and} \quad f''(x) = (2 + 4x^2)e^{x^2}$$

and for $0 \leq x \leq 2$,

$$|f''(x)| \leq [2 + 4(2)^2]e^{(2)^2} = 18e^4$$

Hence,

$$|E_8| \leq \frac{18e^4}{96} \simeq 10.2372$$

(b) Simpson's rule:

$$\int_0^2 e^{x^2}\,dx \simeq \frac{0.25}{3}(1 + 4e^{(0.25)^2} + 2e^{(0.4)^2} + 4e^{(0.75)^2}$$
$$+ 2e^{(1)^2} + 4e^{(1.25)^2} + 2e^{(1.5)^2} + 4e^{(1.75)^2} + e^{(2)^2}) \simeq 16.5386$$

The error estimate is $|E_n| \leq \frac{M(b-a)^5}{180n^4}$. For $n = 8$, $a = 0$, and $b = 2$,

$$|E_8| \leq \frac{M(2-0)^5}{180(8)^4} = \frac{M}{23,040}$$

where M is the maximum value of $|f^{(4)}(x)|$ on $0 \leq x \leq 2$. Now,

$$f^{(3)}(x) = (8x^3 + 12x)e^{x^2} \quad \text{and} \quad f^{(4)}(x) = (16x^4 + 48x^2 + 12)e^{x^2}$$

and for $0 \leq x \leq 2$,

$$|f^{(4)}(x)| \leq [16(2)^4 + 48(2)^2 + 12]e^{(2)^2} = 460e^4$$

Hence,

$$|E_8| \leq \frac{460e^4}{23,040} \simeq 1.0901$$

33. The integral to be approximated is $\int_1^3 \sqrt{x}\, dx$. The derivatives of $f(x) = \sqrt{x}$ are

$$f'(x) = \frac{1}{2}x^{-1/2}, \quad f''(x) = -\frac{1}{4}x^{-3/2}, \quad f^{(3)}(x) = \frac{3}{8}x^{-5/2}, \quad f^{(4)}(x) = -\frac{15}{16}x^{-7/2}$$

(a) For the trapezoidal rule, $|E_n| \leq \frac{M(b-a)^3}{12n^2}$, where M is the maximum value of $|f''(x)|$ on $1 \leq x \leq 3$. Since

$$|f''(x)| \leq \frac{1}{4}(1)^{-3/2} = \frac{1}{4}$$

on $1 \leq x \leq 3$,

$$|E_n| \leq \frac{1}{4}\left[\frac{(3-1)^3}{12n^2}\right] = \frac{1}{6n^2}$$

which is less than 0.00005 if

$$1 < 6(0.00005)n^2 \quad \text{or} \quad n > \sqrt{\frac{1}{6(0.00005)}} \simeq 57.7$$

Hence, 58 subintervals should be used.

(b) For Simpson's rule, $|E_n| \leq \frac{M(b-a)^5}{180n^4}$, where M is the maximum value of $|f^{(4)}(x)|$ on $1 \leq x \leq 3$. Since

$$|f^{(4)}(x)| = \left|-\frac{15}{16}x^{-7/2}\right| \leq \frac{15}{16}(1)^{-7/2} = \frac{15}{16}$$

on $1 \leq x \leq 3$,

$$|E_n| \leq \frac{15}{16}\left[\frac{(3-1)^5}{180n^4}\right] = \frac{1}{6n^4}$$

which is less than 0.00005 if

$$1 < 6(0.00005)n^4 \quad \text{or} \quad n > \left[\frac{1}{6(0.00005)}\right]^{1/4} \simeq 7.6$$

Hence, 8 subintervals should be used.

34. The integral to be approximated is $\int_{0.5}^{1} e^{x^2} dx$. The derivatives of $f(x) = e^{x^2}$ are

$$f'(x) = 2xe^{x^2}, \quad f''(x) = (4x^2 + 2)e^{x^2},$$
$$f^{(3)}(x) = (8x^3 + 12x)e^{x^2}, \text{ and } f^{(4)}(x) = (16x^4 + 48x^2 + 12)e^{x^2}$$

(a) For the trapezoidal rule, $|E_n| \le \frac{M(b-a)^3}{12n^2}$, where M is the maximum value of $|f''(x)|$ on $0.5 \le x \le 1$. Since

$$|f''(x)| \le [4(1)^2 + 2]e^1 = 6e$$

on $0.5 \le x \le 1$,

$$|E_n| \le \frac{6e(1 - 0.5)^3}{12n^2} = \frac{e}{16n^2}$$

which is less than 0.00005 if

$$e < 16(0.00005)n^2 \text{ or } n > \sqrt{\frac{e}{16(0.00005)}} \simeq 58.3$$

Hence, 59 subintervals should be used.

(b) For Simpson's rule, $|E_n| \le \frac{M(b-a)^5}{180n^4}$, where M is the maximum value of $|f^{(4)}(x)|$ on $0.5 \le x \le 1$. Since

$$|f^{(4)}(x)| \le [16(1)^4 + 48(1)^2 + 12]e^1 = 76e$$

on $0.5 \le x \le 1$,

$$|E_n| \le \frac{76e(1 - 0.5)^5}{180n^4} = \frac{19e}{1,440n^4}$$

which is less than 0.00005 if

$$19e < 1,440(0.00005)n^4 \text{ or } n > \left[\frac{19e}{1,440(0.00005)}\right]^{1/4} \simeq 5.2$$

Hence, 6 subintervals should be used.

where C is the constant e^{-C_1}, and the absolute value can be dropped since $400 - Q > 0$. Since the tank initially held 200 gallons containing 3 pounds of salt per gallon for a total of 600 pounds of salt,

$$600 = Q(0) = 400 - C \qquad \text{or} \qquad C = -200$$

Hence,

$$Q(t) = 400 - (-200e^{-t/40}) = 400 + 200e^{-t/40}$$

As t increases without bound, $e^{-t/40}$ approaches zero, and so $Q(t)$ approaches 400. Moreover, the vertical intercept is $Q(0) = 600$. The graph is sketched below.

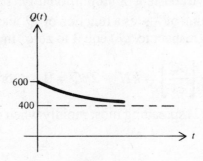

Graph for Problem 25.

27. Let t denote time, Q the number of residents who have been infected, and B the total number of susceptible residents. The differential equation describing the spread of the epidemic is $\frac{dQ}{dt} = kQ(B-Q)$, where k is a positive constant of proportionality. Notice that this is the same differential equation as that in Example 2.4 of the text. In that example, the solution was found to be

$$Q(t) = \frac{B}{1 + Ae^{-Bkt}}$$

Since there are $2,000$ susceptible residents, $B = 2,000$. Since 500 residents had the disease initially, $Q(0) = 500$ and so

$$500 = \frac{2,000}{1 + Ae^0} \qquad 500 + 500A = 2,000 \qquad \text{or} \qquad A = 3$$

Hence,

$$Q(t) = \frac{2,000}{1 + 3e^{-2,000kt}}$$

To find k, use the fact that $Q(1) = 855$ to get

$$855 = \frac{2,000}{1 + 3e^{-2,000k}} \qquad 855 + 2565e^{-2,000k} = 2,000$$

$$2,565e^{-2,000k} = 1,145 \qquad -2,000k = \ln\frac{1,145}{2,565} \simeq -0.8$$

Hence, the function giving the number of residents who have been infected after t weeks is (approximately)

$$Q(t) = \frac{2,000}{1 + 3e^{-0.8t}}$$

29. Let t denote time, Q the number of residents that have been infected, and N the total number of residents. The differential equation describing the spread of the epidemic is

$$\frac{dQ}{dt} = kQ(N - Q) = kQN - kQ^2$$

where k is a positive constant of proportionality. To find when the epidemic is spreading most rapidly, think of $\frac{dQ}{dt}$ as a function of Q and maximize this function by setting its derivative (with respect to Q) equal to zero. In particular,

$$\frac{d}{dQ}\left[\frac{dQ}{dt}\right] = kN - 2kQ = 0 \quad \text{or} \quad Q = \frac{N}{2}$$

Thus, the epidemic is spreading most rapidly when one half of the residents have been infected.

Chapter 7, Review Problems

1.   If $\frac{dy}{dx} = x^3 - 3x^2 + 5$,

$$y = \int (x^3 - 3x^2 + 5)\, dx = \frac{1}{4}x^4 - x^3 + 5x + C$$

2.   Seprate the variables of $\frac{dy}{dx} = 0.02y$ and integrate to get

$$\int \frac{1}{y}\, dy = \int 0.02\, dx \qquad\qquad \ln|y| = 0.02x + C_1$$

$$|y| = e^{0.02x + C_1} = e^{C_1} e^{0.02x} \qquad \text{or} \qquad y = Ce^{0.02x}$$

where $C = \pm e^{C_1}$.

3.   Separate the variables of $\frac{dy}{dx} = k(80 - y)$ and integrate to get

$$\int \frac{1}{80 - y}\, dy = \int k\, dx \qquad\qquad -\ln|80 - y| = kx + C_1$$

$$\ln|80 - y| = -kx - C_1 \qquad\qquad |80 - y| = e^{-kx - C_1} = e^{-C_1} e^{-kx}$$

$$80 - y = Ce^{-kx} \qquad \text{or} \qquad y = 80 - Ce^{-kx}$$

where $C = \pm e^{-C_1}$.

4.   Separate the variables of $\frac{dy}{dx} = y(1 - y)$ and integrate to get

$$\int \frac{1}{y(1 - y)}\, dy = \int 1\, dx$$

Notice that

$$\frac{1}{y(1 - y)} = \frac{1}{y} + \frac{1}{1 - y}$$

Hence,

$$\int \frac{1}{y(1 - y)}\, dy = \int \left[\frac{1}{y} + \frac{1}{1 - y} \right] dy = \ln|y| - \ln|1 - y|$$

and so

$$\ln|y| - \ln|1 - y| = x + C_1 \qquad \ln\left|\frac{y}{1-y}\right| = x + C_1$$

$$\left|\frac{y}{1-y}\right| = e^{x+C_1} = e^{C_1}e^x$$

$$\frac{y}{1-y} = C_2e^x \qquad \text{(where } C_2 = \pm e^{C_1}\text{)}$$

$$y = C_2e^x - C_2ye^x \qquad (1 + C_2e^x)y = C_2e^x$$

$$y = \frac{C_2e^x}{1 + C_2e^x} = \frac{1}{1/C_2e^x + 1} = \frac{1}{1 + Ce^{-x}}$$

where $C = \frac{1}{C_2}$.

5. Separate the variables of $\frac{dy}{dx} = e^{x+y} = (e^x)(e^y)$ and integrate to get

$$\int e^{-y}\, dy = \int e^x\, dx$$

$$-e^{-y} = e^x + C_1 \qquad \text{or} \qquad e^x + e^{-y} + C_1 = 0$$

If $C_1 = -C$ then the solution can be rewritten as

$$e^x + e^{-y} = C$$

In this case, it is better to leave solution as it is; instead of solving for y.

6. Separate the variables of $\frac{dy}{dx} = ye^{-2x}$ and integrate to get

$$\int \frac{dy}{y} = \int e^{-2x}\, dx \qquad \ln|y| = -\frac{1}{2}e^{-2x} + C_1$$

$$|y| = e^{-\frac{1}{2}e^{-2x}+C_1} \qquad |y| = e^{C_1}e^{-\frac{1}{2}e^{-2x}}$$

$$\text{or} \quad y = Ce^{-\frac{1}{2}e^{-2x}} \qquad \text{where} \qquad C = \pm e^{C_1}$$

7. If $\frac{dy}{dx} = 5x^4 - 3x^2 - 2$,

$$y = \int (5x^4 - 3x^2 - 2)\, dx = x^5 - x^3 - 2x + C$$

Since $y = 4$ when $x = 1$,

$$4 = (1)^5 - (1)^3 - 2(1) + C \qquad \text{or} \qquad C = 6$$

Hence, $y = x^5 - x^3 - 2x + 6$.

8. Separate the variables of $\frac{dy}{dx} = 0.06y$ and integrate to get

$$\int \frac{1}{y}\, dy = \int 0.06\, dx \qquad\qquad \ln|y| = 0.06x + C_1$$

$$|y| = e^{0.06x+C_1} = e^{C_1} e^{0.06x} \qquad \text{or} \qquad y = Ce^{0.06x}$$

where $C = \pm e^{C_1}$. Since $y = 100$ when $x = 0$,

$$100 = Ce^0 \qquad \text{or} \qquad C = 100$$

Hence, $y = 100e^{0.06x}$.

9. Separate the variables of $\frac{dy}{dx} = 3 - y$ and integrate to get

$$\int \frac{1}{3-y}\, dy = \int 1\, dx \qquad\qquad -\ln|3-y| = x + C_1$$

$$\ln|3-y| = -x - C_1 \qquad\qquad |3-y| = e^{-x-C_1} = e^{-C_1} e^{-x}$$

$$3 - y = Ce^{-x} \qquad \text{or} \qquad y = 3 - Ce^{-x}$$

where $C = \pm e^{-C_1}$. Since $y = 2$ when $x = 0$,

$$2 = 3 - C \qquad \text{or} \qquad C = 1$$

Hence, $y = 3 - e^{-x}$.

10. If $\frac{d^2y}{dx^2} = 2$, then

$$\frac{dy}{dx} = \int \frac{d^2y}{dx^2}\, dx = \int 2\, dx = 2x + C_1$$

Since $\frac{dy}{dx} = 3$ when $x = 0$,

$$3 - 2(0) + C_1 \qquad \text{or} \qquad C_1 = 3$$

Hence, $\frac{dy}{dx} = 2x + 3$, and

$$y = \int (2x + 3)\, dx = x^2 + 3x + C$$

Since $y = 5$ when $x = 0$

$$5 = (0)^2 + 3(0) + C \qquad \text{or} \qquad C = 5$$

Hence $y = x^2 + 3x + 5$.

11. Let $V(t)$ denote the value of the machine after t years. The rate of change of V is $\frac{dV}{dt} = k(V - 5,000)$, where k is a positive constant of proportionality. Separate the variables and integrate to get

$$\int \frac{1}{V - 5,000} \, dV = \int k \, dt \qquad \ln(V - 5,000) = kt + C_1$$

$$V - 5,000 = e^{kt+C_1} = e^{C_1}e^{kt} \qquad \text{or} \qquad V(t) = 5,000 + Ce^{kt}$$

where $C = e^{C_1}$ and the absolute values can be dropped since $V - 5,000 > 0$. Since the machine was originally worth \$40,000,

$$40,000 = V(0) = 5,000 + C \qquad \text{or} \qquad C = 35,000$$

Hence, $V(t) = 5,000 + 35,000e^{kt}$. Since the machine was worth \$30,000 after 4 years,

$$30,000 = V(4) = 5,000 + 35,000e^{4k}$$

$$35,000e^{4k} = 25,000 \qquad \text{or} \qquad e^{4k} = \frac{25,000}{35,000} = \frac{5}{7}$$

The value of the machine after 8 years is

$$V(8) = 5,000 + 35,000e^{8k} = 5,000 + 35,000[e^{4k}]^2$$

$$= 5,000 + 35,000(\tfrac{5}{7})^2 \simeq \$22,857$$

12. Let $R(t)$ denote the total revenue generated during the next t months and $P(t)$ the price of oil t months from now. Then, $P(t) = 24 + 0.08t$ and

$$\frac{dR}{dt} = \begin{bmatrix} \text{dollars per} \\ \text{month} \end{bmatrix} = \begin{bmatrix} \text{dollars per} \\ \text{barrel} \end{bmatrix} \begin{bmatrix} \text{barrels per} \\ \text{month} \end{bmatrix}$$

$$= P(t)(600) = 600(24 + 0.08t)$$

The general solution of this differential equation is

$$R(t) = 600 \int (24 + 0.08t) \, dt = 600(24t + 0.04t^2) + C$$

Since $R(0) = 0$, it follows that $C = 0$ and the appropriate particular solution is

$$R(t) = 600(24t + 0.04t^2)$$

Since the well will run dry at 36 months, the total future revenue will be

$$R(36) = 600[24(36) + 0.04(36)^2] \simeq \$549,504$$

13. Let $Q(t)$ denote the number of pounds of salt in the tank after t minutes. Then $\frac{dQ}{dt}$ is the rate of change of salt with respect to time (measured in pounds per minute). Thus,

$$\frac{dQ}{dt} = \text{(rate at which salt enters)} - \text{(rate at which salt leaves)}$$

$$= \frac{\text{pounds entering}}{\text{gallon}} \frac{\text{gallons entering}}{\text{minute}} - \frac{\text{pounds leaving}}{\text{gallon}} \frac{\text{gallons leaving}}{\text{minute}}$$

Now,

$$\frac{\text{pounds leaving}}{\text{gallon}} = \frac{\text{pounds of salt in tank}}{\text{gallons of salt in tank}} = \frac{Q}{200 - t}$$

since the tank loses 1 gallon of brine per minute. Hence,

$$\frac{dQ}{dt} = 0(4) - \frac{Q}{200 - t}(5) = \frac{5Q}{t - 200}$$

Separate the variables and integrate to get

$$\int \frac{1}{Q}\, dQ = \int \frac{5}{5 - 200}\, dt \qquad \ln|Q| = 5\ln|t - 200| + C_1$$

$$\ln|Q| = \ln|t - 200|^5 + C_1 \qquad Q = e^{C_1} e^{\ln(t-200)^5}$$

$$Q(t) = C(t - 200)^5$$

where $C = e^{C_1}$, and the absolute values are not needed since Q and $t - 200$ are both positive in the context of this problem. Since there are initially 600 pounds of salt in the tank (3 pounds of salt per gallon times 200 gallons),

$$600 = Q(0) = C(-200)^5 \qquad \text{or} \qquad C = \frac{600}{(-200)^5}$$

Hence,

$$Q(t) = \frac{600}{(-200)^5}(t - 200)^5 = 600\left[\frac{t - 200}{-200}\right]^5 = 600\left[1 - \frac{t}{200}\right]^5$$

The amount of salt in the tank after 100 minutes is

$$Q(100) = 600\left[1 - \frac{100}{200}\right]^5 = 18.75 \text{ pounds}.$$

14. Let $Q(t)$ denote the population in millions t years after 1980. The differential equation describing the population growth is $\frac{dQ}{dt} = kQ(10 - Q)$, where k is a positive constant of proportionality. this differential equation (with $10 = B$) was solved in Example 2.4. Its solution is

$$Q(t) = \frac{10}{1 + Ae^{-10kt}}$$

Since the population was 4 million in 1980, $Q(0) = 4$ and so

$$4 = \frac{10}{1 + A} \qquad 4 + 4A = 10 \qquad \text{or} \qquad A = 1.5$$

Hence, $Q(t) = \frac{10}{1 + 1.5e^{-10kt}}$. Since the population was 4.74 million in 1985, $Q(5) =$ 4.74 and

$$4.74 = \frac{10}{1 + 1.5e^{-50k}} \qquad 4.74 + 7.11e^{-50k} = 10$$

$$7.11e^{-50k} = 5.26 \qquad -50k = \ln\frac{5.26}{7.11} \qquad -10k = \frac{1}{5}\ln\frac{5.26}{7.11} \simeq -0.06$$

Hence, the population function is (approximately)

$$Q(t) = \frac{10}{1 + 1.5e^{-0.06t}}$$

15. Let $Q(t)$ denote the amount of new currency in circulation at time t. The $\frac{dQ}{dt}$ is the rate of change of the new currency with respect to time (measured in dollars per day). Thus,

$$\frac{dQ}{dt} = (\text{rate at which new currency enters}) - (\text{rate at which new currency leaves})$$

Now, the rate at which new currency enters is 18 million per day. The rate at which new currency leaves is

$$\left(\frac{\text{new currency at time } t}{\text{total currency}}\right)(\text{rate at which new currency enters})$$

$$= (\frac{Q(t)}{5000}(18) \text{ million per day}$$

Putting it all together,

$$\frac{dQ}{dt} = 18 - \frac{18Q}{5000}$$

$$\frac{dQ}{dt} = 18(1 - \frac{Q}{5000})$$

To solve this differential equation, separate variables to obtain

$$\frac{dQ}{1 - \frac{Q}{5000}} = 18\, dt$$

and so

$$\int \frac{dQ}{1 - \frac{Q}{5000}} = 18 \int dt$$

$$-5000 \ln \left| 1 - \frac{Q}{5000} \right| = 18t + C$$

When $t = 0$, $Q(0) = 0$ which yields

$$-5000 \ln \left| 1 - \frac{0}{5000} \right| = 18(0) + C \qquad \text{or} \qquad C = 0$$

Therefore, the solution becomes

$$-5000 \ln \left| 1 - \frac{Q}{5000} \right| = 18t \qquad \qquad \ln \left| 1 - \frac{Q}{5000} \right| = \frac{-18t}{5000}$$

Since Q is a part of 5000, then $1 - \frac{Q}{5000} > 0$ and so

$$\ln\left(1 - \frac{Q}{5000}\right) = -\frac{18t}{5000} \qquad \qquad 1 - \frac{Q}{5000} = e^{-\frac{18}{5000}t}$$

Now to find T so that $Q(T) = .9(5000)$ substitute into the last solution

$$1 - \frac{.9(5000)}{5000} = e^{-\frac{18}{5000}T}$$

$$\frac{1}{10} = e^{-\frac{18}{5000}T}$$

or

$$\ln \frac{1}{10} = -\frac{18}{5000}T$$

$$\ln 1 - \ln 10 = -\frac{18}{5000}T$$

$$-\ln 10 = -\frac{18}{5000}T$$

or $T = \frac{5000}{18} \ln 10 \approx 640$ days ≈ 1.75 years.

CHAPTER 8

LIMITS AT INFINITY AND IMPROPER INTEGRALS

---

1 LIMITS AT INFINITY AND L'HÔPITAL'S RULE

2 IMPROPER INTEGRALS

3 PROBABILITY DENSITY FUNCTIONS

REVIEW PROBLEMS

Chapter 8, Section 1

1. $\lim_{x \to \infty} (x^3 - 4x^2 - 4) = \lim_{x \to \infty} x^3 = \infty$

3. $\lim_{x \to \infty} (1 - 2x)(x + 5) = \lim_{x \to \infty} (5 - 9x - 2x^2) = \lim_{x \to \infty} (-2x^2) = -\infty$

5. Divide numerator and denominator by x^2 to get

$$\lim_{x \to \infty} \frac{x^2 - 2x + 3}{2x^2 + 5x + 1} = \lim_{x \to \infty} \frac{1 - 2/x + 3/x^2}{2 + 5/x + 1/x^2} = \frac{1}{2}$$

7. Divide numerator and denominator by x to get

$$\lim_{x \to \infty} \frac{2x + 1}{3x^2 + 2x - 7} = \lim_{x \to \infty} \frac{2 + 1/x}{3x + 2 - 7/x} = \frac{2}{\infty} = 0$$

9. Divide numerator and denominator by x to get

$$\lim_{x \to \infty} \frac{3x^2 - 6x + 2}{2x - 9} = \lim_{x \to \infty} \frac{3x - 6 + 2/x}{2 - 9x} = \frac{\infty}{2} = \infty$$

11. $\lim_{x \to \infty} (2 - 3e^x) = 2 - 3 \lim_{x \to \infty} e^x = 2 - \infty = -\infty$

13. Since $\lim_{x \to \infty} e^{-8x} = 0$, it follows that

$$\lim_{x \to \infty} \frac{3}{2 + 5e^{-8x}} = \frac{3}{2 + 0} = \frac{3}{2}$$

15. Since the limit is of the form ∞/∞, L'Hôpital's rule gives

$$\lim_{x \to \infty} \frac{e^{2x}}{3x} = \lim_{x \to \infty} \frac{2e^{2x}}{3} = \frac{\infty}{3} = \infty$$

17. $\lim\limits_{x\to\infty} \frac{2x}{3e^{-x}} = \frac{2}{3} \lim\limits_{x\to\infty} xe^x = \frac{2}{3}(\infty)(\infty) = \infty$

19. Since the limit is of the form ∞/∞, L'Hôpital's rule (applied twice) gives

$$\lim_{x\to\infty} \frac{x^2}{e^x} = \lim_{x\to\infty} \frac{2x}{e^x} = \lim_{x\to\infty} \frac{2}{e^x} = 0$$

21. Since the limit is of the form ∞/∞, L'Hôpital's rule gives

$$\lim_{x\to\infty} \frac{\sqrt{x}}{e^x} = \lim_{x\to\infty} \frac{1/2\sqrt{x}}{e^x} = \lim_{x\to\infty} \frac{1}{2\sqrt{x}e^x} = \frac{1}{\infty} = 0$$

23. Since the limit is of the form ∞/∞, L'Hôpital's rule gives

$$\lim_{x\to\infty} \frac{\ln x}{x} = \lim_{x\to\infty} \frac{1/x}{1} = \lim_{x\to\infty} \frac{1}{x} = \frac{1}{\infty} = 0$$

25. Since the limit is of the form ∞/∞, L'Hôpital's rule (applied twice) gives

$$\lim_{x\to\infty} \frac{\ln(x^2+1)}{x} = \lim_{x\to\infty} \frac{2x/(x^2+1)}{1} = \lim_{x\to\infty} \frac{2x}{x^2+1} = \lim_{x\to\infty} \frac{2}{2x} = 0$$

27. Since the limit is of the form ∞/∞, L'Hôpital's rule (applied twice) gives

$$\lim_{x\to\infty} \frac{\ln(2x+1)}{\ln(3x-1)} = \lim_{x\to\infty} \frac{2/(2x+1)}{3/(3x-1)} = \lim_{x\to\infty} \frac{6x-2}{6x+3} = \lim_{x\to\infty} \frac{6}{6} = 1$$

29. Since the limit is of the form ∞/∞, L'Hôpital's rule (applied twice) gives

$$\lim_{x\to\infty} \frac{e^x}{x \ln x} = \lim_{x\to\infty} \frac{e^x}{1 + \ln x} = \lim_{x\to\infty} \frac{e^x}{1/x} = \lim_{x\to\infty} xe^x = \infty$$

31. Since the limit is of the form ∞/∞, L'Hôpital's rule (applied twice) gives

$$\lim_{x\to\infty} \frac{x}{e^{\sqrt{x}}} = \lim_{x\to\infty} \frac{1}{(1/(2\sqrt{x}))e^{\sqrt{x}}} = \lim_{x\to\infty} \frac{2\sqrt{x}}{e^{\sqrt{x}}}$$

$$= \lim_{x\to\infty} \frac{2/(2\sqrt{x})}{(1/(2\sqrt{x}))e^{\sqrt{x}}} = \lim_{x\to\infty} \frac{2}{e^{\sqrt{x}}} = 0$$

33. The limit is of the form 0∞. Rewrite the product as a quotient and apply L'Hôpital's rule to get

$$\lim_{x\to\infty} e^{2x} \ln x = \lim_{x\to\infty} \frac{\ln x}{e^{2x}} = \lim_{x\to\infty} \frac{1/x}{2e^{2x}} = \lim_{x\to\infty} \frac{1}{2xe^{2x}} = 0$$

35. Since the limit is of the form ∞^0, let $y = x^{2/x}$ and take logarithms to get

$$\ln y = \ln x^{2/x} = \frac{2 \ln x}{x}$$

Since the limit of $\ln y$ is of the form ∞/∞, apply L'Hôpital's rule to get

$$\lim_{x \to \infty} \ln y = \lim_{x \to \infty} \frac{2 \ln x}{x} = \lim_{x \to \infty} \frac{2/x}{1} = 0$$

Since $\ln y \to 0$, it follows that $y \to e^0 = 1$.

37. Since the limit is of the form 1^∞, let $y = (1 + \frac{2}{x})^x$ and take logarithms to get

$$\ln y = \ln(1 + \frac{2}{x})^x = x \ln(1 + \frac{2}{x}) = \frac{\ln(1 + 2/x)}{1/x}$$

Since the limit of $\ln y$ is of the form $0/0$, apply L'Hôpital's rule to get

$$\lim_{x \to \infty} \ln y = \lim_{x \to \infty} \frac{\ln(1 + 2/x)}{1/x} = \lim_{x \to \infty} \frac{\left[\frac{1}{1+2/x}\right](-2/x^2)}{-1/x^2}$$

$$= \lim_{x \to \infty} \frac{2}{1 + 2/x} = \frac{2}{1 + 0} = 2$$

Since $\ln y \to 2$, it follows that $y \to e^2$.

39. If p is negative, then $p = -n$, where n is some positive number, and so

$$\lim_{x \to \infty} x^p e^{-kx} = \lim_{x \to \infty} \frac{1}{x^n e^{kx}} = \frac{1}{\infty} = 0$$

If $p = 0$, then

$$\lim_{x \to \infty} x^p e^{-kx} = \lim_{x \to \infty} \frac{1}{e^{kx}} = \frac{1}{\infty} = 0$$

If $p > 0$, then

$$\lim_{x \to \infty} x^p e^{-kx} = \lim_{x \to \infty} \frac{x^p}{e^{kx}} = \frac{\infty}{\infty}$$

Each time L'Hôpital's rule is applied, the power of x in the numerator decreases by 1 until eventually it becomes zero or negative. When that occurs, the limit is like one of the two cases already considered and hence is equal to zero.

Chapter 8, Section 2

1. $\int_1^\infty \frac{1}{x^3}\,dx = \lim_{N\to\infty} \int_1^N \frac{1}{x^3}\,dx = \lim_{N\to\infty} -\frac{1}{2x^2}\Big|_1^N = \lim_{N\to\infty}\left[-\frac{1}{2N^2} + \frac{1}{2}\right] = \frac{1}{2}$

3. $\int_1^\infty \frac{1}{\sqrt{x}}\,dx = \lim_{N\to\infty} \int_1^N x^{-1/2}\,dx = \lim_{N\to\infty} 2x^{1/2}\Big|_1^N = \lim_{N\to\infty}(2N^{1/2} - 2) = \infty$

5. $\int_3^\infty \frac{1}{2x-1} = \lim_{N\to\infty} \int_3^N \frac{1}{2x-1}\,dx = \frac{1}{2}\lim_{N\to\infty} \ln|2x-1|\Big|_3^N$

    $= \frac{1}{2}\lim_{N\to\infty}[\ln(2N-1) - \ln 5] = \infty$

7. $\int_3^\infty \frac{1}{(2x-1)^2}\,dx = \lim_{N\to\infty} \int_3^N (2x-1)^{-2}\,dx$

    $= \frac{1}{2}\lim_{N\to\infty} -\frac{1}{2x-1}\Big|_3^N = -\frac{1}{2}\lim_{N\to\infty}\left[\frac{1}{2N-1} - \frac{1}{5}\right] = \frac{1}{10}$

9. $\int_0^\infty 5e^{-2x}\,dx = 5\lim_{N\to\infty} \int_0^N e^{-2x}\,dx = -\frac{5}{2}\lim_{N\to\infty} e^{-2x}\Big|_0^N$

    $= -\frac{5}{2}\lim_{N\to\infty}(e^{-2N} - 1) = \frac{5}{2}$

11. $\int_1^\infty \frac{x^2}{(x_2^3)^2}\,dx = \lim_{N\to\infty} \frac{1}{3}\int_1^N 3x^2(x^3+2)^{-2}\,dx$

    $= \frac{1}{3}\lim_{N\to\infty} -\frac{1}{x^3+2}\Big|_1^N = \frac{1}{3}\lim_{N\to\infty}\left[-\frac{1}{N^3+2} + \frac{1}{3}\right] = \frac{1}{9}$

13. $\int_1^\infty \frac{x^2}{\sqrt{x^3+2}}\,dx = \lim_{N\to\infty} \frac{1}{3}\int_1^N 3x^2(x^3+2)^{-1/2}\,dx$

    $= \frac{1}{3}\lim_{N\to\infty} 2(x^3+2)^{1/2}\Big|_1^N = \frac{1}{3}\lim_{N\to\infty}[2(N^3+2)^{1/2} - 2(3)^{1/2}] = \infty$

15. $\int_1^\infty \frac{e^{-\sqrt{x}}}{\sqrt{x}}\,dx = \lim_{N\to\infty} 2\int_1^N e^{-\sqrt{x}}\left[\frac{1}{2\sqrt{x}}\right]\,dx$

    $= -2\lim_{N\to\infty} e^{-\sqrt{x}}\Big|_1^N = -2\lim_{N\to\infty}\left[e^{-\sqrt{N}} - e^{-1}\right] = \frac{2}{e}$

17. $\int_0^\infty 2xe^{-3x}\,dx = \lim_{N\to\infty} \int_0^N 2xe^{-3x}\,dx$

    $= \lim_{N\to\infty}\left[-\frac{2}{3}xe^{-3x}\Big|_0^N + \frac{2}{3}\int_0^N e^{-3x}\,dx\right]$ (by parts)

    $= \lim_{N\to\infty}\left(-\frac{2}{3}xe^{-3x} - \frac{2}{9}e^{-3x}\right)\Big|_0^N = \lim_{N\to\infty} -\frac{2}{3}e^{-3x}\left(x + \frac{1}{3}\right)\Big|_0^N$

$$= \lim_{N \to \infty} \left[-\tfrac{2}{3} e^{-3N}(N + \tfrac{1}{3}) + \tfrac{2}{3}(\tfrac{1}{3}) \right] = \tfrac{2}{9}$$

19. $\int_0^\infty 5x e^{10-x}\, dx = 5e^{10} \lim_{N \to \infty} \int_0^N x e^{-x}\, dx$

$$= 5e^{10} \lim_{N \to \infty} \left[-x e^{-x} \Big|_0^N + \int_0^N e^{-x}\, dx \right] \qquad \text{(by parts)}$$

$$= 5e^{10} \lim_{N \to \infty} (-x e^{-x} - e^{-x}) \Big|_0^N$$

$$= 5e^{10} \lim_{N \to \infty} [(-N e^{-N} - e^{-N}) - (0 - 1)] = 5e^{10}$$

21. $\int_2^\infty \frac{1}{x \ln x}\, dx = \lim_{N \to \infty} \int_0^N \frac{1}{\ln x}(\frac{1}{x})\, dx$

$$= \lim_{N \to \infty} \ln |\ln x| \Big|_2^N \qquad \text{(by substitution with } u = \ln x\text{)}$$

$$= \lim_{N \to \infty} [\ln |\ln N| - \ln |\ln 2|] = \infty$$

23. $\int_0^\infty x^2 e^{-x}\, dx = \lim_{N \to \infty} \int_0^N x^2 e^{-x}\, dx$

$$= \lim_{N \to \infty} \left[-x^2 e^{-x} \Big|_0^N + 2 \int_0^N x e^{-x}\, dx \right] \qquad \text{(by parts)}$$

$$= \lim_{N \to \infty} \left[(-x^2 e^{-x} - 2x e^{-x}) \Big|_0^N + \int_0^N 2 e^{-x}\, dx \right] \qquad \text{(by parts again)}$$

$$= \lim_{N \to \infty} (-x^2 - 2x - 2) e^{-x} \Big|_0^N$$

$$= \lim_{N \to \infty} [(-N^2 - 2N - 2) e^{-N} - (-2)] = 2$$

25.   To find the present value of the investment of $2,400 per year for N years, divide the N-year interval $0 \le t \le N$ into n equal subintervals and length Δt years, and let t_j denote the beginning of the jth subinterval. Then, during the jth subinterval,

$$\text{Amount generated} \simeq 2,400 \Delta t$$

and

$$\text{Present value} \simeq 2,400 e^{-0.12 t_j} \Delta t$$

Hence, the present value of an N-year investment is

$$\lim_{n \to \infty} \sum_{j=1}^{n} 2,400 e^{-0.12t_j} \Delta t = \int_0^N 2,400 e^{-0.12t} \, dt$$

To find the present value P of the total investment, let $N \to \infty$ to get

$$P = \lim_{N \to \infty} \int_0^N 2,400 e^{-0.12t} \, dt = -\frac{2,400}{0.12} \lim_{N \to \infty} e^{-0.12t} \Big|_0^N$$

$$= -\frac{2,400}{0.12} \lim_{N \to \infty} [e^{-0.12N} - 1] = \frac{2,400}{0.12} = \$20,000$$

27. To find the present value of an apratment complex generating $f(t) = 10,000 + 500t$ dollars per year for N years, divide the N-year time interval $0 \le t \le N$ into n equal subintervals of length Δt years, and let t_j denote the beginning of the jth subinterval. Then, during the jth subinterval,

$$\text{Amount generated} \simeq f(t_j)\,\Delta t$$

and, at an interest rate of 10 percent,

$$\text{Present value} \simeq f(t_j)e^{-0.1t_j}\Delta t$$

Hence, the present value of the apartment complex over an N-year period is

$$\lim_{n \to \infty} \sum_{j=1}^{n} f(t_j)e^{-0.1t_j}\Delta t = \int_0^N f(t)e^{-0.1t} \, dt$$

To find the present value P of the total income, let $N \to \infty$ to get

$$P = \lim_{N \to \infty} \int_0^N (10,000 + 500t)e^{-0.1t} \, dt$$

$$= \lim_{N \to \infty} \left[-10(10,000 + 500t)e^{-0.1t}\Big|_0^N - \int_0^N -5,000e^{-0.1t} \, dt \right]$$

$$= \lim_{N \to \infty} (-100,000 - 5,000t - 50,000)e^{-0.1t}\Big|_0^N$$

$$= \lim_{N \to \infty} (-150,000 - 5,000t)e^{-0.1t}\Big|_0^N$$

$$= -5,000 \lim_{N \to \infty} [(30 + N)e^{-0.1N} - 30] = -5,000(-30) = \$150,000$$

29. To find the present value of an investment generating $f(t) = A + Bt$ dollars per year for N years, divide the N-year time interval $0 \le t \le N$ into n equal subintervals of length Δt years and let t_j denote the beginning of the jth subinterval. Then, during the jth subinterval,

$$\text{Amount generated} \simeq f(t_j)\,\Delta t$$

and

$$\text{Present value} \simeq f(t_j)e^{-rt_j}\,\Delta t$$

Hence, the present value of an N-year investment is

$$\lim_{n \to \infty} \sum_{j=1}^{n} f(t_j)e^{-rt_j}\,\Delta t = \int_{0}^{N} (A + Bt)e^{-rt}\,dt$$

To find the present value P of the total investment, let $N \to \infty$ to get

$$P = \lim_{N \to \infty} \int_{0}^{N} (A + Bt)e^{-rt}\,dt$$

$$= \lim_{N \to \infty} \left[-\frac{1}{r}(A + Bt)e^{-rt} \Big|_{0}^{N} + \frac{B}{r}\int_{0}^{N} e^{-rt}\,dt \right] \qquad \text{(by parts)}$$

$$= \lim_{N \to \infty} \left[-\frac{1}{r}(A + Bt)e^{-rt} - \frac{B}{r^2}e^{-rt} \right]\Big|_{0}^{N}$$

$$= -\frac{1}{r} \lim_{N \to \infty} (A + Bt) + \frac{B}{r})e^{-rt} \Big|_{0}^{N}$$

$$= -\frac{1}{r} \lim_{N \to \infty} \left[(A + BN + \frac{B}{r})e^{-rN} - (A + \frac{B}{r}) \right] = \frac{1}{r}(A + \frac{B}{r}) = \frac{A}{r} + \frac{B}{r^2}$$

31. To find the number of patients after N months, divide the N-month time interval $0 \leq t \leq N$ into n equal subintervals of length Δt months and let t_j denote the beginning of the jth subinterval. Then, the number of people starting treatment during the jth subinterval is approximately $10\,\Delta t$. Of these, the number still receiving treatment at time $t = N$ (that is, $N - t_j$ months later) is approximately $10f(N - t_j)\,\Delta t$. Hence, the number of patients receiving treatment at time $t = N$ is

$$\lim_{n \to \infty} \sum_{j=1}^{n} 10f(N - t_j)\,\Delta t = \int_{0}^{N} 10f(N - t)\,dt$$

and the number P of patients receiving treatment in the long run is

$$P = \lim_{N \to \infty} 10 \int_0^N e^{-(N-t)/20} \, dt$$

$$= \lim_{N \to \infty} 10 e^{-N/20} \int_0^N e^{t/20} \, dt = \lim_{N \to \infty} 200 e^{-N/20} e^{t/20} \Big|_0^N$$

$$= \lim_{N \to \infty} 200 e^{-N/20} (e^{N/20} - 1) = \lim_{N \to \infty} 200(1 - e^{-N/20}) = 200 \text{ patients}$$

33. To find the number of units of the drug in the patient's body after N hours, divide the
     N-hour time interval $0 \le t \le N$ into n equal subintervals of length Δt hours, and
     let t_j denote the beginning of the jth subinterval. Then, during the jth subinterval,
     approximately $5\Delta t$ units of the drug are received. Of these, the number remaining at
     time $t = N$ (that is, $N - t_j$ hours later) is approximately $5f(N - t_j)\Delta t$. Hence, the
     number of units of the drug in the patient's body at time $t = N$ is

$$\lim_{n \to \infty} \sum_{j=1}^{n} 5f(N - t_j) \Delta t = \int_0^N 5f(N - t) \, dt$$

and the number Q of units in the patient's body in the long run is

$$Q = \lim_{N \to \infty} \int_0^N 5f(N - t) dt = \lim_{N \to \infty} \int_0^N 5e^{-(N-t)/10} \, dt$$

$$= \lim_{N \to \infty} 5 e^{-N/10} \int_0^N e^{t/10} \, dt = \lim_{N \to \infty} 50 e^{-N/10} (e^{t/10}) \Big|_0^N$$

$$= \lim_{N \to \infty} 50 e^{-N/10} (e^{N/10} - 1) = \lim_{N \to \infty} 50(1 - e^{-N/10}) = 50 \text{ units}$$

Chapter 8, Section 3

1. (a) $P(2 \le x \le t) = \int_2^5 f(x)\,dx = \int_2^5 \frac{1}{3}\,dx = \frac{1}{3}x\big|_2^5 = 1$

    (b) $P(3 \le x \le 4) = \int_3^4 f(x)\,dx = \int_3^4 \frac{1}{3}\,dx = \frac{1}{3}x\big|_3^4 = \frac{1}{3}$

    (c) $P(x \ge 4) = \int_4^\infty f(x)\,dx = \int_4^5 \frac{1}{3}\,dx = \frac{1}{3}x\big|_4^5 = \frac{1}{3}$

3. (a) $P(0 \le x \le 4) = \int_0^4 f(x)\,dx = \int_0^4 \frac{1}{8}(4-x)\,dx$

$$= \frac{1}{8}(4x - \frac{1}{2}x^2)\big|_0^4 = \frac{1}{8}\left[16 - \frac{16}{2}\right] = 1$$

    (b) $P(2 \le x \le 3) = \int_2^3 f(x)\,dx = \int_2^3 \frac{1}{8}(4-x)\,dx$

$$= \frac{1}{8}(4x - \frac{1}{2}x^2)\big|_2^3 = \frac{1}{8}\left[(12 - \frac{9}{2}) - (8 - \frac{4}{2})\right] = \frac{3}{16}$$

    (c) $P(x \ge 1) = \int_1^\infty f(x)\,dx = \int_1^4 \frac{1}{8}(4-x)\,dx$

$$= \frac{1}{8}(4x - \frac{1}{2}x^2)\big|_1^4 = \frac{1}{8}\left[(16 - \frac{16}{2}) - (4 - \frac{1}{2})\right] = \frac{9}{16}$$

5. (a) $P(1 \le x < \infty) = \int_1^\infty f(x)\,dx = \int_1^\infty \frac{3}{x^4}\,dx = \lim_{N\to\infty} \int_1^N \frac{3}{x^4}\,dx$

$$= \lim_{N\to\infty} -\frac{1}{x^3}\Big|_1^N = \lim_{N\to\infty}\left[-\frac{1}{N^3} + 1\right] = 1$$

    (b) $P(1 \le x \le 2) = \int_1^2 f(x)\,dx = \int_1^2 \frac{3}{x^4}\,dx = -\frac{1}{x^3}\Big|_1^2 = \frac{7}{8}$

    (c) $P(x \ge 2) = \int_2^\infty f(x)\,dx = \lim_{N\to\infty} \int_2^N \frac{3}{x^4}\,dx$

$$= \lim_{N\to\infty} -\frac{1}{x^3}\Big|_2^N = \lim_{N\to\infty}\left[-\frac{1}{N^3} + \frac{1}{8}\right] = \frac{1}{8}$$

7. (a) $P(x \ge 0) = \int_0^\infty f(x)\,dx = \lim_{N\to\infty} \int_0^N 2xe^{-x^2}\,dx$

$$= \lim_{N\to\infty} -e^{-x^2}\Big|_0^N = \lim_{N\to\infty}[-e^{-N^2} + 1] = 1$$

    (b) $P(1 \le x \le 2) = \int_1^2 f(x)\,dx = \int_1^2 2xe^{-x^2}\,dx$

$$= -e^{-x^2}\Big|_1^2 = -e^{-4} + e^{-1} \simeq 0.3496$$

(c) $P(x \leq 2) = \int_{-\infty}^{2} f(x)\,dx = \int_{0}^{2} 2xe^{-x^2}\,dx$

$$= -e^{-x^2}\Big|_{0}^{2} = -e^{-4} + 1 \simeq 0.9817$$

9.   Let x denote the time (in seconds) you must wait. The uniform density function for x is

$$f(x) = \begin{cases} \frac{1}{45} & \text{if } 0 \leq x \leq 45 \\ 0 & \text{otherwise} \end{cases}$$

Hence, the probability that the light turns green within 15 seconds is

$$P(0 \leq x \leq 15) = \int_{0}^{15} \frac{1}{45}\,dx = \frac{x}{45}\Big|_{0}^{15} = \frac{1}{3}$$

11.   Let x denote the number of minutes since the start of the movie at the time of your arrival. The uniform density function for x is

$$f(x) = \begin{cases} \frac{1}{120} & \text{if } 0 \leq x \leq 120 \\ 0 & \text{otherwise} \end{cases}$$

Hence, the probability that you arrive within 10 minutes (before or after) of the start of a movie is

$$P(0 \leq x \leq 10) + P(110 \leq x \leq 120) = 2P(0 \leq x \leq 10)$$

$$= 2\int_{0}^{10} \frac{1}{120}\,dx = \frac{x}{60}\Big|_{0}^{10} = \frac{1}{6}$$

13. $P(x \geq 8) = \int_{8}^{\infty} f(x)\,dx = \lim_{N \to \infty} \int_{8}^{N} \frac{1}{4}e^{-x/4}\,dx$

$$= \lim_{N \to \infty} -e^{-x/4}\Big|_{8}^{N} = \lim_{N \to \infty} [-e^{-N/4} + e^{-2}] = e^{-2} \simeq 0.1353$$

15.   For the probability density function

$$f(x) = \begin{cases} \frac{1}{3} & \text{if } 2 \leq x \leq 5 \\ 0 & \text{otherwise} \end{cases}$$

$$E(x) = \int_{-\infty}^{\infty} xf(x)\,dx = \int_{2}^{5} \frac{x}{3}\,dx = \frac{1}{6}x^2\Big|_{2}^{5} = \frac{1}{6}(25 - 4) = \frac{7}{2}$$

and

$$Var(x) = \int_{-\infty}^{\infty} x^2 f(x)\, dx - [E(x)]^2 = \int_{2}^{5} \frac{1}{3} x^2\, dx - \left[\frac{7}{2}\right]^2$$

$$= \frac{1}{9} x^3 \Big|_{2}^{5} - \frac{49}{4} = \frac{1}{9}(125 - 8) - \frac{49}{4} = \frac{3}{4}$$

17. For the probability density function

$$f(x) = \begin{cases} \frac{1}{8}(4 - x) & \text{if } 0 \le x \le 4 \\ 0 & \text{otherwise} \end{cases}$$

$$E(x) = \int_{-\infty}^{\infty} x f(x)\, dx = \int_{0}^{4} \frac{1}{8} x(4 - x)\, dx = \frac{1}{8} \int_{0}^{4} (4x - x^2)\, dx$$

$$= \frac{1}{8}\left(2x^2 - \frac{1}{3}x^3\right)\Big|_{0}^{4} = \frac{1}{8}\left[32 - \frac{64}{3}\right] = \frac{4}{3}$$

and

$$Var(x) = \int_{-\infty}^{\infty} x^2 f(x)\, dx - [E(x)]^2 = \int_{0}^{4} \frac{1}{8} x^2(4 - x)\, dx - \left[\frac{4}{3}\right]^2$$

$$= \frac{1}{8} \int_{0}^{4} (4x^2 - x^3)\, dx - \frac{16}{9} = \frac{1}{8}\left(\frac{4}{3}x^3 - \frac{1}{4}x^4\right)\Big|_{0}^{4} - \frac{16}{9}$$

$$= \frac{1}{8}\left[\frac{256}{3} - \frac{256}{4}\right] - \frac{16}{9} = \frac{8}{9}$$

19. For the probability density function

$$f(x) = \begin{cases} \frac{1}{10} e^{-x/10} & \text{if } x \ge 0 \\ 0 & \text{if } x < 0 \end{cases}$$

$$E(x) = \int_{-\infty}^{\infty} x f(x)\, dx = \lim_{N \to \infty} \int_0^N \frac{1}{10} x e^{-x/10}\, dx$$

$$= \frac{1}{10} \lim_{N \to \infty} (-10 x e^{-x/10} - 100 e^{-x/10}) \Big|_0^N \quad \text{(by parts)}$$

$$= \frac{1}{10} \lim_{N \to \infty} [(-10 N e^{-N/10} - 100 e^{-N/10}) - (-100)] = 10$$

and

$$Var(x) = \int_{-\infty}^{\infty} x^2 f(x)\, dx - [E(x)]^2 = \lim_{N \to \infty} \int_0^N \frac{1}{10} x^2 e^{-x/10}\, dx - 100$$

$$= \frac{1}{10} \lim_{N \to \infty} (-10 x^2 - 200 x - 2{,}000) e^{-x/10} \Big|_0^N - 100 \qquad \text{(by parts twice)}$$

$$= \frac{1}{10} \lim_{N \to \infty} [(-10 N^2 - 200 N - 2{,}000) e^{-N/10} - (-2{,}000)] - 100 = 100$$

21. Let x denote your waiting time (in minutes). The probability density function for x is

$$f(x) = \begin{cases} \frac{1}{20} & \text{if } 0 \leq x \leq 20 \\ 0 & \text{otherwise} \end{cases}$$

Hence, the average waiting time is

$$E(x) = \int_{-\infty}^{\infty} x f(x)\, dx = \int_0^{20} \frac{x}{20}\, dx = \frac{1}{40} x^2 \Big|_0^{20} = 10 \text{ minutes}$$

23. From Problem 13, the probability density function is

$$f(x) = \begin{cases} \frac{1}{4} e^{-x/4} & \text{if } x \geq 0 \\ 0 & \text{if } x < 0 \end{cases}$$

where x is the waiting time (in minutes). Hence, the average waiting time is

$$E(x) = \int_{-\infty}^{\infty} x f(x)\, dx = \lim_{N \to \infty} \int_{0}^{N} \frac{1}{4} x e^{-x/4}\, dx$$

$$= \frac{1}{4} \lim_{N \to \infty} (-4x e^{-x/4} - 16 e^{-x/4}) \Big|_{0}^{N}$$

$$= \frac{1}{4} \lim_{N \to \infty} [(-4N e^{-N/4} - 16 e^{-N/4} - (-16)] = 4 \text{ minutes}$$

25. For the probability density function

$$f(x) = \begin{cases} \frac{1}{B-A} & \text{if } A \leq x \leq B \\ 0 & \text{otherwise} \end{cases}$$

it was shown in Problem 24 that $E(x) = \frac{A+B}{2}$. Hence,

$$Var(x) = \int_{-\infty}^{\infty} x^2 f(x)\, dx - [E(x)]^2 = \int_{A}^{B} x^2 \left[\frac{1}{B-A} \right] dx - \left[\frac{A+B}{2} \right]^2$$

$$= \frac{1}{B-A} \int_{A}^{B} x^2\, dx - \left[\frac{A+B}{2} \right]^2$$

$$= \frac{1}{B-A} \frac{1}{3} x^3 \Big|_{A}^{B} - \left[\frac{A+B}{2} \right]^2 = \frac{B^3 - A^3}{3(B-A)} - \left[\frac{A+B}{2} \right]^2$$

$$= \frac{(B-A)(B^2 + AB + A^2)}{3(B-A)} - \frac{(A+B)^2}{4}$$

$$= \frac{B^2 + AB + A^2}{3} - \frac{A^2 + 2AB + B^2}{4} = \frac{B^2 - 2AB + A^2}{12} = \frac{(B-A)^2}{12}$$

27. For the exponential probability density function

$$f(x) = \begin{cases} k e^{-kx} & \text{if } x \geq 0 \\ 0 & \text{otherwise} \end{cases}$$

it was shown in Problem 26 that $E(x) = \frac{1}{k}$. Hence,

$$Var(x) = \int_{-\infty}^{\infty} x^2 f(x)\, dx - [E(x)]^2 = \lim_{N \to \infty} \int_0^N kx^2 e^{-kx}\, dx - \frac{1}{k^2}$$

$$= k \lim_{N \to \infty} \left(-\frac{x^2}{k} - \frac{2x}{k^2} - \frac{2}{k^3} \right) e^{-kx} \bigg|_0^N - \frac{1}{k^2} \qquad \text{(by parts twice)}$$

$$= k \lim_{N \to \infty} \left[\left(-\frac{N^2}{k} - \frac{2N}{k^2} - \frac{2}{k^3} \right) e^{-kN} - \left(-\frac{2}{k^3} \right) \right] - \frac{1}{k^2}$$

$$= k \left(\frac{2}{k^3} \right) - \frac{1}{k^2} = \frac{1}{k^2}$$

Chapter 8, Review Problems

1. $\displaystyle\lim_{x\to\infty}(x^2+1)(2-x^4)=\lim_{x\to\infty}(-x^6)=-\infty$

2. $\displaystyle\lim_{x\to\infty}\frac{1+x^2-3x^3}{2x^3+5}=\lim_{x\to\infty}\frac{1/x^3+1/x-3}{2+\frac{5}{x^3}}=-\frac{3}{2}$

3. $\displaystyle\lim_{x\to\infty}\frac{(x+1)(2-x)}{x+3}=\lim_{x\to\infty}\frac{2+x-x^2}{x+3}=\lim_{n\to\infty}\frac{2/x+1-x}{1+3/x}=-\infty$

4. $\displaystyle\lim_{x\to\infty}\frac{2}{3+4e^{-5x}}=\frac{2}{3+0}=\frac{2}{3}$

5.   Since the limit is of the form ∞/∞, L'Hôpital's rule (applied twice) gives

$$\lim_{x\to\infty}\frac{x^2+1}{e^{2x}}=\lim_{x\to\infty}\frac{2x}{2e^{2x}}=\lim_{x\to\infty}\frac{1}{2e^{2x}}=0$$

6. $\displaystyle\lim_{x\to.\infty}\frac{x^2-1}{e^{-2x}}=\lim_{x\to\infty}(x^2-1)e^{2x}=\infty$

7. $\displaystyle\lim_{x\to\infty}\frac{1+e^{-x}}{e^{-x}}=\lim_{x\to\infty}(e^x+1)=\infty$

8.   Since the limit is of the form ∞/∞, L'Hôpital's rule gives

$$\lim_{x\to\infty}\frac{\ln(3x+1)}{\ln(x^2-4)}=\lim_{n\to\infty}\frac{3/(3x+1)}{2x/(x^2-4)}=\lim_{x\to\infty}\frac{3x^2-12}{6x^2-2x}=\frac{3}{6}=\frac{1}{2}$$

9.   Since the limit is of the form ∞/∞, L'Hôpital's rule (applied twice) gives

$$\lim_{x\to\infty}\frac{e^{\sqrt{x}}}{2x}=\lim_{x\to\infty}\frac{e^{\sqrt{x}}[1/2\sqrt{x}]}{2}=\lim_{x\to\infty}\frac{e^{\sqrt{x}}}{4\sqrt{x}}$$

$$=\lim_{x\to\infty}\frac{e^{\sqrt{x}}[1/2\sqrt{x}]}{4[1/2\sqrt{x}]}=\lim_{x\to\infty}\frac{e^{\sqrt{x}}}{4}=\infty$$

10.   Since the limit is of the form $\infty 0$, rewrite the product as

$$x^2(e^{1/x}-1)=\frac{e^{1/x}-1}{1/x^2}$$

which is a quotient with limit of the form $0/0$. By L'Hôpital's rule,

$$\lim_{x\to\infty}\frac{e^{1/x}-1}{1/x^2}=\lim_{x\to\infty}\frac{e^{1/x}(-1x^2)}{-2/x^3}=\lim_{x\to\infty}\frac{xe^{1/x}}{2}=\infty$$

11.   Since the limit is of the form ∞^0, let $y=x^{1/x^2}$ and take logarithms to get

$$\lim_{x\to\infty}\ln y=\lim_{x\to\infty}\frac{\ln x}{x^2}=\lim_{x\to\infty}\frac{1/x}{2x}=\lim_{x\to\infty}\frac{1}{2x^2}=0$$

Since $\ln y \to 0$, it follows that $y \to e^0 = 1$.

12.   Since the limit is of the form 1^∞, let $y = (1 + \frac{2}{x})^{3x}$ and take logarithms to get

$$\ln y = \ln(1 + \frac{2}{x})^{3x} = 3x \ln(1 + \frac{2}{x}) = \frac{\ln(1 + 2/x)}{1/3x}$$

The limit of this quotient is of the form $0/0$, and so, by L'Hôpital's rule,

$$\lim_{x \to \infty} \ln y = \lim_{x \to \infty} \frac{\ln(1 + 2/x)}{1/3x} = \lim_{x \to \infty} \frac{\left[\frac{1}{1+2/x}\right](-2/x^2)}{-1/3x^2}$$

$$= \lim_{x \to \infty} \frac{6}{1 + 2/x} = 6$$

Since $\ln y \to 6$, it follows that $y \to e^6$.

13. $\int_0^\infty \frac{1}{\sqrt[3]{1+2x}} \, dx = \lim_{N \to \infty} \int_0^N (1 + 2x)^{-1/3} \, dx$

$$= \lim_{N \to \infty} \left[\tfrac{3}{4}(1 + 2x)^{2/3}\right]_0^N = \tfrac{3}{4} \lim_{N \to \infty} [(1 + 2N)^{2/3} - 1] = \infty$$

14. $\int_0^\infty (1 + 2x)^{-3/2} \, dx = \lim_{N \to \infty} \int_0^N (1 + 2x)^{-3/2} \, dx$

$$= \lim_{N \to \infty} \left[-(1 + 2x)^{-1/2}\right]_0^N = \lim_{N \to \infty} [-(1 + 2N)^{-1/2} + 1] = 1$$

15. $\int_0^\infty \frac{3x}{x^2+1} \, dx = 3 \lim_{N \to \infty} \int_0^N \frac{x}{x^2+1} \, dx$

$$= 3 \lim_{N \to \infty} \left[\tfrac{1}{2} \ln(x^2 + 1)\right]_0^N = \tfrac{3}{2} \lim_{N \to \infty} [\ln(N^2 + 1) - \ln 1] = \infty$$

16. $\int_0^\infty 3e^{-5x} \, dx = 3 \lim_{N \to \infty} \int_0^N e^{-5x} \, dx$

$$= -\tfrac{3}{5} \lim_{N \to \infty} e^{-5x} \Big|_0^N = -\tfrac{3}{5} \lim_{N \to \infty} [e^{-5N} - 1] = \tfrac{3}{5}$$

17. $\int_0^\infty xe^{-2x} \, dx = \lim_{N \to \infty} \int_0^N xe^{-2x} \, dx = \lim_{N \to \infty} \left[-\tfrac{1}{2}xe^{-2x}\Big|_0^N + \tfrac{1}{2} \int_0^N e^{-2x} \, dx\right]$

$$= \lim_{N \to \infty} \left[-\tfrac{1}{2}xe^{-2x} - \tfrac{1}{4}e^{-2x}\right]\Big|_0^N = \lim_{N \to \infty} \left[(-\tfrac{1}{2}Ne^{-2N} - \tfrac{1}{4}e^{-2N}) + \tfrac{1}{4}\right] = \tfrac{1}{4}$$

18. $\int_0^\infty 2x^2 e^{-x^3} \, dx = 2 \lim_{N \to \infty} \int_0^N x^2 e^{-x^3} \, dx$

Since $\ln y \to 0$, it follows that $y \to e^0 = 1$.

12. Since the limit is of the form 1^∞, let $y = (1 + \frac{2}{x})^{3x}$ and take logarithms to get

$$\ln y = \ln(1 + \frac{2}{x})^{3x} = 3x \ln(1 + \frac{2}{x}) = \frac{\ln(1 + 2/x)}{1/3x}$$

The limit of this quotient is of the form $0/0$, and so, by L'Hôpital's rule,

$$\lim_{x \to \infty} \ln y = \lim_{x \to \infty} \frac{\ln(1 + 2/x)}{1/3x} = \lim_{x \to \infty} \frac{\left[\frac{1}{1+2/x}\right](-2/x^2)}{-1/3x^2}$$

$$= \lim_{x \to \infty} \frac{6}{1 + 2/x} = 6$$

Since $\ln y \to 6$, it follows that $y \to e^6$.

13. $\int_0^\infty \frac{1}{\sqrt[3]{1+2x}} \, dx = \lim_{N \to \infty} \int_0^N (1 + 2x)^{-1/3} \, dx$

$$= \lim_{N \to \infty} \left[\frac{3}{4}(1 + 2x)^{2/3}\right]_0^N = \frac{3}{4} \lim_{N \to \infty} [(1 + 2N)^{2/3} - 1] = \infty$$

14. $\int_0^\infty (1 + 2x)^{-3/2} \, dx = \lim_{N \to \infty} \int_0^N (1 + 2x)^{-3/2} \, dx$

$$= \lim_{N \to \infty} \left[-(1 + 2x)^{-1/2}\right]_0^N = \lim_{N \to \infty} [-(1 + 2N)^{-1/2} + 1] = 1$$

15. $\int_0^\infty \frac{3x}{x^2+1} \, dx = 3 \lim_{N \to \infty} \int_0^N \frac{x}{x^2+1} \, dx$

$$= 3 \lim_{N \to \infty} \left[\frac{1}{2} \ln(x^2 + 1)\right]_0^N = \frac{3}{2} \lim_{N \to \infty} [\ln(N^2 + 1) - \ln 1] = \infty$$

16. $\int_0^\infty 3e^{-5x} \, dx = 3 \lim_{N \to \infty} \int_0^N e^{-5x} \, dx$

$$= -\frac{3}{5} \lim_{N \to \infty} e^{-5x} \Big|_0^N = -\frac{3}{5} \lim_{N \to \infty} [e^{-5N} - 1] = \frac{3}{5}$$

17. $\int_0^\infty xe^{-2x} \, dx = \lim_{N \to \infty} \int_0^N xe^{-2x} \, dx = \lim_{N \to \infty} \left[-\frac{1}{2}xe^{-2x}\Big|_0^N + \frac{1}{2} \int_0^N e^{-2x} \, dx\right]$

$$= \lim_{N \to \infty} \left[-\frac{1}{2}xe^{-2x} - \frac{1}{4}e^{-2x}\right]\Big|_0^N = \lim_{N \to \infty} \left[\left(-\frac{1}{2}Ne^{-2N} - \frac{1}{4}e^{-2N}\right) + \frac{1}{4}\right] = \frac{1}{4}$$

18. $\int_0^\infty 2x^2 e^{-x^3} \, dx = 2 \lim_{N \to \infty} \int_0^N x^2 e^{-x^3} \, dx$

$$= 2 \lim_{N \to \infty} -\tfrac{1}{3} e^{-x^3} \Big|_0^N = -\tfrac{2}{3} \lim_{N \to \infty} [e^{-N^3} - 1] = \tfrac{2}{3}$$

19. $\int_0^\infty x^2 e^{-2x}\, dx = \lim\limits_{N \to \infty} \int_0^N x^2 e^{-2x}\, dx$

$$= \lim_{N \to \infty} \left[-\tfrac{1}{2} x^2 e^{-2x} \Big|_0^N + \int_0^N x e^{-2x}\, dx \right]$$

$$= \lim_{N \to \infty} \left[-\tfrac{1}{2} x^2 e^{-2x} - \tfrac{1}{2} x e^{-2x} - \tfrac{1}{4} e^{-2x} \right] \Big|_0^N$$

$$= \lim_{N \to \infty} \left[\left(-\tfrac{1}{2} N^2 e^{-2N} - \tfrac{1}{2} N e^{-2N} - \tfrac{1}{4} e^{-2N} \right) + \tfrac{1}{4} \right] = \tfrac{1}{4}$$

20. $\int_2^\infty \frac{1}{x(\ln x)^2}\, dx = \lim\limits_{N \to \infty} \int_2^N \frac{1}{x(\ln x)^2}\, dx$

$$= \lim_{N \to \infty} -\frac{1}{\ln x} \Big|_0^N = \lim_{N \to \infty} \left[-\frac{1}{\ln N} + \frac{1}{\ln 2} \right] = \frac{1}{\ln 2}$$

21. $\int_0^\infty x^5 e^{-x^3}\, dx = \lim\limits_{N \to \infty} \int_0^N x^3 (x^2 e^{-x^3})\, dx$

$$= \lim_{N \to \infty} \left[-\tfrac{1}{3} x^3 e^{-x^3} \Big|_0^N + \int_0^N x^2 e^{-x^3}\, dx \right]$$

$$= \lim_{N \to \infty} \left[-\tfrac{1}{3} x^3 e^{-x^3} - \tfrac{1}{3} e^{-x^3} \right] \Big|_0^N = \lim_{N \to \infty} \left[\left(-\tfrac{1}{3} N^3 e^{-N^3} - \tfrac{1}{3} e^{-N^3} \right) + \tfrac{1}{3} \right] = \tfrac{1}{3}$$

22. It was determined in Problem 23 of Section 7.2 that the number of subscribers in N years will be

$$P_0 f(N) + \int_0^N r(t) f(N - t)\, dt$$

where $P_0 = 20,000$ is the current number of subscribers, $f(t) = e^{-t/10}$ is the fraction of subscribers remaining at least t years, and $r(t) = 1,000$ is the rate at which new subscriptions are sold. Hence, the number of subscribers in the long run is

$$\lim_{N \to \infty} \left[(20,000e^{-N/10}) + \int_0^N 1,000e^{-(N-t)/10} \, dt \right]$$

$$= \lim_{N \to \infty} \left[20,000e^{-N/10} + 1,000e^{-N/10} \int_0^N e^{t/10} \, dt \right]$$

$$= 0 + 10,000 \lim_{N \to \infty} (e^{-N/10})(e^{t/10}) \Big|_0^N$$

$$= 10,000 \lim_{N \to \infty} e^{-N/10}(e^{N/10} - 1)$$

$$= 10,000 \lim_{N \to \infty} (1 - e^{-N/10}) = 10,000$$

23. To find the present value of the investment of N years, divide the interval $0 \le t \le N$ into n subintervals of length Δt years, and let t_j denote the beginning of the jth subinterval. Then during the jth subinterval,

$$\text{Amount generated} \simeq f(t_j) \, \Delta t$$

and

$$\text{Present value} \simeq f(t_j)e^{-0.1t_j} \, \Delta t$$

Hence, the present value of the N-year investment is

$$\lim_{n \to \infty} \sum_{j=1}^n f(t_j)e^{-0.1t_j} \, \Delta t = \int_0^N f(t)e^{-0.1t} \, dt$$

Hence, the present value P of the total investment is

$$P = \lim_{N \to \infty} \int_0^N f(t)e^{-0.1t} \, dt = \lim_{N \to \infty} \int_0^N (8,000 + 400t)e^{-0.1t} \, dt$$

$$= \lim_{N \to \infty} \left[-10(8,000 + 400t)e^{-0.1t} \Big|_0^N + 4,000 \int_0^N e^{-0.1t} \, dt \right]$$

$$= \lim_{N \to \infty} [-10(8,000 + 400t)e^{-0.1t} - 40,000e^{-0.1t}] \Big|_0^N$$

$$= \lim_{N \to \infty} (-120,000 - 4,000t)e^{-0.1t} \Big|_0^N$$

$$= \lim_{N \to \infty} [(-120,000 - 4,000N)e^{-0.1N} + 120,000] = \$120,000$$

24. As in Problem 22, in N years, the population of the city will be

$$P_0 f(N) + \int_0^N r(t) f(N - t)\, dt$$

where now $P_0 = 100,000$ is the current population, $f(t) = e^{-t/20}$ is the fraction of the residents remaining for at least t years, and $r(t) = 100t$ is the rate of new arrivals. Hence, in the long run, the number of residents will be

$$\lim_{N \to \infty} \left[100,000e^{-N/20} + \int_0^N 100te^{-(N-t)/20}\, dt \right]$$

$$= 0 + \lim_{N \to \infty} 100e^{-N/20} \int_0^N te^{t/20}\, dt$$

$$= \lim_{N \to \infty} 100e^{-N/20}[20te^{t/20} - 400e^{t/20}] \Big|_0^N$$

$$= \lim_{N \to \infty} 100e^{-N/20}[(20Ne^{N/20} - 400e^{N/20}) + 400]$$

$$= \lim_{N \to \infty} 100(20N - 400 + 400e^{-N/20}) = \infty$$

That is, the population will increase without bound.

25. (a) $P(1 \le x \le 4) = \int_1^4 f(x)\, dx = \int_1^4 \frac{1}{3}\, dx = \frac{1}{3}x \Big|_1^4 = 1$

(b) $P(2 \le x \le 3) = \int_2^3 f(x)\, dx = \int_2^3 \frac{1}{3}\, dx = \frac{1}{3}x \Big|_2^3 = \frac{1}{3}$

(c) $P(x \le 2) = \int_{-\infty}^2 f(x)\, dx = \int_1^2 \frac{1}{3}\, dx = \frac{1}{3}x \Big|_1^2 = \frac{1}{3}$

26. (a) $P(0 \le x \le 3) = \int_0^3 f(x)\, dx = \int_0^3 \frac{2}{9}(3 - x)\, dx$

$$= \frac{2}{9}(3x - \frac{1}{2}x^2) \Big|_0^3 = \frac{2}{9} \left[(9 - \frac{9}{2}) - 0 \right] = 1$$

(b) $P(1 \le x \le 2) = \int_1^2 f(x)\, dx = \int_1^2 \frac{2}{9}(3 - x)\, dx$

$$= \frac{2}{9}(3x - \frac{1}{2}x^2) \Big|_1^2 = \frac{2}{9} \left[(6 - \frac{4}{2}) - (3 - \frac{1}{2}) \right] = \frac{1}{3}$$

27. (a) $P(x \geq 0) = \int_0^\infty f(x)\, dx = \lim_{N \to \infty} \int_0^N 0.2 e^{-0.2x}\, dx$

$$= \lim_{N \to \infty} -e^{-0.2x}\Big|_0^N = \lim_{N \to \infty} [-e^{-0.2N} + 1] = 1$$

(b) $P(1 \leq x \leq 4) = \int_1^4 0.2 e^{-0.2x}\, dx = -e^{-0.2x}\Big|_1^4$

$$= -e^{-0.8} + e^{-0.2} \simeq 0.3694$$

(c) $P(x \geq 5) = \int_5^\infty f(x)\, dx = \lim_{N \to \infty} \int_5^N 0.2 e^{-0.2x}\, dx$

$$= \lim_{N \to \infty} -e^{-0.2x}\Big|_0^N = \lim_{N \to \infty} [-e^{-0.2N} + e^{-1}] \simeq 0.3679$$

28. For the probability density function

$$f(x) = \begin{cases} \frac{1}{3} & \text{if } 1 \leq x \leq 4 \\ 0 & \text{otherwise} \end{cases}$$

$$E(x) = \int_{-\infty}^\infty x f(x)\, dx = \int_1^4 \frac{1}{3} x\, dx = \frac{1}{6} x^2 \Big|_1^4 = \frac{1}{6}(16 - 1) = \frac{5}{2}$$

and

$$Var(x) = \int_{-\infty}^\infty x^2 f(x)\, dx - [E(x)]^2 = \int_1^4 \frac{1}{3} x^2\, dx - \left[\frac{5}{2}\right]^2$$

$$= \frac{1}{9} x^3 \Big|_1^4 - \frac{25}{4} = \frac{1}{9}(64 - 1) - \frac{25}{4} = \frac{3}{4}$$

29. For the probability density function

$$f(x) = \begin{cases} \frac{2}{9}(3 - x) & \text{if } 0 \leq x \leq 3 \\ 0 & \text{otherwise} \end{cases}$$

$$E(x) = \int_{-\infty}^\infty x f(x)\, dx = \int_0^3 \frac{2}{9} x(3 - x)\, dx = \frac{2}{9} \int_0^3 (3x - x^2)\, dx$$

$$= \frac{2}{9}\left(\frac{3}{2} x^2 - \frac{1}{3} x^3\right)\Big|_0^3 = \frac{2}{9}\left[\frac{27}{2} - \frac{27}{3}\right] = 1$$

and

$$Var(x) = \int_{-\infty}^{\infty} x^2 f(x) \, dx - [E(x)]^2 = \int_0^3 \frac{2}{9} x^2 (3-x) \, dx - 1$$

$$= \frac{2}{9} \int_0^3 (3x^2 - x^3) \, dx - 1 = \frac{2}{9} (x^3 - \frac{1}{4} x^4) \Big|_0^3 - 1$$

$$= \frac{2}{9} \left[27 - \frac{81}{4} \right] - 1 = \frac{1}{2}$$

30. For the probability density function

$$f(x) = \begin{cases} 0.2e^{-0.2x} & \text{if } x \geq 0 \\ 0 & \text{if } x < 0 \end{cases}$$

$$E(x) = \int_{-\infty}^{\infty} x f(x) \, dx = \lim_{N \to \infty} \int_0^N 0.2x e^{-0.2x} \, dx$$

$$= \lim_{N \to \infty} \left[-x e^{-0.2x} \Big|_0^N + \int_0^N e^{-0.2x} \, dx \right]$$

$$= \lim_{N \to \infty} -(x+5)e^{-0.2x} \Big|_0^N = \lim_{N \to \infty} [-(N+5)e^{-0.2N} + 5] = 5$$

and

$$Var(x) = \int_{-\infty}^{\infty} x^2 f(x) \, dx - [E(x)]^2 = \lim_{N \to \infty} \int_0^N 0.2x^2 e^{-0.2x} \, dx - 25$$

$$= \lim_{N \to \infty} \left[-x^2 e^{-0.2x} \Big|_0^N + \int_0^N 2x e^{-02.x} \, dx \right] - 25$$

$$= \lim_{N \to \infty} \left[(-x^2 e^{-0.2x} - 10x e^{-0.2x}) \Big|_0^N + \int_0^N 10 e^{-0.2x} \, dx \right] - 25$$

$$= \lim_{N \to \infty} -(x^2 + 10x + 50)e^{-0.2x} \Big|_0^N - 25$$

$$= \lim_{N \to \infty} [-(N^2 + 10N + 50)e^{-0.2N} + 50] - 25 = 25$$

31. Let x denote the number of minutes between your arrival and the next batch of cookies. Then x is unifomly distributed with probability density function

$$f(x) = \begin{cases} \frac{1}{24} & \text{if } 0 \le x \le 45 \\ 0 & \text{otherwise} \end{cases}$$

The probability that you arrive within 5 minutes (before or after) of the cookie is

$$P(0 \le x \le 5) + P(40 \le x \le 45) = 2P(0 \le x \le 5) = 2\int_0^5 \frac{1}{45}\, dx = \frac{2}{45}x \Big|_0^5 = \frac{2}{9}$$

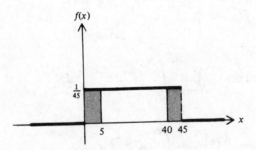

Density function for Problem 31.

32. Let x denote the time (in minutes) between the arrivals of successive cars. Then the probability density function for x is

$$f(x) = \begin{cases} 0.5e^{-0.5x} & \text{if } x \ge 0 \\ 0 & \text{if } x < 0 \end{cases}$$

(a) The probability that two cars will arrive at least 6 minutes apart is

$$P(x \ge 6) = \int_6^\infty f(x)\, dx = \lim_{N \to \infty} \int_6^N 0.5e^{-0.5x}\, dx$$

$$= \lim_{N \to \infty} -e^{-0.5x} \Big|_6^N = \lim_{N \to \infty} [-e^{-0.5N} + e^{-3}] \simeq 0.0498$$

(b) The average time interval between successive arrivals is

$$E(x) = \int_{-\infty}^{\infty} x f(x)\, dx = \lim_{N \to \infty} \int_{0}^{N} 0.5x e^{-0.5x}\, dx$$

$$= \lim_{N \to \infty} \left[-xe^{-0.5x} \Big|_{0}^{N} + \int_{0}^{N} e^{-0.5x}\, dx \right]$$

$$= \lim_{N \to \infty} (-x - 2)e^{-0.5x} \Big|_{0}^{N}$$

$$= \lim_{N \to \infty} [(-N - 2)e^{-0.5N} + 2] = 2 \text{ minutes}$$

CHAPTER 9

FUNCTIONS OF SEVERAL VARIABLES

---

1 FUNCTIONS OF SEVERAL VARIABLES

2 PARTIAL DERIVATIVES

3 THE CHAIN RULE AND THE TOTAL DIFFERENTIAL

4 ISOQUANTS: APPLICATIONS TO ECONOMICS

5 RELATIVE MAXIMA AND MINIMA

6 LAGRANGE MULTIPLIERS

7 THE METHOD OF LEAST SQUARES

REVIEW PROBLEMS

Chapter 9, Section 1

1. If $f(x, y) = (x - 1)^2 + 2xy^3$, then the domain of f consists of all ordered pairs (x, y) of real numbers. Moreover,

$$f(2, -1) = (2 - 1)^2 + 2(2)(-1)^3 = -3$$

and

$$f(1, 2) = (1 - 1)^2 + 2(1)(2)^3 = 16$$

3. If $f(x, y) = \sqrt{y^2 - x^2}$, then the domain of f consists of all ordered pairs (x, y) of real numbers for which $y^2 - x^2 \geq 0$, or, equivalently, for which $|y| \geq |x|$. Moreover,

$$f(4, 5) = \sqrt{25 - 16} = \sqrt{9} = 3$$

and

$$f(-1, 2) = \sqrt{4 - 1} = \sqrt{3} \simeq 1.732$$

5. If $f(x, y) = \frac{x}{\ln y}$, then the domain of f consists of all ordered pairs (x, y) of real numbers for which $y > 0$ (since $\ln y$ is defined only for positive values of y) and $y \neq 1$ (since $\ln 1 = 0$). Moreover,

$$f(-1, e^3) = \frac{-1}{\ln e^3} = -\frac{1}{3}$$

and

$$f(\ln 9, e^2) = \frac{\ln 9}{\ln e^2} = \frac{1}{2} \ln 9 = \ln 9^{1/2} = \ln 3 \simeq 1.099$$

7. (a) Since $Q(x, y) = 10x^2 y$ and $x = 20$ and $y = 40$,

$$Q(20, 40) = 10(20)^2(40) = 160,000 \text{ units}$$

(b) With one more skilled worker, $x = 21$ and the additional output is

$$Q(21, 40) - Q(20, 40) = 10(21)^2(40) - 160,000 = 16,400 \text{ units}$$

(c) With one more unskilled worker, $y = 41$ and the additional output is

$$Q(20, 41) - Q(20, 40) = 10(20)^2(41) - 160,000 = 4,000 \text{ units}$$

(d) With one more skilled worker and one more unskilled worker, $x = 21$ and $y = 41$, and the additional output is

$$Q(21, 41) - Q(20, 40) = 10(21)^2(41) - 160,000 = 20,810 \text{ units}$$

9. (a) Let R denote the total monthly revenue. Then,

$$R = (\text{revenue from first brand}) + (\text{revenue from second brand})$$

$$= x_1 D_1(x_1, x_2) + x_2 D_2(x_1, x_2)$$

Hence,

$$R(x_1, x_2) = x_1(200 - 10x_1 + 20x_2) + x_2(100 + 5x_1 - 10x_2)$$

$$= 200x_1 - 10x_1^2 + 20x_1 x_2 + 100x_2 + 5x_1 x_2 - 10x_2^2$$

$$= 200x_1 - 10x_1^2 + 25x_1 x_2 + 100x_2 - 10x_2^2$$

(b) If $x_1 = 6$ and $x_2 = 5$, then

$$R(6, 5) = 200(6) - 10(6)^2 + 25(6)(5) + 100(5) - 10(5)^2 = \$1,840$$

11. If $Q(K, L) = AK^\alpha L^{1-\alpha}$, then

$$Q(mK, mL) = A(mK)^\alpha (mL)^{1-\alpha} = Am^\alpha K^\alpha m^{1-\alpha} L^{1-\alpha}$$

$$= m^{(\alpha + 1 - \alpha)} AK^\alpha L^{1-\alpha} = mAK^\alpha L^{1-\alpha} = mQ(K, L)$$

Chapter 9, Section 2

1. If $f(x, y) = 2xy^5 + 3x^2y + x^2$, then

$$f_x(x, y) = 2y^5 + 6xy + 2x \qquad \text{and} \qquad f_y(x, y) = 10xy^4 + 3x^2$$

3. If $z = (3x + 2y)^5$, then

$$\frac{\partial z}{\partial x} = 5(3x + 2y)^4(3) = 15(3x + 2y)^4$$

and

$$\frac{\partial z}{\partial y} = 5(3x + 2y)^4(2) = 10(3x + 2y)^4$$

5. If $f(x, y) = \frac{3y}{2x} = \frac{3}{2}xy^{-1}$, then

$$f_x(x, y) = -\frac{3}{2}yx^{-2} = -\frac{3y}{2x^2} \qquad \text{and} \qquad f_y(x, y) = \frac{3}{2}x^{-1} = \frac{3}{2x}$$

7. If $z = xe^{xy}$, then

$$\frac{\partial z}{\partial x} = x(ye^{xy}) + e^{xy} = (xy + 1)e^{xy} \qquad \text{and} \qquad \frac{\partial z}{\partial y} = x^2e^{xy}$$

9. If $f(x, y) = \frac{2x + 3y}{y - x}$, then

$$f_x(x, y) = \frac{(y - x)(2) - (2x + 3y)(-1)}{(y - x)^2} = \frac{5y}{(y - x)^2}$$

and

$$f_y(x, y) = \frac{(y - x)(3) - (2x + 3y)(1)}{(y - x)^2} = \frac{-5x}{(y - x)^2}$$

11. If $z = x \ln y$, then

$$\frac{\partial z}{\partial x} = \ln y \qquad \text{and} \qquad \frac{\partial z}{\partial y} = \frac{x}{y}$$

13. Since $Q = 60K^{1/2}L^{1/3}$, the partial derivative

$$Q_K = \frac{\partial Q}{\partial K} = 30L^{1/3}K^{-1/2} = \frac{30L^{1/3}}{K^{1/2}}$$

is the rate of change of output with respect to the capital investment. For any values of K and L, this is an approximation to the additional number of units that will be produced each week if the capital investment is increased from K to $K + 1$ while the size of the labor force is not changed. In particular, if the capital investment K is increased from 900 (thousand) to 901 (thousand) and the size of the labor force is $L = 1,000$, the resulting change in output is

$$\Delta Q = Q_K(900, 1000) = \frac{30(1,000)^{1/3}}{(900)^{1/2}} = \frac{30(10)}{30} = 10 \text{ units}$$

15. The profit is

$$P(x, y) = (x - 30)(70 - 5x + 4y) + (y - 40)(80 + 6x - 7y) \text{ cents}$$

An approximation to the change in profit that will result if y is increased by 1 cent while x is held fixed is the partial derivative

$$P_y(x, y) = (x - 30)(4) + (70 - 5x + 4y)(0)$$

$$+ (y - 40)(-7) + (80 + 6x - 7y)(1)$$

$$= 10x - 14y + 240$$

If the current prices are $x = 50$ and $y = 52$, the change in profit is

$$\Delta P = P_y(50, 52) = 10(50) - 14(52) + 240 = 12 \text{ cents}$$

17. (a) Sugar and artificial sweetner are substitute commodities since an increase in the price of one would cause an increase in the demand for the other. Similarly, butter and margine are substitute commodities.

(b) If two commodities are substitutes and Q_1 is the demand for the first commodity and Q_2 is the demand for the second commodity, then

$$\frac{\partial Q_1}{\partial p_2} \geq 0 \qquad \text{and} \qquad \frac{\partial Q_2}{\partial p_1} \geq 0$$

since the demand for each is an increasing function of the price of the other.

(c) If $Q_1 = 3,000 + \frac{400}{p_1 + 3} + 50p_2$, then $\frac{\partial Q_1}{\partial p_2} = 50 \geq 0$, and if $Q_2 = 2,000 - 100p_1 + \frac{500}{p_2 + 4}$, then $\frac{\partial Q_2}{\partial p_1} = -100 < 0$. Hence the commodities are not substitutes.

19. If $f(x, y) = 5x^4 y^3 + 2xy$, then

$$f_x = 5(4x^3)y^3 + 2(1)y = 20x^3 y^3 + 2y$$

and

$$f_y = 5x^4(3y^2) + 2x(1) = 15x^4 y^2 + 2x$$

Hence,

$$f_{xx} = \frac{\partial}{\partial x}(f_x) = 20(3x^2)y^3 + 0 = 60x^2 y^3$$

$$f_{yy} = \frac{\partial}{\partial y}(f_y) = 15x^4(2y) + 0 = 30x^4 y$$

$$f_{xy} = \frac{\partial}{\partial y}(f_x) = 20x^3(3y^2) + 2(1) = 60x^3 y^2 + 2$$

$$f_{yx} = \frac{\partial}{\partial x}(f_y) = 15(4x^3)y^2 + 2(1) = 60x^3 y^2 + 2$$

21. If $f(x, y) = e^{x^2 y}$, then

$$f_x = 2xye^{x^2 y} \qquad \text{and} \qquad f_y = x^2 e^{x^2 y}$$

Hence,

$$f_{xx} = \frac{\partial}{\partial x}(f_x) = 2xy(e^{x^2 y})(2xy) + e^{x^2 y}(2y) = 2y(2x^2 y + 1)e^{x^2 y}$$

$$f_{yy} = \frac{\partial}{\partial y}(f_y) = x^2(e^{x^2 y})(x^2) = x^4 e^{x^2 y}$$

$$f_{xy} = \frac{\partial}{\partial y}(f_x) = 2xy(e^{x^2 y})(x^2) + e^{x^2 y}(2x) = 2x(x^2 y + 1)e^{x^2 y}$$

$$f_{yx} = \frac{\partial}{\partial x}(f_y) = x^2(e^{x^2 y})(2xy) + e^{x^2 y}(2x) = 2x(x^2 y + 1)e^{x^2 y}$$

23. If $f(x, y) = \sqrt{x^2 + y^2}$, then

$$f_x = \frac{1}{2}(x^2 + y^2)^{-1/2}(2x) = x(x^2 + y^2)^{-1/2}$$

$$f_y = \frac{1}{2}(x^2 + y^2)^{-1/2}(2y) = y(x^2 + y^2)^{-1/2}$$

Hence, by the product rule,

$$f_{xx} = x(-\frac{1}{2})(x^2 + y^2)^{-3/2}(2x) + (x^2 + y^2)^{-1/2}$$

$$= -\frac{x^2}{(x^2 + y^2)^{3/2}} + \frac{x^2 + y^2}{(x^2 + y^2)^{3/2}} = \frac{y^2}{(x^2 + y^2)^{3/2}}$$

By the same calculation with x and y reversed,

$$f_{yy} = \frac{x^2}{(x^2 + y^2)^{3/2}}$$

in addition,

$$f_{xy} = x(-\frac{1}{2})(x^2 + y^2)^{-3/2}(2y) = -\frac{xy}{(x^2 + y^2)^{3/2}}$$

and, by a similar calculation,

$$f_{yx} = -\frac{xy}{(x^2 + y^2)^{3/2}}$$

25. If $\frac{\partial^2 Q}{\partial K^2}$ is negative, the marginal product of capital decreases as K increases. This implies that for a fixed level of labor, the effect on output of the addition of $ 1,000 of capital is greater when capital investment is small than when capital investment is large. If $\frac{\partial^2 Q}{\partial K^2}$ is positive, it follows that for a fixed level of labor, the effect on output of the addition of $ 1,000 of capital is greater when capital investment is large than when it is small.

27. (a) According to the law of diminishing returns, $\frac{\partial Q}{\partial L}$ is increased if $L < L_0$ and $\frac{\partial Q}{\partial L}$ is decreasing if $L > L_0$. Rephrased in terms of the derivative of $\frac{\partial Q}{\partial L}$, the law states that

$$\frac{\partial^2 Q}{\partial L^2} > 0 \text{ if } L < L_0 \qquad \text{and} \qquad \frac{\partial^2 Q}{\partial L^2} < 0 \text{ if } L > L_0$$

     (b) A typical graph of Q as a function of L that reflects this situation is shown below. When the size of the labor force is very small, (inadequate for the utilization of existing equipment, for example), the addition of one worker will have little impact on output. When the size of the labor force is very large (close to the saturation level), the addition of one worker will again have little impact on output.

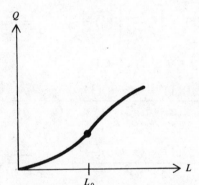

Production function for Problem 27.

Chapter 9, Section 3

1. If $z = x + 2y$, $x = 3t$, and $y = 2t + 1$, then

$$\frac{\partial z}{\partial x} = 1, \qquad \frac{\partial z}{\partial y} = 2, \qquad \frac{dx}{dt} = 3, \quad \text{and} \quad \frac{dy}{dt} = 2$$

Hence, by the chain rule,

$$\frac{dz}{dt} = \frac{\partial z}{\partial x}\frac{dx}{dt} + \frac{\partial z}{\partial y}\frac{dy}{dt} = 1(3) + 2(2) = 7$$

3. If $z = \frac{x}{y}$, $x = t^2$, and $y = 3t$, then

$$\frac{\partial z}{\partial x} = \frac{1}{y}, \qquad \frac{\partial z}{\partial y} = -\frac{x}{y^2}, \qquad \frac{dx}{dt} = 2t, \quad \text{and} \quad \frac{dy}{dt} = 3$$

Hence, by the chain rule,

$$\frac{dz}{dt} = \frac{\partial z}{\partial x}\frac{dx}{dt} + \frac{\partial z}{\partial y}\frac{dy}{dt} = \left[\frac{1}{y}\right](2t) + \left[-\frac{x}{y^2}\right](3)$$

$$= \left[\frac{1}{3t}\right](2t) - \left[\frac{t^2}{(3t)^2}\right](3) = \frac{2}{3} - \frac{3}{9} = \frac{1}{3}$$

5. If $z = \frac{x+y}{x-y}$, $x = t^3 + 1$, and $y = 1 - t^3$, then

$$\frac{\partial z}{\partial x} = \frac{(x-y)(1) - (x+y)(1)}{(x-y)^2} = \frac{-2y}{(x-y)^2}$$

$$\frac{\partial z}{\partial y} = \frac{(x-y)(1) - (x+y)(-1)}{(x-y)^2} = \frac{2x}{(x-y)^2}$$

$$\frac{dx}{dt} = 3t^2 \qquad \text{and} \qquad \frac{dy}{dt} = -3t^2$$

Hence, by the chain rule,

$$\frac{dz}{dt} = \frac{\partial z}{\partial x}\frac{dx}{dt} + \frac{\partial z}{\partial y}\frac{dy}{dt} = \left[\frac{-2y}{(x-y)^2}\right](3t^2) + \left[\frac{2x}{(x-y)^2}\right](-3t^2)$$

$$= \frac{-2(1-t^3)(3t^2) + 2(t^3+1)(-3t^2)}{[(t^3+1) - (1-t^3)]^2}$$

$$= \frac{-6t^2 + 6t^5 - 6t^5 - 6t^2}{(2t^3)^2} = -\frac{12t^2}{4t^6} = -\frac{3}{t^4}$$

7. If $z = (x - y^2)^3$, $x = t^2$, and $y = 2t$, then

$$\frac{\partial z}{\partial x} = 3(x - y^2)^2, \quad \frac{\partial z}{\partial y} = -6y(x - y^2)^2, \quad \frac{dx}{dt} = 2t, \quad \text{and} \quad \frac{dy}{dt} = 2$$

Hence, by the chain rule,

$$\frac{dz}{dt} = \frac{\partial z}{\partial x}\frac{dx}{dt} + \frac{\partial z}{\partial y}\frac{dy}{dt} = 3(x - y^2)^2(2t) - 6y(x - y^2)^2(2)$$

$$= (6t - 12y)(x - y^2)^2 = [6t - 12(2t)][t^2 - (2t)^2]^2$$

$$= -18t(9t^4) = -162t^5$$

9.   If $z = 2x + 3y$, $x = t^2$, and $y = 5t$, then

$$\frac{\partial z}{\partial x} = 2, \quad \frac{\partial z}{\partial y} = 3, \quad \frac{dx}{dt} = 2t, \quad \text{and} \quad \frac{dy}{dt} = 5$$

Hence, by the chain rule,

$$\frac{dz}{dt} = \frac{\partial z}{\partial x}\frac{dx}{dt} + \frac{\partial z}{\partial y}\frac{dy}{dt} = 2(2t) + 3(5) = 4t + 15$$

and when $t = 2$,

$$\frac{dz}{dt} = 2(4) + 3(5) = 23$$

11.   If $z = \frac{3x}{y}$, $x = t$, and $y = t^2$, then

$$\frac{\partial z}{\partial x} = \frac{3}{y}, \quad \frac{\partial z}{\partial y} = -\frac{3x}{y^2}, \quad \frac{dx}{dt} = 1, \quad \text{and} \quad \frac{dy}{dt} = 2t$$

Hence, by the chain rule,

$$\frac{dz}{dt} = \frac{\partial z}{\partial x}\frac{dx}{dt} + \frac{\partial z}{\partial y}\frac{dy}{dt} = \left[\frac{3}{y}\right](1) + \left[-\frac{3x}{y^2}\right](2t) = \frac{3}{y} - \frac{6xt}{y^2}$$

When $t = 3$, $x = 3$ and $y = 3^2 = 9$, and so

$$\frac{dz}{dt} = \frac{3}{9} - \frac{6(3)(3)}{81} = \frac{1}{3} - \frac{2}{3} = -\frac{1}{3}$$

13.   If $z = xy$, $x = e^{2t}$, and $y = e^{3t}$, then

$$\frac{\partial z}{\partial x} = y, \quad \frac{\partial z}{\partial y} = x, \quad \frac{dx}{dt} = 2e^{2t}, \quad \text{and} \quad \frac{dy}{dt} = 3e^{3t}$$

Hence, by the chain rule,

$$\frac{dz}{dt} = \frac{\partial z}{\partial x}\frac{dx}{dt} + \frac{\partial z}{\partial y}\frac{dy}{dt} = y(2e^{2t}) + x(3e^{3t})$$

When $t = 0$, $x = 1$ and $y = 1$, and so

$$\frac{dz}{dt} = 1(2) + 1(3) = 5$$

15. The demand for bicycles is $z = f(x,y) = 200 - 24\sqrt{x} + 4(0.1y + 5)^{3/2}$, where $x = 129 + 5t$ is the price of bicycles and $y = 80 + 10\sqrt{3t}$ is the price of gasoline. Then,

$$\frac{\partial z}{\partial x} = -\frac{12}{x^{1/2}}, \quad \frac{\partial z}{\partial y} = 0.6(0.1y + 5)^{1/2}, \quad \frac{dx}{dt} = 5 \quad \text{and} \quad \frac{dy}{dt} = \frac{15}{(3t)^{1/2}}$$

Hence, by the chain rule, the rate of change of demans with respect to time is

$$\frac{dz}{dt} = \frac{\partial z}{\partial x}\frac{dx}{dt} + \frac{\partial z}{\partial y}\frac{dy}{dt} = \left[-\frac{12}{x^{1/2}}\right](5) + 0.6(0.1y + 5)^{1/2}\left[\frac{15}{(3t)^{1/2}}\right]$$

$$= -\frac{60}{x^{1/2}} + 9(0.1y + 5)^{1/2}\left[\frac{1}{(3t)^{1/2}}\right]$$

When $t = 3$, $x = 144$ and $y = 110$, and so

$$\frac{dz}{dt} = -\frac{60}{12} + 9(11 + 5)^{1/2}\left(\frac{1}{3}\right) = -5 + 12 = 7 \text{ bicycles per month}$$

17. Since the output is $Q(x,y) = 0.08x^2 + 0.12xy + 0.03y^2$, the change in output is

$$\Delta Q \simeq dQ = \frac{\partial Q}{\partial x}\Delta x + \frac{\partial Q}{\partial y}\Delta y = (0.16x + 0.12y)\Delta x + (0.12x + 0.06y)\Delta y$$

If $x = 80$, $y = 200$, $\Delta x = \frac{1}{2}$, and $\Delta y = 2$,

$$\Delta Q = [0.16(80) + +0.12(200)]\left(\frac{1}{2}\right) + [0.12(80) + 0.06(200)](2) = 61.60 \text{ units}$$

19. Since profit is

$$P(x,y) = (x - 30)(70 - 5x + 4y) + (y - 40)(80 + 6x - 7y)$$

$$= -20x - 5x^2 + 10xy - 5,300 + 240y - 7y^2$$

the change in profit is

$$\Delta P \simeq dP = \frac{\partial P}{\partial x}\Delta x + \frac{\partial P}{\partial y}\Delta y = (-20 - 10x + 10y)\Delta x + (10x + 240 - 14y)\Delta y$$

If $x = 50$, $y = 52$, $\Delta x = 1$, and $\Delta y = 2$,

$$\Delta P \simeq (-20 - 500 + 520)(1) + (500 + 240 - 728)(2) = 24 \text{ cents}$$

21. The area of the inner recangle is $A(x, y) = xy$. The area of the concrete is the change in area of ΔA due to an increase in x of $\Delta x = 1.6$ and a change in y of $\Delta y = 1.6$. Hence,

$$\text{Area of concrete} = \Delta A \simeq dA = \frac{\partial A}{\partial x}\Delta x + \frac{\partial A}{\partial y}\Delta y = y\Delta x + x\Delta y$$

and when $x = 40$, $y = 30$, $\Delta x = 1.6$, and $\Delta y = 1.6$,

$$\text{Area of concrete} = 30(1.6) + 40(1.6) = 112 \text{ square yards}$$

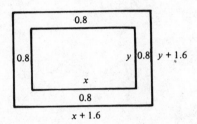

Rectangular garden for Problem 21.

23. The volume of a cylinder is $V(r, h) = \pi r^2 h$. Apply the approximation formula for percentage change with $\Delta r = 0.01r$ and $\Delta h = -0.015h$ to get

$$\text{Percentage change in } V \simeq 100\frac{(\partial V/\partial r)\Delta r + (\partial V/\partial h)\Delta h}{V}$$

$$= 100\frac{2\pi r h(0.01r) + \pi r^2(-0.015h)}{\pi r^2 h}$$

$$= 100\frac{0.02\pi r^2 h - 0.015\pi r^2 h}{\pi r^2 h}$$

$$= 100\frac{0.005\pi r^2 h}{\pi r^2 h}$$

$$= 0.5 \text{ percent}$$

25. The output function is $Q(K, L) = AK^\alpha L^{1-\alpha}$. Apply the approximation formula for percentage change with $\Delta K = 0.01K$ and $\Delta L = 0.01L$ to get

Percentage change in $Q \simeq 100 \dfrac{(\partial Q/\partial K)\Delta K + (\partial Q/\partial L)\Delta L}{Q}$

$$= 100 \frac{A\alpha K^{\alpha-1}L^{1-\alpha}(0.01K) + A(1-\alpha)K^{\alpha}L^{1-\alpha-1}(0.01L)}{AK^{\alpha}L^{1-\alpha}}$$

$$= 100 \frac{(0.01)\alpha AK^{\alpha}L^{1-\alpha} + (0.01)(1-\alpha)AK^{\alpha}L^{1-\alpha}}{AK^{\alpha}L^{1-\alpha}}$$

$$= 100 \frac{0.01 AK^{\alpha}L^{1-\alpha}}{AK^{\alpha}L^{1-\alpha}}$$

$$= 1 \text{ percent}$$

27. If $z = f(x, y)$, where $x = at$ and $y = bt$, then

$$\frac{\partial z}{\partial x} = f_x, \qquad \frac{\partial z}{\partial y} = f_y, \qquad \frac{dx}{dt} = a, \quad \text{and} \quad \frac{dy}{dt} = b$$

Hence, by the chain rule,

$$\frac{dz}{dt} = \frac{\partial z}{\partial x}\frac{dx}{dt} + \frac{\partial z}{\partial y}\frac{dy}{dt} = af_x + bf_y$$

Now,

$$\frac{d^2 z}{dt^2} = \frac{d}{dt}\left(\frac{dz}{dt}\right) = \frac{d}{dt}(af_x + bf_y) = a\frac{d}{dt}(f_x) + b\frac{d}{dt}(f_y)$$

But f_x and f_y are both functions of x and y, where $x = at$ and $y = bt$. Hence the chain rule must be applied again to get

$$\frac{d^2 z}{dt^2} = a\left[\frac{\partial}{\partial x}(f_x)\frac{dx}{dt} + \frac{\partial}{\partial y}(f_x)\frac{dy}{dt}\right] + b\left[\frac{\partial}{\partial x}(f_y)\frac{dx}{dt} + \frac{\partial}{\partial y}(f_y)\frac{dy}{dt}\right]$$

$$= a[af_{xx} + bf_{xy}] + b[af_{yx} + fb_{yy}]$$

$$= a^2 f_{xx} + abf_{xy} + abf_{yx} + b^2 f_{yy}$$

$$= a^2 f_{xx} + 2abf_{xy} + b^2 f_{yy}$$

Chapter 9, Section 4

1.   If $f(x,y) = x + 2y$, then when $f = 1$,

$$x + 2y = 1 \qquad \text{or} \qquad y = -\frac{1}{2}x + \frac{1}{2}$$

(a straight line with slope $-\frac{1}{2}$ and y intercept $\frac{1}{2}$); when $f = 2$,

$$x + 2y = 2 \qquad \text{or} \qquad y = -\frac{1}{2}x + 1$$

(a straight line with slope $-\frac{1}{2}$ and y intercept 1); and when $f = 3$,

$$x + 2y = 3 \qquad \text{or} \qquad y = -\frac{1}{2}x + \frac{3}{2}$$

(a straight line with slope $-\frac{1}{2}$ and y intercept $\frac{3}{2}$). The graphs of these level curves are three parallel lines.

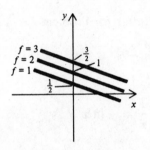

Level curves for Problem 1.   Level curves for Problem 3.

3.   If $f(x,y) = x^2 - 4x - y$, then when $f = -4$,

$$x^2 - 4x - y = -4 \qquad \text{or} \qquad y = x^2 - 4x + 4 = (x - 2)^2$$

(a parabola opening upward with vertex at $(2,0)$) and when $f = 5$,

$$x^2 - 4x - y = 5 \qquad \text{or} \qquad y = x^2 - 4x - 5 = (x - 5)(x + 1)$$

(a parabola opening upward with vertex at $(2,-9)$).

5.   If $f(x,y) = xy$, then when $f = 1$,

$$xy = 1 \qquad \text{or} \qquad y = \frac{1}{x}$$

(a hyperbola in the first and third quadrants); when $f = -1$,

$$xy = -1 \qquad \text{or} \qquad y = -\frac{1}{x}$$

(a hyperbola in the second and fourth quatrands); when $f = 2$,

$$xy = 2 \qquad \text{or} \qquad y = \frac{2}{x}$$

(a hyperbola in the first and third quadrants); and when $f = -2$,

$$xy = -2 \qquad \text{or} \qquad y = -\frac{2}{x}$$

(a hyperbola in the second and fourth quadrants).

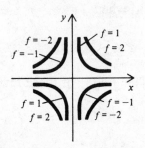

Level curves for Problem 5.

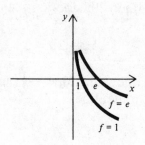

Level curves for Problem 7.

7.   If $f(x, y) = xe^y$, then when $f = 1$,

$$xe^y = 1 \qquad e^y = \frac{1}{x} \qquad \text{or} \qquad y = \ln\frac{1}{x} = -\ln x$$

and when $f = e$,

$$xe^y = e \qquad e^y = \frac{e}{x} \qquad \text{or} \qquad y = \ln\frac{e}{x} = \ln e - \ln x = 1 - \ln x$$

9.   <u>Formula</u>: If $f(x, y) = \frac{x}{y}$, then

$$f_x = \frac{1}{y} \qquad f_y = -\frac{x}{y^2} \qquad \text{and} \qquad \frac{dy}{dx} = -\frac{f_x}{f_y} = -\frac{1/y}{-x/y^2} = \frac{y}{x}$$

<u>Implicit differentiation</u>: If $\frac{x}{y} = C$, then $Cy - x = 0$ and implicit differentiaon with y regarded as a function of x gives

$$C\frac{dy}{dx} - 1 = 0 \qquad \text{or} \qquad \frac{dy}{dx} = \frac{1}{C} = \frac{1}{x/y} = \frac{y}{x} \qquad (\text{since } C = \frac{x}{y})$$

11.   <u>Formula</u>: If $f(x, y) = x^2 y + 2y^3 - 3x - 2y$, then

$$f_x = 2xy - 3 \qquad\qquad f_y = x^2 + 6y^2 - 2$$

and

$$\frac{dy}{dx} = -\frac{f_x}{f_y} = -\frac{2xy - 3}{x^2 + 6y^2 - 2} = \frac{3 - 2xy}{x^2 + 6y^2 - 2}$$

Implicit differentiation: If $x^2 y + 2y^3 - 3x - 2y = C$, then implicit differentiation with y regarded as a function of x gives

$$x^2 \frac{dy}{dx} + y(2x) + 2\left[3y^2 \frac{dy}{dx}\right] - 3 - 2\frac{dy}{dx} = 0$$

$$(x^2 + 6y^2 - 2)\frac{dy}{dx} = 3 - 2xy \quad \text{or} \quad \frac{dy}{dx} = \frac{3 - 2xy}{x^2 + 6y^2 - 2}$$

13.   Formula: If $f(x, y) = x \ln y$, then

$$f_x = \ln y, \quad f_y = \frac{x}{y}, \quad \text{and} \quad \frac{dy}{dx} = -\frac{f_x}{f_y} = -\frac{\ln x}{x/y} = -\frac{y \ln y}{x}$$

Implicit differentiation: If $x \ln y = C$, then implicit differentiation and y regarded as a function of x gives

$$x\left[\frac{1}{y}\frac{dy}{dx}\right] + (\ln y)(1) = 0 \quad \frac{x}{y}\frac{dy}{dx} = -\ln y \quad \text{or} \quad \frac{dy}{dx} = -\frac{y \ln y}{x}$$

15.   If $f(x, y) = x^2 + xy + y^3$, then

$$f_x = 2x + y \quad f_y = x + 3y^2 \quad \text{and} \quad \frac{dy}{dx} = -\frac{f_x}{f_y} = -\frac{2x + y}{x + 3y^2}$$

When $x = 0$ and $f = 8$, it follows that $8 = y^3$ or $y = 2$. The slope of the tangent at the point $(0, 2)$ is the corresponding value of $\frac{dy}{dx}$, that is,

$$\text{Slope} = \frac{dy}{dx} = -\frac{2(0) + 2}{0 + 3(4)} = -\frac{1}{6}$$

17.   If $f(x, y) = (x^2 + y)^3$, then

$$f_x = 6x(x^2 + y)^2 \qquad f_y = 3(x^2 + y)^2$$

and so

$$\frac{dy}{dx} = -\frac{f_x}{f_y} - \frac{6x(x^2 + y)^2}{3(x^2 + y)^2} = -2x$$

When $x = -1$ and $f = 8$, it follows that $8 = (1 + y)^3$ or $y = 1$. The slope of the tangent at the point $(-1, 1)$ is the corresponding value of $\frac{dy}{dx}$, that is

$$\text{Slope} = \frac{dy}{dx} = -2(-1) = 2$$

19.   Since output is $f(x, y) = 10xy^{1/2}$ and currently $x = 30$ and $y = 36$, the current level of output is $f(30, 36) = 1,800$ units. The combinations of x and y for which output will remain at this level are the coordinates of the points that lie on the constant production curve $f(x, y) = 1,800$. For any value of x, the slope of this curve is an estimate of the change in y (unskilled labor) that should be made to offset a 1-unit increase in x (skilled labor) so that the level of output will remain the same. That is,

$$\text{Change in unskilled labor} \simeq \frac{dy}{dx} = -\frac{f_x}{f_y} = -\frac{10y^{1/2}}{5xy^{-1/2}} = -\frac{2y}{x}$$

and when $x = 30$ and $y = 36$,

$$\text{Change in unskilled labor} \simeq -\frac{2(36)}{30} = -2.4$$

That is, unskilled labor should be reduced by approximately 2.4 hours.

21. Since output is $Q = 200K^{1/2}L^{1/3}$ and currently $K = 60$ (thousand) and $L = 10,000$, the present output is $Q(60, 10000)$. The combinations of K and L for which output will remain at this level are the coordinates of the points that lie on the corresponding constant-production curves. To estimate the change in L that should be made to offset an increase in K of ΔK so that the level of output will remain the same, use the approximation formula

$$\Delta L \simeq \frac{dL}{dK}\Delta K = -\frac{Q_K}{Q_L}\Delta K = -\frac{200\left(\frac{1}{2}\right)K^{-1/2}L^{1/3}}{200\left(\frac{1}{3}\right)K^{1/2}L^{-2/3}}\Delta K = -\frac{3L}{2K}\Delta K$$

When $K = 60$, $L = 10,000$, and $\Delta K = 0.5$ (thousand)

$$\Delta L \simeq \frac{dL}{dK}\Delta K = -\frac{3(10,000)}{2(60)}(0.5) = -250(0.5) = -125$$

That, is the labor force should be reduced by approximately 125 worker-hours.

23. The utility function is $U(x, y) = 2x^3y^2$ and the current level of utility is $U(5, 4) = 4,000$. The goal is to estimate the change Δy corresponding to a change of $\Delta x = -1$ on the corresponding indifference curve. From

$$\Delta y \simeq \frac{dy}{dx}\Delta x = -\frac{dy}{dx} = \frac{U_x}{U_y} = \frac{6x^2y^2}{4x^3y} = \frac{3y}{2x}$$

with $x = 5$ and $y = 4$ you get

$$\Delta y \simeq \frac{3(4)}{2(5)} = 1.2 \text{ units}$$

That is, the consumer could substitute approximately 1.2 units of the second commodity without affecting total utility.

Chapter 9, Section 5

1.   If $f(x, y) = 5 - x^2 - y^2$, then

$$f_x = -2x \qquad \text{and} \qquad f_y = -2y$$

which are both equal to zero only if $x = 0$ and $y = 0$. Hence, $(0, 0)$ is the only critical point. Since $f_{xx} = -2$, $f_{yy} = -2$, and $f_{xy} = 0$,

$$D(x, y) = f_{xx}f_{yy} - (f_{xy})^2 = (-2)(-2) - 0 = 4$$

Since

$$D(0, 0) = 4 > 0 \qquad \text{and} \qquad f_{xx}(0, 0) = -2 < 0$$

it follows that f has a relative maximum at $(0, 0)$.

3.   If $f(x, y) = xy$, then

$$f_x = y \qquad \text{and} \qquad f_y = x$$

which are both zero only when $x = 0$ and $y = 0$. Hence, $(0, 0)$ is the only critical point. Since $f_{xx} = 0$, $f_{yy} = 0$, and $f_{xy} = 1$,

$$D(x, y) = f_{xx}f_{yy} - (f_{xy})^2 = (0)(0) - 1 = -1$$

Since $D(0, 0) = -1 < 0$, it follows that f has a saddle point at $(0, 0)$.

5.   If $f(x, y) = 2x^3 + y^3 + 3x^2 - 3y - 12x - 4$, then

$$f_x = 6x^2 + 6x - 12 = 6(x + 2)(x - 1) \quad \text{and} \quad f_y = 3y^2 - 3 = 3(y + 1)(y - 1)$$

from which it is clear that $f_x = 0$ when $x = -2$ and $x = 1$, and $f_y = 0$ when $y = -1$ and $y = 1$. Hence, the critical points of f are $(-2, -1)$, $(-2, 1)$, $(1, -1)$, and $(1, 1)$. The second-order partial derivatives are

$$f_{xx} = 12x + 6, \qquad f_{yy} = 6y, \quad \text{and} \quad f_{xy} = 0$$

Hence,

$$D(x, y) = f_{xx}f_{yy} - (f_{xy})^2 = (12x + 6)(6y) - 0 = 36y(2x + 1)$$

Since

$$D(-2, -1) = 36(-1)(-3) = 108 > 0 \quad \text{and} \quad f_{xx}(-2, -1) = -24 + 6 = -18 < 0$$

f has a relative maximum at $(-2, -1)$; since

$$D(-2, 1) = 36(1)(-3) = -108 < 0$$

f has a saddle point at $(-2, 1)$; since

$$D(1, -1) = 36(-1)(3) = -108 < 0$$

f has a saddle point at $(1, -1)$; and since

$$D(1, 1) = 36(1)(3) = 108 > 0 \quad \text{and} \quad f_{xx}(1, 1) = 12 + 6 = 18 > 0$$

f has a realtive minimum at $(1, 1)$.

7.   If $f(x, y) = x^3 + y^2 - 6xy + 9x + 5y + 2$, then

$$f_x = 3x^2 - 6y + 9 \qquad \text{and} \qquad f_y = 2y - 6x + 5$$

Setting $f_x = 0$ and $f_y = 0$ gives

$$3x^2 - 6y + 9 = 0 \qquad \text{and} \qquad 2y - 6x + 5 = 0$$

or

$$x^2 - 2y + 3 = 0 \qquad \text{and} \qquad -6x + 2y + 5 = 0$$

Adding the two equations yields

$$x^2 - 6x + 8 = 0 \qquad (x - 4)(x - 2) = 0 \quad \text{or} \quad x = 4 \text{ and } x = 2$$

If $x = 4$, the second equation gives $-24 + 2y + 5 = 0$ or $y = \frac{19}{2}$, and if $x = 2$, the second equation gives $-12 + 2y = 0$ or $y = \frac{7}{2}$. Hence, the critical points are $(4, \frac{19}{2})$ and $(2, \frac{7}{2})$. Since

$$f_{xx} = 6x, \qquad f_{yy} = 2, \quad \text{and} \quad f_{xy} = -6$$

$$D(x, y) = f_{xx}f_{yy} - (f_{xy})^2 = (6x)(2) - (-6)^2 = 12x - 36 = 12(x - 3)$$

Since

$$D(4, \frac{19}{2}) = 12(4 - 3) = 12 > 0 \qquad \text{and} \qquad f_{xx}(4, \frac{19}{2}) = 24 > 0$$

f has a relative minimum at $(4, \frac{19}{2})$, and since

$$D(2, \frac{7}{2}) = 12(2 - 3) = -12 < 0$$

f has a saddle point at $(2, \frac{7}{2})$.

9.   Since the first type of system sells for x hundred dollars per system and costs $\$1,000$ (or 10 hundred dollars) per system, the corresponding profit per system is $(x - 10)$. The demand for the first system is $40 - 8x + 5y$. Hence, the profit from the sale of the first type of system is

$$P_1 = (x - 10)(40 - 8x + 5y)$$

The second type of system sells for y hundred dollars per system, and costs $\$3,000$ (or 30 hundred dollars) per system. Since the demand for the second system is $50 + 9x - 7y$, the profit from the sale of this type of system is

$$P_2 = (y - 30)(50 + 9x - 7y)$$

Hence, the total profit is

$$P(x, y) = P_1 + P_2 = (x - 10)(40 - 8x + 5y) + (y - 30)(50 + 9x - 7y)$$

$$= -150x - 8x^2 + 14xy - 1,900 + 210y - 7y^2$$

The first-order partial derivatives are

$$P_x = -150 - 16x + 14y \qquad \text{and} \qquad P_y = 14x + 210 - 14y$$

which are equal to zero when

$$-150 - 16x + 14y = 0 \qquad \text{and} \qquad 210 + 14x - 14y = 0$$

Adding these two equations yields

$$60 - 2x = 0 \qquad \text{or} \qquad x = 30$$

When $x = 30$, the first equation gives

$$-150 - 16(30) + 14y = 0, \qquad 14y = 630, \qquad \text{or} \quad y = 45$$

Hence, $(30, 45)$ is the only critical point. The second-order partial derivatives are

$$P_{xx} = -16, \qquad P_{yy} = -14, \qquad \text{and} \quad P_{xy} = 14$$

Hence,

$$D(x, y) = P_{xx}P_{yy} - (P_{xy})^2 = (-16)(-14) - (14)^2 = 28 > 0$$

for all x and y. Thus, f has a relative extremum at $(30, 45)$. Moreover, since $f_{xx} = -16 < 0$, this relative extremum is a relative maximum. Assuming that the relative maximum and absolute maximum are the same, it follows that to maximize profit, the telephone company should price the first system at $3,000$ ($x = 30$ hundred dollars) and the second at $4,500$ ($y = 45$ hundred dollars).

11. For the domestic market, x machines will sell for $60 - \frac{x}{5} + \frac{y}{20}$ thousand dollars apiece and the cost of producing each machine is 10 thousand dollars. Thus, the profit from the domestic market is

$$P_1 = (\text{number of machines})(\text{profit per machine})$$

$$= x \left[60 - \frac{x}{5} + \frac{y}{20} - 10 \right] = x \left[50 - \frac{x}{5} + \frac{y}{20} \right]$$

For the foreign market, y machines will sell for $50 - \frac{y}{10} + \frac{x}{20}$ thousand dollars apiece and the cost of producing each machine is 10 thousand dollars. Thus, the profit from the foreign market is

$$P_2 = y \left[50 - \frac{y}{10} + \frac{x}{20} - 10 \right] = y \left[40 - \frac{y}{10} + \frac{x}{20} \right]$$

The total profit is

$$P(x,y) = P_1 + P_2 = x\left[50 - \frac{x}{5} + \frac{y}{20}\right] + y\left[40 - \frac{y}{10} + \frac{x}{20}\right]$$

$$= 50x - \frac{x^2}{5} + \frac{xy}{10} + 40y - \frac{y^2}{10}$$

The first-order partial derivatves are

$$P_x = 50 - \frac{2x}{5} + \frac{y}{10} \qquad \text{and} \qquad P_y = \frac{x}{10} + 40 - \frac{y}{5}$$

which are equal to zero when

$$50 - \frac{2x}{5} + \frac{y}{10} = 0 \qquad \text{or} \qquad \frac{x}{10} + 40 - \frac{y}{5} = 0$$

or, equivalently, when

$$-8x + 2y = -1,000 \qquad \text{and} \qquad x - 2y = -400$$

Adding these two equations yields

$$-7x = -1,400 \qquad \text{or} \qquad x = 200$$

and substituting $x = 200$ into the second equation yields $200 - 2y = -400$ or $y = 300$. The second-order partial derivatives are

$$P_{xx} = -\frac{2}{5}, \qquad P_{yy} = -\frac{1}{5}, \qquad \text{and} \qquad P_{xy} = \frac{1}{10}$$

Hence,

$$D(x,y) = P_{xx}P_{yy} - (P_{xy})^2 = \frac{2}{25} - \frac{1}{100} = \frac{3}{100} > 0$$

Hence, $(200, 300)$ is a relative extremum. Moreover, since $P_{xx} = -\frac{2}{5} < 0$, this relative extremum is a relative maximum. Assuming that the relative maximum and absolute maximum are the same, it follows that to generate the largest possible profit, 200 machines should be supplied to the domestic market and 300 to the foreign market.

13.   (a)   From the fact that the values of C on the level curves of $f = C$ decrease as the curves approach the origin, it follows that f has a relative minimum at $(0,0)$.

      (b)   The values of C on the level curves $f = C$ in the first and third quadrants decrease as the curves approach the origin, while the values of C on the level curves in the second and fourth quadrants increase as the curves approach the origin. Thus, f has a saddle point at $(0,0)$.

Chapter 9, Section 6

1. For $f(x, y) = xy$ subject ot the constraint that $g(x, y) = x + y = 1$, the partial derivatives are

$$f_x = y \qquad f_y = x \qquad g_x = 1 \quad \text{and} \quad g_y = 1$$

Hence, the three Lagrange equations are

$$y = \lambda \qquad x = \lambda \quad \text{and} \quad x + y = 1$$

From the first two equations, $x = y$ which, when substituted into the third equation gives $2x = 1$ or $x = \frac{1}{2}$. Since $x = y$, the corresponding value for y is $y = \frac{1}{2}$. Thus, the constrained maximum is $f(\frac{1}{2}, \frac{1}{2}) = \frac{1}{4}$.

3. For $f(x, y) = x^2 + y^2$ subject to the constraint that $g(x, y) = xy = 1$, the partial derivatives are

$$f_x = 2x \qquad f_y = 2y \qquad g_x = y \quad \text{and} \quad g_y = x$$

Hence, the three Lagrange equations are

$$2x = \lambda y \qquad 2y = \lambda x \quad \text{and} \quad xy = 1$$

Multiply the first equation by y and the second by x to get

$$2xy = \lambda y^2 \qquad \text{and} \qquad 2xy = \lambda x^2$$

and set the two expressions for $2xy$ equal to each other to get

$$\lambda y^2 = \lambda x^2 \qquad y^2 = x^2 \quad \text{or} \quad x = \pm y$$

(Note that another solution of the equation $\lambda y^2 = \lambda x^2$ is $\lambda = 0$, which implies that x and y are zero, which is not consistent with the third equations.) If $y = x$, the third equation becomes $x^2 = 1$, which implies that $x = \pm 1$ and $y = \pm 1$. If $y = -x$, the third equation becomes $-x^2 = 1$, which has no solutions. Thus, the two points at which the constrained extrema can occur are $(1, 1)$ and $(-1, -1)$. Since $f(1, 1) = 2$ and $f(-1, -1) = 2$, it follows that the minimum value is 2 and is attained at the two points $(1, 1)$ and $(-1, -1)$.

5. For $f(x, y) = x^2 - y^2$ subject to the constraint $g(x, y) = x^2 + y^2 = 4$, the partial derivatives are

$$f_x = 2x \qquad f_y = -2y \qquad g_x = 2x \quad \text{and} \quad g_y = 2y$$

Hence, the three Lagrange equations are

$$2x = 2\lambda x \qquad -2y = 2\lambda y \quad \text{and} \quad x^2 + y^2 = 4$$

From the first equation, either $\lambda = 1$ or $x = 0$. If $x = 0$, the third equation becomes $y^2 = 4$ or $y = \pm 2$. From the second equation, either $\lambda = -1$ or $y = 0$. If $y = 0$, the third equation becomes $x^2 = 4$ or $x = \pm 2$. If neither x nor y is zero, the first equation implies $\lambda = 1$ while the second implies $\lambda = -1$, which is impossible. Hence, the only points at which the constrained extrema can occur are $(0, -2)$, $(0, 2)$, $(-2, 0)$, and $(2, 0)$. Since

$$f(0,-2) = -4, \quad f(0,2) = -4, \quad f(-2,0) = 4, \quad \text{and} \quad f(2,0) = 4$$

it follows that the constrained maximum is -4 and is attained at the two points $(0,-2)$ and $(0,2)$.

7. Let f denote the amount of fencing needed to enclose the pasture. Then, $f(x,y) = x + 2y$. The goal is to minimize this function subject to the constraint that the area $g(x,y) = xy = 3,200$. The partial derivatives are

$$f_x = 1 \qquad f_y = 2 \qquad g_x = y \quad \text{and} \quad g_y = x$$

and hence the three Lagrange equations are

$$1 = \lambda y \qquad 2 = \lambda x \quad \text{and} \quad xy = 3,200$$

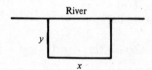

River

y

x

Rectangular pasture for Problem 7.

From the first equation, $\lambda = \frac{1}{y}$ (or $y = 0$, which is impossible by the third equation). From the second equation, $\lambda = \frac{2}{x}$ (or $x = 0$, which is impossible by the third equation). Setting the two expressions for λ equal to each other gives $\frac{1}{y} = \frac{2}{x}$ or $x = 2y$, and substituting this into the third equation yields

$$2y^2 = 3,200 \qquad y^2 = 1,600 \quad \text{or} \quad y = \pm 40$$

Only the positive value is meaningful in the context of this problem. Hence, $y = 40$, and (since $x = 2y$), $x = 80$. That is, to minimize the amount of fencing, the dimensions of the field should be 40 meters by 80 meters.

9. Let f denote the volume of the parcel. Then $f(x,y) = x^2 y$. The girth $4x$ plus the length y can be at most 72 inches. Thus the goal is to maximize the function $f(x,y)$ subject to the constraint that $g(x,y) = 4x + y = 72$. The partial derivatives are

$$f_x = 2xy \qquad f_y = x^2 \qquad g_x = 4 \quad \text{and} \quad g_y = 1$$

and hence the three Lagrange equations are

$$2xy = 4\lambda \qquad x^2 = \lambda \quad \text{and} \quad 4x + y = 72$$

From the first equation, $\lambda = \frac{1}{2}xy$, which, combined with the second equation, gives $\frac{1}{2}xy = x^2$ or $y = 2x$. (Another solution is $x = 0$, which is impossible in the context of this problem.) Substituting $y = 2x$ into the third equation gives $6x = 72$ or $x = 12$, and, since $y = 2x$, the corresponding value of y is $y = 24$. Hence, the largest volume is $f(12, 24) = (12)^2(24) = 3,456$ cubic inches.

11. Let f denote the cost of constructing the cylindrical can. The area of the top is πx^2, the area of the base is πx^2, and the area of the cardboard side is $2\pi xy$, where x is the radius and y the height. Let k denote the cost per square inch of constructing the cardboard side. Then, $2k$ is the cost per square inch of constructing the top and bottom. The goal is to minimize the total cost

$$f(x,y) = 2k(2\pi x^2) + k(2\pi xy) = 4k\pi x^2 + 2k\pi xy$$

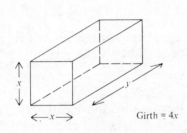

Parcel for Problem 9.

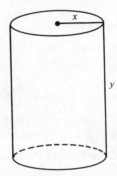

Cylindrical can for Problem 11.

subject to the constraint that the volume is to be $g(x, y) = \pi x^2 y = 4\pi$. The partial derivatives are

$$f_x = 8k\pi x + 2k\pi y \qquad f_y = 2k\pi x \qquad g_x = 2\pi xy \quad \text{and} \quad g_y = \pi x^2$$

and hence the three Lagrange equations are

$$8k\pi x + 2k\pi y = 2\lambda\pi xy \qquad 2k\pi x = \lambda\pi x^2 \quad \text{and} \quad \pi x^2 y = 4\pi$$

From the first equation, $\lambda = \frac{4k}{y} + \frac{k}{x}$ (since $x \neq 0$ and $y \neq 0$), and from the second equation, $\lambda = \frac{2k}{x}$ (since $x \neq 0$). Setting the two expressions for λ equal to each other gives

$$\frac{4k}{y} + \frac{k}{x} = \frac{2k}{x} \qquad 4x + y = 2y \quad \text{or} \quad y = 4x$$

Substituting this in the third equation yields

$$\pi x^2(4x) = 4\pi \qquad \text{or} \qquad x = 1$$

and since $y = 4x$, the corresponding value of y is $y = 4$. Hence, to minimize cost, the radius of the can should be 1 inch and the height should be 4 inches.

13. The goal is to maximize $Q(x, y) = 60x^{1/3}y^{2/3}$ subject to the constraint that $g(x, y) = x + y = 120$ (thousand). The partial derivatives are

$$Q_x = 20x^{-2/3}y^{2/3} \qquad Q_y = 40x^{1/3}y^{-1/3} \qquad g_x = 1 \quad \text{and} \quad g_y = 1$$

and hence the three Lagrange equations are

$$20x^{-2/3}y^{2/3} = \lambda \qquad 40x^{1/3}y^{-1/3} = \lambda \quad \text{and} \quad x + y = 120$$

It follows from the first two equations that

$$20x^{-2/3}y^{2/3} = 40x^{1/3}y^{-1/3} \qquad \text{or} \qquad y = 2x$$

Substituting this into the third equation gives $x + 2x = 120$ or $x = 40$, and (since $y = 2x$), the corresponding value of y is $y = 80$. That is, to generate maximal output, $\$40,000$ should be spent on labor and $\$80,000$ should be spent on equipment.

15. From Problem 13, the Lagrange equations were

$$\lambda = 20x^{-2/3}y^{2/3} \qquad \lambda = 40x^{1/3}y^{-1/3} \quad \text{and} \quad x + y = 120$$

from which it was determined that the maximal output occurs when $x = 40$ and $y = 80$. Substituting these values in the first equation gives

$$\lambda = 20(40)^{-2/3}(80)^{2/3} = 31.75$$

which implies that the maximal output will increase by approximately 31.75 units if the available money is increased by 1 thousand dollars and allocated optimally.

17. Let P denote the profit (in units of $\$1,000$). Then,

$$P = (\text{number of units})(\text{price per unit} - \text{cost per unit})$$

$$- \text{total amount spent on development and promotion}$$

The number of units is $\frac{320}{y+2} + \frac{160x}{x+4}$, where x thousand is spent on development and y thousand is spent on promotion. The price per unit is $\$150$ and the cost per unit is $\$50$, so the price per unit minus the cost per unit is $\$100$, or $\frac{1}{10}$ thousand dollars. Hence,

$$P(x,y) = \frac{1}{10}\left[\frac{320y}{y+2} + \frac{160x}{x+4}\right] - (x+y)$$

$$= \frac{32y}{y+2} + \frac{16x}{x+4} - x - y$$

The partial derivates of P are

$$P_x = \frac{(x+4)(16) - (16x)(1)}{(x+4)^2} - 1 = \frac{64}{(x+4)^2} - 1$$

and

$$P_y = \frac{(y+2)(32) - (32y)(1)}{(y+2)^2} - 1 = \frac{64}{(y+2)^2} - 1$$

(a)   To maximize profit when unlimited funds are available is to maximize $P(x, y)$ without constraints. To do this, find the critical points by seting P_x and P_y equal to zero, that is,

$$0 = P_x = \frac{64}{(x+4)^2} - 1 \qquad 64 = (x+4)^2 \qquad x + 4 = 8 \quad \text{or} \quad x = 4$$

and

$$0 = P_y = \frac{64}{(y+2)^2} - 1 \qquad 64 = (y+2)^2 \qquad y + 2 = 8 \quad \text{or} \quad y = 6$$

Thus, $\$4,000$ should be spent on development and $\$6,000$ should be spent on promotion to maximize profit.

(b)   If there were a restriction on the amount spent on development and promotion, then the constraint would be

$$g(x, y) = x + y = k, \quad \text{for some positive constant } k$$

The corresponding Lagrange equations would be

$$\frac{64}{(x+4)^2} - 1 = \lambda \qquad \frac{64}{(y+2)^2} - 1 = \lambda \quad \text{and} \quad x + y = k$$

To get the answer in part (a), λ must be equal to zero. To see this from another point of view, recall that $\lambda = \frac{dM}{dk}$, where M is the maximum profit if k thousand dollars is available. This maximum profit will be greatest when its derivative $\frac{dM}{dk}$ is zero, that is, when $\lambda = 0$.

(c)   Beginning with the Lagrange equations from part (b), set λ equal to zero to get

$$\frac{64}{(x+4)^2} - 1 = 0, \qquad x + 4 = 8, \quad \text{or} \quad x = 4$$

and

$$\frac{64}{(y+2)^2} - 1 = 0, \qquad y + 2 = 8, \quad \text{or} \quad y = 6$$

as in part (a).

19.   The goal is to maximize the utility function $U(x, y) = x^{\alpha} y^{\beta}$ subject to the budgetary constraint that $ax + by = c$. The three Lagrange equations are

$$\alpha x^{\alpha-1} y^{\beta} = a\lambda \qquad \beta x^{\alpha} y^{\beta-1} = b\lambda \quad \text{and} \quad ax + by = c$$

From the first two equations,

$$\frac{\alpha x^{\alpha-1} y^{\beta}}{a} = \frac{\beta x^{\alpha} y^{\beta-1}}{b} \qquad \frac{\alpha y}{a} = \frac{\beta x}{b} \quad \text{or} \quad y = \frac{a\beta x}{b\alpha}$$

Substituting this now the third equation gives.

$$ax + b\left[\frac{a\beta x}{ba}\right] = c \qquad \left[a + \frac{a\beta}{\alpha}\right]x = c \qquad \left[\frac{a\alpha + a\beta}{\alpha}\right]x = c$$

$$x = \frac{c\alpha}{a\alpha + a\beta} = \frac{c\alpha}{a(\alpha + \beta)} = \frac{c\alpha}{a} \qquad \text{(since } \alpha + \beta = 1)$$

Finally, since $y = \frac{a\beta x}{b\alpha}$, it follows that

$$y = \frac{a\beta}{b\alpha}\left[\frac{c\alpha}{a}\right] = \frac{c\beta}{b}$$

Chapter 9, Section 7

1. The sum $S(m, b)$ of the squares of the vertical distances from the three given points is

$$S(m, b) = d_1^2 + d_2^2 + d_3^2 = (b - 1)^2 + (2m + b - 3)^2 + (4m + b - 2)^2$$

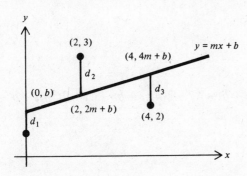

Graph for Problem 1.

To minimize $S(m, b)$, set the partial derivatives $\frac{\partial S}{\partial m}$ and $\frac{\partial S}{\partial b}$ equal to zero to get

$$\frac{\partial S}{\partial m} = 2(2m + b - 3)(2) + 2(4m + b - 2)(4) = 40m + 12b - 28 = 0$$

and

$$\frac{\partial S}{\partial b} = 2(b + 1) + 2(2m + b - 3) + 2(4m + b - 2) = 12m + 6b - 12 = 0$$

and solve the resulting simplified equations

$$10m + 3b = 7 \qquad \text{and} \qquad 6m + 3b = 6$$

to get $m = \frac{1}{4}$ and $b = \frac{3}{2}$. Hence, the equation of the least-squares line is

$$y = \frac{1}{4}x + \frac{3}{2}$$

3. The sum $S(m, b)$ of the squares of the vertical distances from the four given points is

$$S(m, b) = (m + b - 2)^2 + (2m + b - 4)^2 + (4m + b - 4)^2 + (5m + b - 2)^2$$

To minimize $S(m, b)$, set the partial derivates $\frac{\partial S}{\partial m}$ and $\frac{\partial S}{\partial b}$ equal to zero to get

$$\frac{\partial S}{\partial m} = 2(m + b - 2) + 2(2m + b - 4)(2) + 2(4m + b - 4)(4) + 2(5m + b - 2)(5)$$

$$= 92m + 24b - 72 = 0$$

and

$$\frac{\partial S}{\partial b} = 2(m + b - 2) + 2(2m + b - 4) + 2(4m + b - 4) + 2(5m + b - 2)$$

$$= 24m + 8b - 24 = 0$$

and solve the resulting simplified equations

$$23m + 6b = 18 \qquad \text{and} \qquad 3m + b = 3$$

to get $m = 0$ and $b = 3$. Hence, the equation of the least-squares line is $y = 3$.

5.

x	y	xy	x^2
1	2	2	1
2	2	4	4
2	3	6	4
5	5	25	25
$\Sigma x = 10$	$\Sigma y = 12$	$\Sigma xy = 37$	$\Sigma x^2 = 34$

From the formulas

$$m = \frac{n\Sigma xy - \Sigma x \Sigma y}{n\Sigma x^2 - (\Sigma x)^2} \qquad \text{and} \qquad b = \frac{\Sigma x^2 \Sigma y - \Sigma x \Sigma xy}{n\Sigma x^2 - (\Sigma x)^2}$$

with $n = 4$,

$$m = \frac{4(37) - 10(12)}{4(34) - (10)^2} = \frac{7}{9} \qquad \text{and} \qquad b = \frac{34(12) - 10(37)}{4(34) - (10)^2} = \frac{19}{18}$$

Hence, the equation of the least-squares line is

$$y = \frac{7}{9}x + \frac{19}{18} = 0.78x + 1.06$$

7.

x	y	xy	x^2
-2	5	-10	4
0	4	0	0
2	3	6	4
4	2	8	16
6	1	6	36

$$\Sigma x = 10 \quad \Sigma y = 15 \quad \Sigma xy = 10 \quad \Sigma x^2 = 60$$

From the formulas

$$m = \frac{n\Sigma xy - \Sigma x \Sigma y}{n\Sigma x^2 - (\Sigma x)^2} \quad \text{and} \quad b = \frac{\Sigma x^2 \Sigma y - \Sigma x \Sigma xy}{n\Sigma x^2 - (\Sigma x)^2}$$

with $n = 5$,

$$m = \frac{5(10) - 10(15)}{5(60) - (10)^2} = -\frac{100}{200} = -\frac{1}{2}$$

and

$$b = \frac{60(15) - 10(10)}{5(60) - (10)^2} = \frac{800}{200} = 4$$

Hence, the equation of the least-squares line is $y = -\frac{1}{2}x + 4$.

9.   (a)   Let x denote the number of catalogs requested and y the number of applications received (both in units of $1,000$). The given points (x, y) are plotted on the accompanying graph.

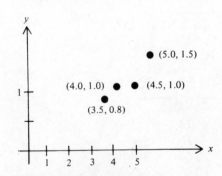

Graph for Problem 9.

(b)

x	y	xy	x^2
4.5	1.0	4.5	20.25
3.5	0.8	2.8	12.25
4.0	1.0	4.0	16.00
5.0	1.5	7.5	25.00
17.0	4.3	18.8	73.50

From the formulas

$$m = \frac{n\Sigma xy - \Sigma x \Sigma y}{n\Sigma x^2 - (\Sigma x)^2} \quad \text{and} \quad b = \frac{\Sigma x^2 \Sigma y - \Sigma x \Sigma xy}{n\Sigma x^2 - (\Sigma x)^2}$$

with $n = 4$

$$m = \frac{4(18.8) - 17(4.3)}{4(73.5) - (17)^2} = 0.42$$

and

$$b = \frac{73.5(4.3) - 17(18.8)}{4(73.5) - (17)^2} = -.071$$

Hence, the equation of the least-squares line is $y = 0.42x - 0.71$.

(c)   If $4,800$ catalogs are requested by December 1, $x = 4.8$ and

$$y = 0.42(4.8) - 0.71 = 1.306$$

which means that approximately $1,306$ completed applications will be received by March 1.

11. (a)   Let x denote the number of hours after the polls open and y the corresponding percentage of registered voters that have already cast their ballots. Then

x	2	4	6	8	10
y	12	19	24	30	37

(b)

x	y	xy	x^2
2	12	24	4
4	19	76	16
6	24	144	36
8	30	240	64
10	37	370	100
30	122	854	220

From the formulas

$$m = \frac{n\Sigma xy - \Sigma x \Sigma y}{n\Sigma x^2 - (\Sigma x)^2} \qquad \text{and} \qquad b = \frac{\Sigma x^2 \Sigma y - \Sigma x \Sigma xy}{n\Sigma x^2 - (\Sigma x)^2}$$

with $n = 5$

$$m = \frac{5(854) - 30(122)}{5(220) - (30)^2} = \frac{610}{200} = 3.05$$

and

$$b = \frac{220(122) - 30(854)}{5(220) - (30)^2} = \frac{1,220}{200} = 6.10$$

Hence, the equation of the least-squares line is $y = 3.05x + 6.10$.

(c)   When the polls close at 8:00 p.m., $x = 12$ and so

$$y = 3.05(12) + 6.1 = 42.7$$

which means that approximately 42.7 percent of the registered voters can be expected to vote.

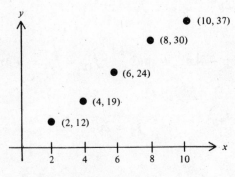

Graph for Problem 11.

Chapter 9, Review Problems

1. (a) If $f(x, y) = 2x^3y + 3xy^2 + \frac{y}{x}$, then

$$f_x = 6x^2y + 3y^2 - \frac{y}{x^2} \quad \text{and} \quad f_y = 2x^3 + 6xy + \frac{1}{x}$$

(b) If $f(x, y) = (xy^2 + 1)^5$, then

$$f_x = 5(xy^2 + 1)^4(y^2) = 5y^2(xy^2 + 1)^4$$

and

$$f_y = 5(xy^2 + 1)^4(2xy) = 10xy(xy^2 + 1)^4$$

(c) If $f(x, y) = xye^{xy}$, then

$$f_x = xye^{xy}(y) + e^{xy}(y) = y(xy + 1)e^{xy}$$

and

$$f_y = xye^{xy}(x) + e^{xy}(x) = x(xy + 1)e^{xy}$$

2. (a) If $f(x, y) = x^2 + y^3 - 2xy^2$, then

$$f_x = 2x - 2y^2 \qquad f_y = 3y^2 - 4xy$$

and so

$$f_{xx} = 2 \qquad f_{yy} = 6y - 4x \quad \text{and} \quad f_{xy} = f_{yx} = -4y$$

(b) If $f(x, y) = e^{x^2+y^2}$, then

$$f_x = 2xe^{x^2+y^2} \qquad f_y = 2ye^{x^2+y^2}$$

and so

$$f_{xx} = 2xe^{x^2+y^2}(2x) + e^{x^2+y^2}(2) = 2(2x^2 + 1)e^{x^2+y^2}$$

$$f_{yy} = 2ye^{x^2+y^2}(2y) + e^{x^2+y^2}(2) = 2(2y^2 + 1)e^{x^2+y^2}$$

and

$$f_{xy} = f_{yx} = 2xe^{x^2+y^2}(2y) = 4xye^{x^2+y^2}$$

(c) If $f(x, y) = x \ln y$, then

$$f_x = \ln y \qquad f_y = \frac{x}{y}$$

and so

$$f_{xx} = 0 \qquad f_{yy} = -\frac{x}{y^2} \quad \text{and} \quad f_{xy} = f_{yx} = \frac{1}{y}$$

3.   Let $Q = 40K^{1/3}L^{1/2}$ denote the daily output, where K denotes the capital invest-ment and L the size of the labor force. The marginal product of capital is

$$\frac{\partial Q}{\partial K} = \frac{40}{3}K^{-2/3}L^{1/2} = \frac{40L^{1/2}}{3K^{2/3}}$$

which is approximately the change ΔQ in output due to a 1 (thousand dollar) unit increase in capital. When $K = 125$ (thousand) and $L = 900$,

$$\Delta Q \simeq \frac{\partial Q}{\partial K} = \frac{40(900)^{1/2}}{3(125)^{2/3}} = \frac{40(30)}{3(25)} = 16 \text{ units}$$

4.   The marginal product of labor is the partial derivative $\frac{\partial Q}{\partial L}$. To say that this partial derivative increases as K increases is to say that its derivative with respect to K is positive, that is, $\frac{\partial^2 Q}{\partial K \partial L} > 0$.

5.   (a)   If $z = x^3 - 3xy^2$, $x = 2t$, and $y = t^2$, then

$$\frac{\partial z}{\partial x} = 3x^2 - 3y^2 \qquad \frac{\partial z}{\partial y} = -6xy \qquad \frac{dx}{dt} = 2 \quad \text{and} \quad \frac{dy}{dt} = 2t$$

Hence, by the chain rule,

$$\frac{dz}{dt} = \frac{\partial z}{\partial x}\frac{dx}{dt} + \frac{\partial z}{\partial y}\frac{dy}{dt} = (3x^2 - 3y^2)(2) + (-6xy)(2t)$$

$$= [3(2t)^2 - 3(t^2)^2](2) + [-6(2t)(t^2)](2t)$$

$$= 24t^2 - 6t^4 - 24t^4 = 6t^2(4 - 5t^2)$$

(b)   If $z = x \ln y$, $x = 2t$, and $y = e^t$, then

$$\frac{\partial z}{\partial x} = \ln y \qquad \frac{\partial z}{\partial y} = \frac{x}{y} \qquad \frac{dx}{dt} = 2 \quad \text{and} \quad \frac{dy}{dt} = e^t$$

Hence, by the chain rule,

$$\frac{dz}{dt} = \frac{\partial z}{\partial x}\frac{dx}{dt} + \frac{\partial z}{\partial y}\frac{dy}{dt} = (\ln y)(2) + \left[\frac{x}{y}\right](e^t)$$

$$= 2(\ln e^t) + \left[\frac{2t}{e^t}\right](e^t) = 2t + 2t = 4t$$

6.   Since the demand is $Q(x, y) = 240 + 0.1y^2 - 0.2x^2$, the change in demand is

$$\Delta Q \simeq dQ = \frac{\partial Q}{\partial x}\Delta x + \frac{\partial Q}{\partial y}\Delta y = (-0.4x)\Delta x + (0.2y)\Delta y$$

When $x = 45$, $y = 48$, $\Delta x = 2$, and $\Delta y = -1$,

$$\Delta Q \simeq (-0.4)(45)(2) + (0.2)(48)(-1) = -46$$

That is, the demand will drop by approximately 46 cans per week.

7.   Output is $Q(K, L) = 120K^{1/3}L^{2/3}$. Apply the approximation formula for percentage change with $\Delta K = 0.02K$ and $\Delta L = 0.01L$ to get

$$\text{Percentage change in } Q = 100\frac{(\partial Q/\partial K)\Delta K + (\partial Q/\partial L)\Delta L}{Q}$$

$$= 100\frac{(40K^{-2/3}L^{2/3})(0.02K) + (80K^{1/3}L^{-1/3})(0.01L)}{120K^{1/3}L^{2/3}}$$

$$= 100\frac{0.8K^{1/3}L^{2/3} + 0.8K^{1/3}L^{2/3}}{120K^{1/3}L^{2/3}}$$

$$= 100\frac{1.6K^{1/3}L^{2/3}}{120K^{1/3}L^{2/3}}$$

$$= 1.33 \text{ percent}$$

8.   The price of apple pies is $p(x,y) = \frac{1}{2}x^{1/3}y^{1/2}$ dollars per pie. The price of apples t months from now will be $x = 23 + \sqrt{8t}$ cents per pound, and bakers' wages t months from now will be $y = 3.96 + 0.02t$ dollars per hour. The weekly demand for the pies is $Q(p) = \frac{3,600}{p}$. The goal of the problem is to find $\frac{dQ}{dt}$ when $t = 2$. By the one-variable chain rule,

$$\frac{dQ}{dt} = \frac{dQ}{dp}\frac{dp}{dt} = -\frac{3,600}{p^2}\frac{dp}{dt}$$

By the two-variable chain rule,

$$\frac{dp}{dt} = \frac{\partial p}{\partial x}\frac{dx}{dt} + \frac{\partial p}{\partial y}\frac{dy}{dt}$$

$$= \left(\frac{1}{6}x^{-2/3}y^{1/2}\right)\left(\frac{1}{2}\right)(8t)^{-1/2}(8) + \frac{1}{4}x^{1/3}y^{-1/2}(0.02)$$

$$= \frac{2}{3}x^{-2/3}y^{1/2}(8t)^{-1/2} + 0.005x^{1/3}y^{-1/2}$$

When $t = 2$, $p = 3$, $x = 27$, and $y = 4$. Hence, putting it all together,

$$\frac{dQ}{dt} = -\frac{3,600}{9}\left[\frac{2}{3}\left(\frac{1}{9}\right)(2)\left(\frac{1}{4}\right) + 0.005(3)\left(\frac{1}{2}\right)\right] \simeq -17.81$$

That is, the weekly demand for pies will be decreasing at the rate of approximately 18 pies per week.

9.   (a)   If $f(x,y) = x^2 - y$, then when $f = 2$,

$$x^2 - y = 2 \qquad \text{or} \qquad y = x^2 - 2$$

which is a parabola opening upward with vertex at $(0, -2)$, and when $f = -2$,

$$x^2 - y = -2 \qquad \text{or} \qquad y = x^2 + 2$$

which is a parabola opening upward with vertex at $(0, 2)$.

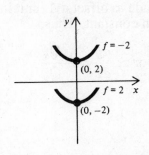

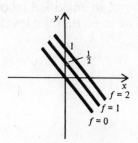

Level curves for Problem 9a.　　　　Level curves for Problem 9b.

(b)　If $f(x, y) = 6x + 2y$, then when $f = 0$,

$$6x + 2y = 0 \qquad \text{or} \qquad y = -3x$$

which is a straight line through the origin with slope -3; when $f = 1$,

$$6x + 2y = 1 \qquad \text{or} \qquad y = -3x + \frac{1}{2}$$

which is a straight line with y intercept $\frac{1}{2}$ and slope -3; and when $f = 2$,

$$6x + 2y = 2 \qquad \text{or} \qquad y = -3x + 1$$

which is a straight line with y intercept 1 and slope -3.

10.　(a)　If $f(x, y) = x^2 - y^3$, then $f_x = 2x$, $f_y = -3y^2$, and so

$$\frac{dy}{dx} = -\frac{f_x}{f_y} = \frac{2x}{-3y^2} = \frac{2x}{3y^2}$$

which is the slope of the tangent line at any point on the level curve. When $x = 1$ and $f = 2$, $2 = 1 - y^3$ or $y = -1$. At the point $(x, y) = (1, -1)$,

$$\text{Slope} = \frac{dy}{dx} = \frac{2(1)}{3(-1)^2} = \frac{2}{3}$$

(b)　If $f(x, y) = xe^y$, then $f_x = e^y$, $f_y = xe^y$, and so

$$\frac{dy}{dx} = -\frac{f_x}{f_y} = -\frac{e^y}{xe^y} = -\frac{1}{x}$$

which is the slope of the tangent line at any point on the level curve. When $x = 2$ and $f = 2e^y$ or $y = 0$. At the point $(x, y) = (2, 0)$,

$$\text{Slope} = \frac{dy}{dx} = -\frac{1}{2}$$

11.   Output is $Q(x, y) = 60x^{1/3}y^{2/3}$, where x denotes the number of skilled workers and y the number of unskilled workers. The combination of x and y for which output will remain at the current level are the coordinates of the points (x, y) that lie on the constant-production curve $Q = k$, where k is the current level of output. For any value of x, the slope of this constant-production curve is an approximation to the change in unskilled labor y that should be made to offset a 1-unit increase in skilled labor x so that the level of output will remain constant. Thus,

$$\Delta y = \text{change in unskilled labor} \simeq \frac{dy}{dx} = -\frac{Q_x}{Q_y}$$

$$= -\frac{20x^{-2/3}y^{2/3}}{40x^{1/3}y^{-1/3}} = -\frac{y}{2x}$$

When $x = 10$ and $y = 40$,

$$\Delta y \simeq \frac{dy}{dx} = -\frac{40}{2(10)} = -2$$

That is, the level of unskilled labor should be decreases by approximately 2 workers.

12.   (a)   If $f(x, y) = x^3 + y^3 + 3x^2 - 18y^2 + 81y + 5$, then

$$f_x = 3x^2 + 6x \qquad \text{and} \qquad f_y = 3y^2 - 36y + 81$$

To find the cricial points, set f_x and f_y equal to zero to get

$$0 = f_x = 3x^2 + 6x = 3x(x + 2) \qquad \text{or} \qquad x = 0 \quad \text{and} \quad x = -2$$

and

$$0 = f_y = 3y^2 - 36y + 81 = 3(y - 3)(y - 9) \quad \text{or} \quad y = 3 \quad \text{and} \quad y = 9$$

Hence, the critical points of f are $(0, 3)$, $(0, 9)$, $(-2, 3)$, and $(-2, 9)$. Since

$$f_{xx} = 6x + 6 \qquad f_{yy} = 6y - 36 \qquad \text{and} \qquad f_{xy} = 0$$

it follows that

$$D(x, y) = (f_{xx})(f_{yy}) - (f_{xy})^2 = (6x + 6)(6y - 36) - 0$$

$$= 36(x + 1)(y - 6)$$

Since

$$D(0, 3) = 36(1)(-3) = -108 < 0$$

f has a saddle point at $(0, 3)$; since

$$D(0, 9) = 36(1)(3) = 108 > 0 \qquad \text{and} \qquad f_{xx}(0, 9) = 6 > 0$$

f has a relative minimum at $(0, 9)$; since

$$D(-2,3) = 36(-1)(-3) = 108 > 0 \qquad \text{and} \qquad f_{xx}(-2,3) = -6 < 0$$

f has a relative maximum at $(-2,3)$; and since

$$D(-2,9) = 36(-1)(3) = -108 < 0$$

f has a saddle point at $(-2,9)$.

(b)   If $f(x,y) = x^2 + y^3 + 6xy - 7x - 6y$, then

$$f_x = 2x + 6y - 7 \qquad \text{and} \qquad f_y = 3y^2 + 6x - 6$$

To find the critical points, set f_x and f_y equal to zero to get

$$2x + 6y + 7 = 0 \qquad \text{and} \qquad 3y^2 + 6x - 6 = 0$$

or

$$2x + 6y - 7 = 0 \qquad \text{and} \qquad 2x + y^2 - 2 = 0$$

Subtracting the two equations gives

$$y^2 - 6y + 5 = 0 \qquad (y-1)(y-5) = 0 \quad \text{or} \quad y = 1 \quad \text{and} \quad y = 5$$

When $y = 1$, the first equation gives

$$2x + 6 - 7 = 0 \qquad \text{or} \qquad x = \frac{1}{2}$$

and when $y = 5$, the first equation gives

$$2x + 30 - 7 = 0 \qquad \text{or} \qquad x = -\frac{23}{2}$$

Hence, the critical points are $\left(\frac{1}{2}, 1\right)$ and $\left(-\frac{23}{2}, 5\right)$. Since

$$f_{xx} = 2 \qquad f_{yy} = 6y \quad \text{and} \quad f_{xy} = 6$$

it follows that

$$D(x,y) = (f_{xx})(f_{yy}) - (f_{xy})^2 = 2(6y) - 36 = 12(y - 3)$$

Since

$$D\left(\frac{1}{2}, 1\right) = 12(-2) = -24 < 0$$

f has a saddle point at $\left(\frac{1}{2}, 1\right)$, and since

$$D\left(-\frac{23}{2}, 5\right) = 12(2) = 24 > 0 \qquad \text{and} \qquad f_{xx} = 2 > 0$$

f has a relative minimum at $\left(-\frac{23}{2}, 5\right)$.

13.   For $f(x,y) = x^2 + 2y^2 + 2x + 3$ subject to the constraint that $g(x,y) = x^2 + y^2 = 4$, the partial derivatives are

$$f_x = 2x + 2 \qquad f_y = 4y \qquad g_x = 2x \quad \text{and} \quad g_y = 2y$$

and hence the three Lagrange equations are

$$2x + 2 = \lambda(2x) \qquad 4y = \lambda(2y) \quad \text{and} \quad x^2 + y^2 = 4$$

From the first equation, $\lambda = 1 + \frac{1}{x}$. From the second equation, $\lambda = 2$ or $y = 0$. If $y = 0$, the third equation gives $x = \pm 2$. If $y \neq 0$, setting the two expressions for λ equal to each other yields

$$1 + \frac{1}{x} = 2 \qquad \text{or} \qquad x = 1$$

and, from the third equation, $y = \pm\sqrt{3}$. Hence, the points at which the constrained extrema can occur are $(-2, 0)$, $(2, 0)$, $(1, \sqrt{3})$, and $(1, -\sqrt{3})$. Since

$$f(-2, 0) = 3 \qquad f(2, 0) = 11 \qquad f(1, \sqrt{3}) = 12 \quad \text{and} \quad f(1, -\sqrt{3}) = 12$$

it follows that the constrained maximum is 12 and is attained at $(1, \sqrt{3})$ and at $(1, -\sqrt{3})$, and that the constrained minimum is 3 and is attained at $(-2, 0)$.

14. Let x denote the length of the rectangle, y the width, and $f(x, y)$ the corresponding area. Then $f(x, y) = xy$. Since the rectangle is to have a fixed perimeter, the goal is to maximize $f(x, y)$ subject to the constraint that $g(x, y) = x + y = k$, for some constant k. The partial derivatives are

$$f_x = y \qquad f_y = x \qquad g_x = 1 \quad \text{and} \quad g_y = 1$$

and the three Lagrange equations are

$$y = \lambda \qquad x = \lambda \quad \text{and} \quad x + y = k$$

From the first two equations, $x = y$, which implies that the rectangle of greatest area is a square.

15. Let P denote profit. Then

$$P = (\text{number of units sold})(\text{price per unit} - \text{cost per unit})$$

$$- \text{total amount spent on development and promotion}$$

The number of units sold is $\frac{250y}{y+2} + \frac{100x}{x+5}$, where x thousand dollars is spent on development and y thousand dollars is spent on promotion. The selling price is \$350 per unit and the cost is \$150 per unit. Hence, the price per unit minus the cost per unit is \$200 or $\frac{1}{5}$ of a thousand dollars. Putting it all together,

$$P(x, y) = \frac{1}{5}\left[\frac{250y}{y+2} + \frac{100x}{x+5}\right] - (x + y) = \frac{50y}{y+2} + \frac{20x}{x+5} - x - y$$

The partial derivatives are

$$P_x = \frac{(x+5)(20) - (20x)(1)}{(x+5)^2} - 1 = \frac{100}{(x+5)^2} - 1$$

and

$$P_y = \frac{(y+2)(50) - (50y)(1)}{(y+2)^2} - 1 = \frac{100}{(y+2)^2} - 1$$

To find the critical points, set P_x and P_y equal to zero to get

$$0 = P_x = \frac{100}{(x+5)^2} - 1 \qquad (x+5)^2 = 100 \qquad x+5 = 10 \quad \text{or} \quad x = 5$$

and

$$0 = P_y = \frac{100}{(y+2)^2} - 1 \qquad (y+2)^2 = 100 \qquad y+2 = 10 \quad \text{or} \quad y = 8$$

Hence, the critical point is $(5, 8)$. The second-order partial derivatives are

$$P_{xx} = -\frac{200}{(x+5)^3} \qquad P_{yy} = -\frac{200}{(y+2)^3} \quad \text{and} \quad P_{xy} = 0$$

Hence,

$$D(x, y) = (P_{xx})(P_{yy}) - (P_{xy})^2 = \frac{40,000}{(x+5)^3(y+2)^3}$$

Since

$$D(5, 8) = \frac{40,000}{(10)^3(10)^3} > 0 \qquad \text{and} \qquad P_{xx}(5, 8) = -\frac{200}{(10)^3} < 0$$

$P(x, y)$ has a relative maximum at $(5, 8)$. Assuming that the absolute maximum and relative maximum are the same, it follows that to maximize profit, $\$5,000$ should be spent on development and $\$8,000$ should be spent on promotion.

16. From Problem 15, the profit function is

$$P(x, y) = \frac{50y}{y+2} + \frac{20x}{x+5} - x - y$$

The constraint is $g(x, y) = x + y = 11$ thousand dollars. Hence, the partial derivatives are

$$P_x = \frac{100}{(x+5)^2} - 1 \qquad P_y = \frac{100}{(y+2)^2} - 1 \qquad g_x = 1 \quad \text{and} \quad g_y = 1$$

and the three Lagrange equations are

$$\frac{100}{(x+5)^2} - 1 = \lambda \qquad \frac{100}{(y+2)^2} - 1 = \lambda \quad \text{and} \quad x + y = 11$$

From the first two equations,

$$\frac{100}{(x+5)^2} = \frac{100}{(y+2)^2} \qquad (x+5)^2 = (y+2)^2 \quad \text{or} \quad y = x + 3$$

and from the third equation,

$$x + (x + 3) = 11 \quad \text{or} \quad x = 4$$

Since $y = x + 3$, the corresponding value of y is $y = 7$. Hence, to maximize profit, $\$4,000$ should be spent on development and $\$7,000$ should be spent on promotion.

17. The increase in maximal profit M resulting from an increase in available money by 1 thousand dollars is

$$\Delta M \simeq \frac{dM}{dk} = \lambda$$

where, from Problem 16, $\lambda = \frac{100}{(x+5)^2} - 1$. Since the optimal allocation of 11 thousand dollars is $x = 4$ and $y = 7$, the increase in maximal profit resulting from the decision to spend 12 thousand dollars is

$$\Delta M \simeq \lambda = \frac{100}{(x+5)^2} - 1 = \frac{100}{(9)^2} - 1 = \frac{100}{81} - 1 = 0.235 \text{ thousand or } \$235$$

18. The sum of the squares of the vertical distances from the four given points to the line is

$$S(m, b) = (m + b - 1)^2 + (m + b - 2)^2 + (3m + b - 2)^2 + (4m + b - 3)^2$$

To minimize $S(m, b)$, set its partial derivatives equal to zero to get

$$\frac{\partial S}{\partial m} = 2(m + b - 1) + 2(m + b - 2) + 2(3m + b - 2)(3) + 2(4m + b - 3)(4)$$

$$= 27m + 9b - 21 = 0$$

and

$$\frac{\partial S}{\partial b} = 2(m + b - 1) + 2(m + b - 2) + 2(3m + b - 2) + 2(4m + b - 3)$$

$$= 18m + 8b - 16 = 0$$

and solve the resulting simplified equations

$$9m + 3b = 7 \quad \text{and} \quad 9m + 4b = 8$$

to get $b = 1$ and $m = \frac{4}{9}$. Hence, the equation of the least-squares line is

$$y = \frac{4}{9}x + 1$$

19. (a) Let x denote the monthly advertising expenditure and y the corresponding monthly sales (both measured in units of $\$1,000$). Then,

x	3	4	7	9	10
y	78	86	138	145	156

(b)

x	y	xy	x^2
3	78	234	9
4	86	344	16
7	138	966	49
9	145	1305	81
10	156	1560	100
33	603	4409	255

Apply the least-squares formulas

$$m = \frac{n\Sigma xy - \Sigma x \Sigma y}{n\Sigma x^2 - (\Sigma x)^2} \qquad \text{and} \qquad b = \frac{\Sigma x^2 \Sigma y - \Sigma x \Sigma xy}{n\Sigma x^2 - (\Sigma x)^2}$$

with $n = 5$ to get

$$m = \frac{5(4,409) - 33(603)}{5(255) - (33)^2} \simeq 11.54$$

and

$$b = \frac{255(603) - 33(4,409)}{5(255) - (33)^2} \simeq 44.45$$

Hence, the equation of the least squares line is

$$y = 11.54x + 44.45$$

(c)   If the monthly advertising expenditure is $\$5,000$, then $x = 5$ and

$$y = 11.54(5) + 44.45 = 102.15$$

which says that monthly sales will be approximately $\$102,150$.

CHAPTER 10

DOUBLE INTEGRALS

---

1 DOUBLE INTEGRALS

2 FINDING LIMITS OF INTEGRATION

3 APPLICATIONS OF DOUBLE INTEGRALS

REVIEW PROBLEMS

Chapter 10, Section 1

1. $\int_2^1 \int_1^2 x^2 y \, dx \, dy = \int_0^1 \left[\int_1^2 x^2 y \, dx \right] dy = \int_0^1 \frac{1}{3}x^3 y \Big|_1^2 \, dy$

$= \int_0^1 \left[\frac{8}{3}y - \frac{1}{3}y \right] dy = \frac{7}{3} \int_0^1 y \, dy = \frac{7}{6}y^2 \Big|_0^1 = \frac{7}{6}$

3. $\int_0^{\ln 2} \int_{-1}^0 2xe^y \, dx \, dy = \int_0^{\ln 2} \left[\int_{-1}^0 2xe^y \, dx \right] dy$

$= \int_0^{\ln 2} x^2 e^y \Big|_{-1}^0 \, dy = \int_0^{\ln 2} -e^y \, dy$

$= -e^y \Big|_0^{\ln 2} = -e^{\ln 2} + 1 = -2 + 1 = -1$

5. $\int_1^3 \int_0^1 \frac{2xy}{x^2+1} \, dx \, dy = \int_1^3 \left[\int_0^1 \frac{2xy}{x^2+1} \, dx \right] dy$

$= \int_1^3 y \ln(x^2 + 1) \Big|_0^1 \, dy = \int_1^3 y(\ln 2 - \ln 1) \, dy$

$= \int_1^3 y \ln 2 \, dy = \ln 2 \left(\frac{1}{2}y^2 \right) \Big|_1^3 = \ln 2 \left[\frac{9}{2} - \frac{1}{2} \right] = 4 \ln 2$

7. $\int_0^4 \int_0^{\sqrt{x}} x^2 y \, dy \, dx = \int_0^4 \left[\int_0^{\sqrt{x}} x^2 y \, dy \right] dx$

$= \int_0^4 \frac{1}{2}x^2 y^2 \Big|_0^{\sqrt{x}} \, dx = \int_0^4 \frac{1}{2}x^3 \, dx = \frac{1}{8}x^4 \Big|_0^4 = 32$

9. $\int_0^1 \int_{y-1}^{1-y} (2x + y) \, dx \, dy = \int_0^1 \left[\int_{y-1}^{1-y} (2x + y) \, dx \right] dy = \int_0^1 (x^2 + xy) \Big|_{y-1}^{1-y} \, dy$

$$= \int_0^1 [[(1-y)^2 + (1-y)y] - [(y-1)^2 + (y-1)y]]\, dy$$

$$= \int_0^1 [(1-y)y - (y-1)y]\, dy = \int_0^1 (-2y^2 + 2y)\, dy$$

$$= -2\left[\tfrac{1}{3}y^3 - \tfrac{1}{2}y^2\right]\big|_0^1 = -2\left[\tfrac{1}{3} - \tfrac{1}{2}\right] = \tfrac{1}{3}$$

11. $\displaystyle \int_0^2 \int_{x^2}^4 xe^y\, dy\, dx = \int_0^2 \left[\int_{x^2}^4 xe^y\, dy\right] dx$

$$= \int_0^2 xe^y\Big|_{x^2}^4\, dx = \int_0^2 [xe^4 - xe^{x^2}]\, dx$$

$$= e^4 \int_0^2 x\, dx - \int_0^2 xe^{x^2}\, dx = e^4 \left(\tfrac{1}{2}x^2\right)\big|_0^2 - \tfrac{1}{2}e^{x^2}\Big|_0^2$$

$$= \tfrac{1}{2}e^4(4-0) - \tfrac{1}{2}(e^4 - 1) = \tfrac{3}{2}e^4 + \tfrac{1}{2}$$

13. $\displaystyle \int_0^1 \int_0^x \frac{2y}{x^3+1}\, dy\, dx = \int_0^1 \left[\int_0^x \frac{2y}{x^3+1}\, dy\right] dx$

$$= \int_0^1 \frac{y^2}{x^3+1}\bigg|_0^x\, dx = \int_0^1 \frac{x^2}{x^3+1}\, dx$$

$$= \tfrac{1}{3}\ln|x^3+1|\big|_0^1 = \tfrac{1}{3}(\ln 2 - \ln 1) = \tfrac{1}{3}\ln 2$$

15. $\displaystyle \int_0^1 \int_0^x xe^y\, dy\, dx = \int_0^1 \left[\int_0^x xe^y\, dx\right] dx = \int_0^1 xe^y\Big|_0^x\, dx$

$$= \int_0^1 (xe^x - x)\, dx = \int_0^1 xe^x\, dx - \int_0^1 x\, dx$$

$$= (xe^x - e^x)\big|_0^1 - \tfrac{1}{2}x^2\big|_0^1 = [(e-e)+1] - \tfrac{1}{2} = \tfrac{1}{2}$$

17. $\displaystyle \int_0^1 \int_y^1 ye^{x+y}\, dx\, dy = \int_0^1 \left[\int_y^1 ye^y e^x\, dx\right] dy$

$$= \int_0^1 ye^y e^x\bigg|_y^1\, dy = \int_0^1 (ye^y e^1 - ye^y e^y)\, dy$$

$$= e\int_0^1 ye^y\, dy - \int_0^1 ye^{2y}\, dy$$

$$= e\,(ye^y - e^y)\big|_0^1 - \left[\tfrac{1}{2}ye^{2y} - \tfrac{1}{4}e^{2y}\right]\big|_0^1$$

$$= e[(e-e)+1] - \left[\left(\tfrac{1}{2}e^2 - \tfrac{1}{4}e^2\right) + \tfrac{1}{4}\right] = e - \tfrac{1}{4}e^2 - \tfrac{1}{4}$$

Chapter 10, Section 2

1. R can be described in terms of vertical cross sections by

$$0 \leq x \leq 3 \quad \text{and} \quad x^2 \leq y \leq 3x$$

and in terms of horizontal cross sections by

$$0 \leq y \leq 9 \quad \text{and} \quad \frac{1}{3}y \leq x \leq \sqrt{y}$$

$$0 \leq y \leq 9 \quad \text{and} \quad \frac{1}{3}y \leq x \leq \sqrt{y}$$

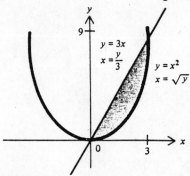

Region for Problem 1.

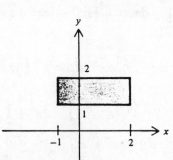

Region for Problem 3.

3. R can be described in terms of vertical cross sections by

$$-1 \leq x \leq 2 \quad \text{and} \quad 1 \leq y \leq 2$$

and in terms of horizontal cross sections by

$$1 \leq y \leq 2 \quad \text{and} \quad -1 \leq x \leq 2$$

5. R can be described in terms of vertical cross section by

$$1 \leq x \leq e \quad \text{and} \quad 0 \leq y \leq \ln x$$

and in terms of horizontal cross sections by

$$0 \leq y \leq 1 \quad \text{and} \quad e^y \leq x \leq e$$

7.   Describe R by $-1 \leq x \leq 2$ and $-1 \leq y \leq 0$. Then,

$$\int_R \int 3xy^2 \, dA = \int_{-1}^{0} \int_{-1}^{2} 3xy^2 \, dx \, dy = \int_{-1}^{0} \frac{3}{2}x^2y^2 \bigg|_{-1}^{2} \, dy$$

$$= \int_{-1}^{0} (6y^2 - \frac{3}{2}y^2) \, dy = \int_{-1}^{0} \frac{9}{2}y^2 \, dy = \frac{3}{2}y^3 \bigg|_{-1}^{0} = 0 - \frac{3}{2}(-1) = \frac{3}{2}$$

9.   Describe R by $0 \leq x \leq 1$ and $0 \leq y \leq x$. Then,

$$\int_R \int xe^y \, dA = \int_0^1 \int_0^x xe^y \, dy \, dx = \int_0^1 xe^y \Big|_0^x \, dx$$

$$= \int_0^1 (xe^x - x) \, dx = \int_0^1 xe^x \, dx - \int_0^1 x \, dx$$

$$= (xe^x - e^x)\Big|_0^1 - \frac{1}{2}x^2 \Big|_0^1 = [(e - e) + 1] - \frac{1}{2} = \frac{1}{2}$$

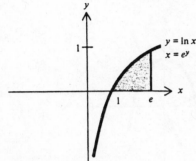

Region for Problem 5.

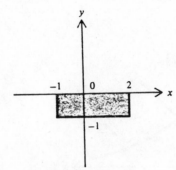

Region for Problem 7.

11. Describe R by $0 \le x \le 2$ and $x^2 \le y \le 2x$. Then,

$$\int_R \int (2y - x) \, dA = \int_0^2 \int_{x^2}^{2x} (2y - x) \, dy \, dx = \int_0^2 (y^2 - xy)\Big|_{x^2}^{2x} \, dx$$

$$= \int_0^2 [(4x^2 - 2x^2) - (x^4 - x^3)] \, dx = \int_0^2 (2x^2 - x^4 + x^3) \, dx$$

$$= \left[\frac{2}{3}x^3 - \frac{1}{5}x^5 + \frac{1}{4}x^4\right]\Big|_0^2 = \frac{16}{3} - \frac{32}{5} + \frac{16}{4} = \frac{44}{15}$$

13. Describe R by $2 \le x \le 4$ and $x \le y \le \frac{16}{x}$. Then,

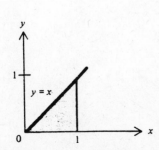

Region for Problem 9.

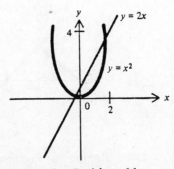

Region for Problem 11.

$$\int_R\int 2\,dA = \int_2^4 \int_x^{16/x} 2\,dy\,dx = \int_2^4 2y\Big|_x^{16/x}\,dx$$

$$= \int_2^4 \left[\frac{32}{x} - 2x\right]\,dx = (32\ln|x| - x^2)\Big|_2^4$$

$$= (32\ln 4 - 16) - (32\ln 2 - 4) = 32\ln\left(\frac{4}{2}\right) - 12 = 32\ln 2 - 12$$

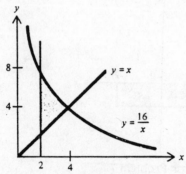

Region for Problem 13.

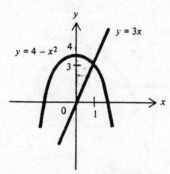

Region for Problem 15.

15. Describe R by $0 \le x \le 1$ and $3x \le y \le 4 - x^2$. Then,

$$\int_R\int 4x\,dA = \int_0^1 \int_{3x}^{4-x^2} 4x\,dy\,dx = \int_0^1 4xy\Big|_{3x}^{4-x^2}\,dx$$

$$= \int_0^1 [4x(4 - x^2) - 12x^2]\,dx = \int_0^1 (16x - 4x^3 - 12x^2)\,dx$$

$$= (8x^2 - x^4 - 4x^3)\Big|_0^1 = 8 - 1 - 4 = 3$$

17. Describe R by $0 \le y \le 1$ and $y - 1 \le x \le 1 - y$. Then,

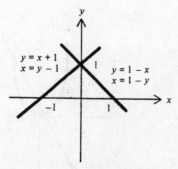

Region for Problem 17.

$$\int_R \int (2x+1)\, dA = \int_0^1 \int_{y-1}^{1-y} (2x+1)\, dx\, dy = \int_0^1 (x^2+1)\Big|_{y-1}^{1-y} dy$$

$$= \int_0^1 [[(1-y)^2 + (1-y)] - [(y-1)^2 + (y-1)]]\, dy$$

$$= \int_0^1 (2-2y)\, dy = (2y - y^2)\Big|_0^1 = 2 - 1 = 1$$

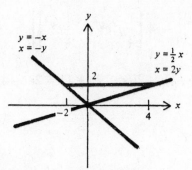

Region for Problem 19.

19. Describer R by $0 \le y \le 2$ and $-y \le x \le 2y$. Then,

$$\int_R \int \frac{1}{y^2+1}\, dA = \int_0^2 \int_{-y}^{2y} \frac{1}{y^2+1}\, dx\, dy = \int_0^2 \frac{x}{y^2+1}\Big|_{-y}^{2y} dy$$

$$= \int_0^2 \left[\frac{2y}{y^2+1} + \frac{y}{y^2+1}\right] dy = 3\int_0^2 \frac{y}{y^2+1}\, dy$$

$$= \frac{3}{2}\ln(y^2+1)\Big|_0^2 = \frac{3}{2}(\ln 5 - \ln 1) = \frac{3}{2}\ln 5$$

21. Describe R by $0 \le y \le 1$ and $y \le x \le y^{1/3}$. Then,

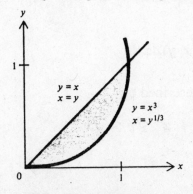

Region for Problem 21.

$$\int_R \int 12x^2 e^{y^2}\, dA = \int_0^1 \int_y^{y^{1/3}} 12x^2 e^{y^2}\, dx\, dy$$

$$= \int_0^1 4x^3 e^{y^2}\Big|_y^{y^{1/3}}\, dy = \int_0^1 (4ye^{y^2} - 4y^3 e^{y^2})\, dy$$

$$= 4\int_0^1 ye^{y^2}\, dy - 4\int_0^1 y^2 [ye^{y^2}]\, dy$$

$$= 2e^{y^2}\Big|_0^1 - 4\left[\frac{1}{2}y^2 e^{y^2} - \frac{1}{2}e^{y^2}\right]\Big|_0^1$$

$$= 2(e-1) - 2[(e-e)+1] = 2e - 4$$

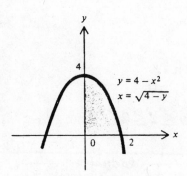

Region for Problem 23.

23. The region for $\int_0^2 \int_0^{4-x^2} f(x,y)\, dy\, dx$ is described by

$$0 \le x \le 2 \quad \text{and} \quad 0 \le y \le 4 - x^2$$

This region can also be described by

$$0 \le y \le 4 \quad \text{and} \quad 0 \le x \le \sqrt{4-y}$$

Hence, an equivalent iterated integral is

$$\int_0^4 \int_0^{\sqrt{4-y}} f(x,y)\, dx\, dy$$

25. The region for $\int_0^1 \int_{x^3}^{\sqrt{x}} f(x,y)\, dy\, dx$ is described by

$$0 \le x \le 1 \quad \text{and} \quad x^3 \le y \le \sqrt{x}$$

This region can also be described by

$$0 \leq y \leq 1 \qquad \text{and} \qquad y^2 \leq x \leq y^{1/3}$$

Hence, an equivalent iterated integral is

$$\int_0^1 \int_{y^2}^{y^{1/3}} f(x,y)\, dx\, dy$$

27.　The region for $\int_1^{e^2} \int_{\ln x}^2 f(x,y)\, dy\, dx$ is described by

$$1 \leq x \leq e^2 \qquad \text{and} \qquad \ln x \leq y \leq 2$$

This region can aslo be described by

$$0 \leq y \leq 2 \qquad \text{and} \qquad 1 \leq x \leq e^y$$

Hence, an equivalent iterated integral is

$$\int_0^2 \int_1^{e^y} f(x,y)\, dx\, dy$$

29.　The region for $\int_{-1}^1 \int_{x^2+1}^2 f(x,y)\, dy\, dx$ is described by

$$-1 \leq x \leq 1 \qquad \text{and} \qquad x^2 + 1 \leq y \leq 2$$

This region can also be described by

$$1 \leq y \leq 2 \qquad \text{and} \qquad -\sqrt{y-1} \leq x \leq \sqrt{y-1}$$

Hence, an equivalent iterated integral is

$$\int_1^2 \int_{-\sqrt{y-1}}^{\sqrt{y-1}} f(x,y)\, dx\, dy$$

31. The region R for $\int_0^1 \int_x^{2-x} f(x,y)\,dy\,dx$ is described by

$$0 \le x \le 1 \quad\text{and}\quad x \le y \le 2 - x$$

This region can also be thought of as $R = R_1 + R_2$, where R_1 is described by

$$0 \le y \le 1 \quad\text{and}\quad 0 \le x \le y$$

and R_2 is described by

$$1 \le y \le 2 \quad\text{and}\quad 0 \le x \le 2 - y$$

Hence, an equivalent iterated integral is

$$\int_0^1 \int_0^y f(x,y)\,dx\,dy + \int_1^2 \int_0^{2-y} f(x,y)\,dx\,dy$$

33. The region R for $\int_{-3}^2 \int_{x^2}^{6-x} f(x,y)\,dy\,dx$ is described by

$$-3 \le x \le 2 \quad\text{and}\quad x^2 \le y \le 6 - x$$

R can also be thought of as $R = R_1 + R_2$, where R_1 is described by

$$0 \le y \le 4 \quad\text{and}\quad -\sqrt{y} \le x \le \sqrt{y}$$

and R_2 is described by

$$4 \le y \le 9 \quad\text{and}\quad -\sqrt{y} \le x \le 6 - y$$

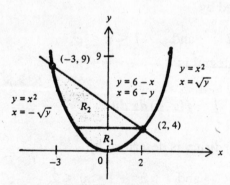

Region for Problem 33.

Hence, an equivalent iterated integral is

$$\int_0^4 \int_{-\sqrt{y}}^{\sqrt{y}} f(x,y)\,dx\,dy + \int_4^9 \int_{-\sqrt{y}}^{6-y} f(x,y)\,dx\,dy$$

Chapter 10, Section 3

1. Describe R by $-4 \leq x \leq 2$ and $0 \leq y \leq x + 4$. Then,

$$\text{Area of } R = \int_R \int 1 \, dA = \int_{-4}^2 \int_0^{x+4} 1 \, dy \, dx = \int_{-4}^2 y \Big|_0^{x+4} dx$$

$$= \int_{-4}^2 (x+4) \, dx = \left(\frac{1}{2}x^2 + 4x\right)\Big|_{-4}^2 = (2+8) - (8-16) = 18$$

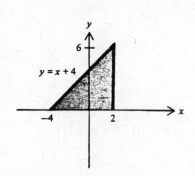

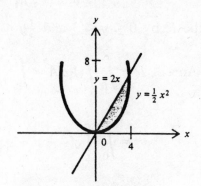

Region for Problem 1.   Region for Problem 3.

3. Describe R by $0 \leq x \leq 4$ and $\frac{1}{2}x^2 \leq y \leq 2x$. Then,

$$\text{Area of } R = \int_R \int 1 \, dA = \int_0^4 \int_{x^2/2}^{2x} 1 \, dy \, dx = \int_0^4 y \Big|_{x^2/2}^{2x} dx$$

$$= \int_0^4 (2x - \frac{1}{2}x^2) \, dx = (x^2 - \frac{1}{6}x^3)\Big|_0^4 = 16 - \frac{64}{6} = \frac{16}{3}$$

5. Describe R by $1 \leq x \leq 3$ and $x^2 - 4x + 3 \leq y \leq 0$. Then,

$$\text{Area of } R = \int_R \int 1 \, dA = \int_1^3 \int_{x^2-4x+3}^0 1 \, dy \, dx$$

$$= \int_1^3 y \Big|_{x^2-4x+3}^0 dx = \int_1^3 (-x^2 + 4x - 3) \, dx$$

$$= \left(-\frac{1}{3}x^3 + 2x^2 - 3x\right)\Big|_1^3 = (-9 + 18 - 12) - \left(-\frac{1}{3} + 2 - 4\right) = \frac{4}{3}$$

7.   Describe R by $1 \le x \le e$ and $0 \le y \le \ln x$. Then,

$$\text{Area of } R = \int_R \int 1 \, dA = \int_1^e \int_0^{\ln x} 1 \, dy \, dx = \int_1^3 y \Big|_0^{\ln x} dx$$

$$= \int_1^e \ln x \, dx = (x \ln x)\Big|_1^e - \int_1^e x \left[\frac{1}{x}\right] dx$$

$$= (x \ln x - x)\Big|_1^e = (e \ln e - e) - (0 - 1) = 1$$

9.   Describe R by $0 \le y \le 3$ and $\frac{1}{3}y \le x \le \sqrt{4-y}$. Then,

$$\text{Area of } R = \int_R \int 1 \, dA = \int_0^3 \int_{y/3}^{\sqrt{4-y}} 1 \, dx \, dy$$

$$= \int_0^3 \left[\sqrt{4-y} - \frac{1}{3}y\right] dy = \left[-\frac{2}{3}(4-y)^{3/2} - \frac{1}{6}y^2\right]\Big|_0^3$$

$$= \left[-\frac{2}{3} - \frac{9}{6}\right] - \left[-\frac{2}{3}(8) - 0\right] = \frac{19}{16}$$

11.   Describe R by $0 \le x \le 1$ and $0 \le y \le 2$. Then,

$$\text{Volume} = \int_R \int f(x,y) \, dA = \int_0^1 \int_0^2 (6 - 2x - y) \, dy \, dx$$

$$= \int_0^1 (6y - 2xy - \frac{1}{2}y^2)\Big|_0^2 dx = \int_0^1 (12 - 4x - 2) \, dx$$

$$= \int_0^1 (10 - 4x) \, dx = (10x - 2x^2)\Big|_0^1 = 10 - 2 = 8$$

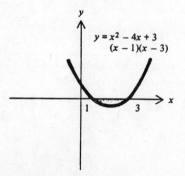

Region for Problem 5.

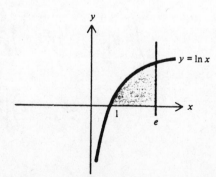

Region for Problem 7.

13.   Describe R by $0 \le y \le 1$ and $0 \le x \le 2y$. Then,

$$\text{Volume} = \int_R\!\!\int f(x,y)\,dA = \int_0^1 \int_0^{2y} e^{y^2}\,dx\,dy = \int_0^1 xe^{y^2}\Big|_0^{2y}\,dy$$

$$= \int_0^1 2ye^{y^2}\,dy = e^{y^2}\Big|_0^1 = e - 1$$

Note that this integration could not have been performed in the opposite order.

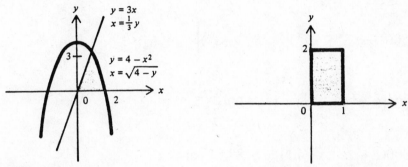

Region for Problem 9.   Region for Problem 11.

15.   Describe R by $0 \le x \le 1$ and $0 \le y \le 1 - x$. Then,

$$\text{Volume} = \int_R\!\!\int f(x,y)\,dA = \int_0^1 \int_0^{1-x} e^{2x+y}\,dy\,dx$$

$$= \int_0^1 \int_0^{1-x} e^{2x}e^y\,dy\,dx = \int_0^1 e^{2x}e^y\Big|_0^{1-x}\,dx$$

$$= \int_0^1 (e^{2x}e^{1-x} - e^{2x}e^0)\,dx = \int_0^1 (e^{x+1} - e^{2x})\,dx$$

$$= \left[e^{x+1} - \frac{1}{2}e^{2x}\right]\Big|_0^1 = \left[e^2 - \frac{1}{2}e^2\right] - \left[e - \frac{1}{2}\right] = \frac{1}{2}e^2 - e + \frac{1}{2}$$

17.   The average carbon monoxide level over the region R is

$$\frac{1}{\text{area of } R} \int_R\!\!\int f(x,y)\,dA = \frac{\int_R\!\int f(x,y)\,dA}{\int_R\!\int 1\,dA}$$

19.   Describe R by $0 \le x \le 1$ and $0 \le y \le x$. The area of R is $\frac{1}{2}$ (base)(height) $= \frac{1}{2}(1)(1) = \frac{1}{2}$. Hence,

$$\text{Average property value} = \frac{1}{\text{area of } R} \int_R \int f(x, y)\, dA$$

$$= 2 \int_0^1 \int_0^x 400 x e^{-y}\, dy\, dx = 2 \int_0^1 -400 x e^{-y}\Big|_0^x dx$$

$$= 2 \int_0^1 (-400 x e^{-x} + 400 x)\, dx = -800 \int_0^1 x e^{-x}\, dx + 800 \int_0^1 x\, dx$$

$$= -800(-x e^{-x} - e^{-x})\Big|_0^1 + 800(\tfrac{1}{2} x^2)\Big|_0^1$$

$$= -800[(-e^{-1} - e^{-1}) + 1] + 800(\tfrac{1}{2})$$

$$= -800(1 - 2e^{-1}) + 400 \simeq \$188.61 \text{ per acre}$$

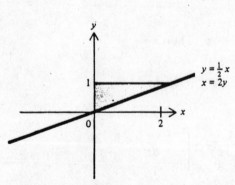

Region for Problem 13.

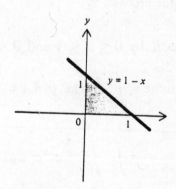

Region for Problem 15.

21. Describe R by $0 \le y \le 2$ and $y \le x \le 4 - y$. The area of R is $\frac{1}{2}$ (base)(height) $= \frac{1}{2}(4)(2) = 4$. Hence,

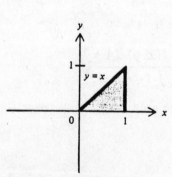

Region for Problem 19.

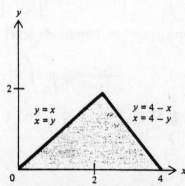

Region for Problem 21.

$$\text{Average value} = \frac{1}{\text{area of } R} \int_R \int f(x,y)\,dA = \frac{1}{4}\int_0^2 \int_y^{4-y} 3y\,dx\,dy$$

$$= \int_0^2 \frac{3}{4}yx\Big|_y^{4-y} dy = \frac{3}{4}\int_0^2 [y(4-y)-y^2]\,dy$$

$$= \frac{3}{4}\int_0^2 (4y-2y^2)\,dy = \frac{3}{4}\left(2y^2 - \frac{2}{3}y^3\right)\Big|_0^2$$

$$= \frac{3}{4}\left[2(4)-\frac{2}{3}(8)\right] = \frac{3}{4}\left[8-\frac{16}{3}\right] = 6-4 = 2$$

23. Describe R by $-2 \le x \le 2$ and $0 \le y \le 4-x^2$. Then,

$$\int_R \int f(x,y)\,dA = \int_{-2}^2 \int_0^{4-y^2} x\,dy\,dx = \int_{-2}^2 xy\Big|_0^{4-x^2} dx$$

$$= \int_{-2}^2 x(4-x^2)\,dx = \int_{-2}^2 (4x-x^3)\,dx$$

$$= (2x^2 - \frac{1}{4}x^4)\Big|_{-2}^2 = \left[8-\frac{16}{4}\right] - \left[8-\frac{16}{4}\right] = 0$$

Hence,

$$\text{Average value} = \frac{1}{\text{area of } R} \int_R \int f(x,y)\,dA = \frac{0}{\text{area of } R} = 0$$

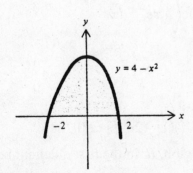

Region for Problem 23.

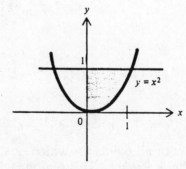

Region for Problem 25.

25. Describe R by $0 \le x \le 1$ and $x^2 \le y \le 1$. Then,

$$\text{Area of } R = \int_R \int 1\, dA = \int_0^1 \int_{x^2}^1 1\, dy\, dx = \int_0^1 y\Big|_{x^2}^1 dx$$

$$= \int_0^1 (1 - x^2)\, dx = \left(x - \frac{1}{3}x^3\right)\Big|_0^1 = 1 - \frac{1}{3} = \frac{2}{3}$$

and

$$\int_R \int f(x, y)\, dA = \int_0^1 \int_{x^2}^1 e^x y^{-1/2}\, dy\, dx$$

$$= \int_0^1 2e^x y^{1/2}\Big|_{x^2}^1 dx = \int_0^1 (2e^x - 2xe^x)\, dx$$

$$= 2\int_0^1 e^x\, dx - 2\int_0^1 xe^x\, dx = 2e^x\big|_0^1 - 2(xe^x - e^x)\big|_0^1$$

$$= (4e^x - 2xe^x)\big|_0^1 = (4e - 2e) - 4 = 2e - 4$$

Hence,

$$\text{Average value} = \frac{1}{\text{area of } R} \int_R \int f(x, y)\, dA = \frac{3}{2}(2e - 4) = 3(e - 2)$$

27. The probability that $0 \le x \le 1$ and $0 \le y \le 2$ is

$$\int_0^1 \int_0^2 xe^{-x}e^{-y}\, dy\, dx = \int_0^1 -xe^{-x}e^{-y}\Big|_0^2 dx$$

$$= \int_0^1 (-xe^{-x}e^{-2} + xe^{-x})\, dx = (1 - e^{-2})\int_0^1 xe^{-x}\, dx$$

$$= (1 - e^{-2})(-xe^{-x} - e^{-x})\Big|_0^1$$

$$= (1 - e^{-2})[(-e^{-1} - e^{-1}) + 1] = (1 - e^{-2})(1 - 2e^{-1}) \simeq 0.2285$$

29. The set of all points for which $x + y \le 1$ is the region R in the first quadrant below the line $x + y = 1$. This region is described by

$$0 \le x \le 1 \qquad \text{and} \qquad 0 \le y \le 1 - x$$

Hence the probability that $x + y \le 1$ is

$$\int_R \int f(x, y)\, dA = \int_0^1 \int_0^{1-x} xe^{-x}e^{-y}\, dy\, dx$$

$$= \int_0^1 -xe^{-x}e^{-y}\Big|_0^{1-x} dx$$

$$= \int_0^1 [-xe^{-x}e^{-(1-x)} + xe^{-x}]\, dx = \int_0^1 (xe^{-x} - xe^{-1})\, dx$$

$$= \int_0^1 xe^{-x}\, dx - e^{-1}\int_0^1 x\, dx = (-xe^{-x} - e^{-x})\Big|_0^1 - e^{-1}(\tfrac{1}{2}x^2)\Big|_0^1$$

$$= [(-e^{-1} - e^{-1}) + 1] - e^{-1}\left[\frac{1}{2}\right] = -\frac{5}{2}e^{-1} + 1 \simeq 0.0803$$

31. The probability that the appliance does not fail during the first year is equal to 1 minus the probability that it does fail during this period. The appliance will fail during the first year if both of the independent components fail, that is, if $0 \le x \le 1$ and $0 \le y \le 1$. Thus the probability that the warranty expires before the appliance becomes unusable is

$$1 - P[0 \le x \le 1 \text{ and } 0 \le y \le 1] = 1 - \int_0^1 \int_0^1 \frac{1}{4}e^{-x/2}e^{-y/2}\, dy\, dx$$

$$= 1 - \int_0^1 \frac{1}{4}e^{-x/2} - 2e^{-y/2}\Big|_0^1 dx$$

$$= 1 - \int_0^1 \frac{1}{4}e^{-x/2}(-2e^{-1/2} + 2)\, dx = 1 - \frac{1}{4}(2 - 2e^{-1/2})\int_0^1 e^{-x/2}\, dx$$

$$= 1 - \frac{1}{2}(1 - e^{-1/2})(-2e^{-x/2})\Big|_0^1 = 1 - \frac{1}{2}(1 - e^{-1/2})(-2e^{-1/2} + 2)$$

$$= 1 - (1 - e^{-1/2})(1 - e^{-1/2}) = 1 - (1 - e^{-1/2})^2 \simeq 0.8452$$

33. The probability that the entire transaction will take more than half an hour is 1 minus the probability that the transaction will take no more than half an hour (30 minutes). The transaction will take no more than 30 minutes if $x + y \le 30$, that is, if (x, y) is in the region R in the first quadrant below the line $x + y = 30$. This region is described by

$$0 \leq x \leq 30 \qquad \text{and} \qquad 0 \leq y \leq 30 - x$$

Hence, the probability that the transaction will take more than half an hour is

$$1 - P[(x, y) \text{ in } R] = 1 - \int_R \int f(x, y)\, dA$$

$$= 1 - \int_0^{30} \int_0^{30-x} \frac{1}{300} e^{-x/30} e^{-y/10}\, dy\, dx$$

$$= 1 - \int_0^{30} \frac{1}{300} e^{-x/30} (-10 e^{-y/10}) \Big|_0^{30-x} dx$$

$$= 1 - \int_0^{30} \frac{1}{300} e^{-x/30} [-10 e^{(x-30)/10} + 10]\, dx$$

$$= 1 - \int_0^{30} \frac{1}{300} e^{-x/30} [10 - 10 e^{-3} e^{x/10}]\, dx$$

$$= 1 - \int_0^{30} [\frac{1}{30} e^{-x/30} - \frac{1}{30} e^{-3} e^{x/15}]\, dx$$

$$= 1 - (-e^{-x/30}) \Big|_0^{30} + \frac{1}{2} e^{-3} (e^{x/15}) \Big|_0^{30}$$

$$= 1 - (-e^{-1} + 1) + \frac{1}{2} e^{-3} (e^2 - 1)$$

$$= e^{-1} + \frac{1}{2} e^{-1} - \frac{1}{2} e^{-3} = \frac{3}{2} e^{-1} - \frac{1}{2} e^{-3} \simeq 0.5269$$

Chapter 10, Review Problems

1. $\int_0^1 \int_{-2}^0 (2x + 3y)\, dy\, dx = \int_0^1 \left[2xy + \frac{3}{2}y^2\right]\Big|_{-2}^0 dx$

    $= \int_0^1 [0 - (-4x + 6)]\, dx = \int_0^1 (4x - 6)\, dx = (2x^2 - 6x)\big|_0^1 = -4$

2. $\int_0^1 \int_0^y x\sqrt{1 - y^3}\, dx\, dy = \int_0^1 \frac{1}{2}x^2\sqrt{1 - y^3}\Big|_0^y dy$

    $= \frac{1}{2}\int_0^1 y^2\sqrt{1 - y^3}\, dy = -\frac{1}{9}(1 - y^3)^{3/2}\big|_0^1 = \frac{1}{9}$

3. $\int_0^1 \int_{-x}^x \frac{6xy^2}{x^5+1}\, dy\, dx = \int_0^1 \frac{2xy^3}{x^5+1}\Big|_{-x}^x dx$

    $= \int_0^1 \left[\frac{2x}{x^5+1}\right][x^3 - (-x^3)]\, dx = 4\int_0^1 \frac{x^4}{x^5+1}\, dx$

    $= \frac{4}{5}\ln|x^5 + 1|\big|_0^1 = \frac{4}{5}(\ln 2 - \ln 1) = \frac{4}{5}\ln 2$

4. $\int_0^1 \int_0^2 e^{-x-y}\, dy\, dx = \int_0^1 \int^2 e^{-x}e^{-y}\, dy\, dx$

    $= \int_0^1 -e^{-x}e^{-y}\Big|_0^2 dx = \int_0^1 (-e^{-x}e^{-2} + e^{-x})\, dx$

    $= (1 - e^{-2})\int_0^1 e^{-x}\, dx = (1 - e^{-2})(-e^{-x})\big|_0^1$

    $= (1 - e^{-2})(-e^{-1} + 1) \simeq 0.5466$

5. $\int_0^1 \int_0^x xe^{2y}\, dy\, dx = \int_0^1 \frac{1}{2}xe^{2y}\Big|_0^x dx$

    $= \int_0^1 \left[\frac{1}{2}xe^{2x} - \frac{1}{2}x\right] dx = \frac{1}{2}\int_0^1 xe^{2x}\, dx - \frac{1}{2}\int_0^1 x\, dx$

    $= \frac{1}{2}\left[\frac{1}{2}xe^{2x}\big|_0^1 - \frac{1}{2}\int_0^1 e^{2x}\, dx\right] - \frac{1}{2}\int_0^1 x\, dx$

    $= \frac{1}{2}\left[\frac{1}{2}xe^{2x} - \frac{1}{4}e^{2x}\right]\big|_0^1 - \frac{1}{4}x^2\big|_0^1$

    $= \frac{1}{2}\left[\left(\frac{1}{2}e^2 - \frac{1}{4}e^2\right) + \frac{1}{4}\right] - \frac{1}{4} = \frac{1}{8}e^2 - \frac{1}{8} = \frac{1}{8}(e^2 - 1)$

6. Describe R by $-1 \le x \le 2$ and $0 \le y \le 3$. Then,

$$\int_R \int 6x^2 y \, dA = \int_{-1}^{2} \int_{0}^{3} 6x^2 y \, dy \, dx = \int_{-1}^{2} 3x^2 y^2 \Big|_{0}^{3} \, dx$$

$$= \int_{-1}^{2} 27x^2 \, dx = 9x^3 \Big|_{-1}^{2} = 9(8 + 1) = 81$$

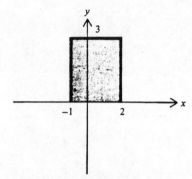

Region for Problem 6.

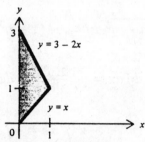

Region for Problem 7.

7.   Describe R by $0 \le x \le 1$ and $x \le y \le 3 - 2x$. Then,

$$\int_R \int (x + 2y) \, dA = \int_{0}^{1} \int_{x}^{3-2x} (x + 2y) \, dy \, dx = \int_{0}^{1} (xy + y^2) \Big|_{x}^{3-2x} \, dx$$

$$= \int_{0}^{1} [3x - 2x^3 + (3 - 2x)^2 - (x^2 + x^2)] \, dx = \int_{0}^{1} (9 - 9x) \, dx$$

$$= \left(9x - \frac{9}{2} x^2 \right) \Big|_{0}^{1} = 9 - \frac{9}{2} = \frac{9}{2}$$

8.   Describe R by $0 \le x \le 4$ and $\frac{1}{2} x \le y \le \sqrt{x}$. Then,

$$\int_R \int 40x^2 y \, dA = \int_{0}^{4} \int_{x/2}^{\sqrt{x}} 40x^2 y \, dy \, dx = \int_{0}^{4} 20x^2 y^2 \Big|_{x/2}^{\sqrt{x}} \, dx$$

$$= \int_{0}^{4} (20x^3 - 5x^4) \, dx = (5x^4 - x^5) \Big|_{0}^{4} = 256$$

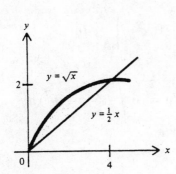

Region for Problem 8.   Region for Problem 9.

9.   Describe R by $0 \leq x \leq 1$ and $x \leq y \leq \sqrt[3]{x}$. Then,

$$\int_R \int 6y^2 e^{x^2} \, dA = \int_0^1 \int_x^{\sqrt[3]{x}} 6y^2 e^{x^2} \, dy \, dx$$

$$= \int_0^1 2y^3 e^{x^2} \Big|_x^{\sqrt[3]{x}} \, dx = \int_0^1 (2x e^{x^2} - 2x^3 e^{x^2}) \, dx$$

$$= \int_0^1 2x e^{x^2} \, dx - \int_0^1 x^2 [2x e^{x^2}] \, dx$$

$$= e^{x^2} \Big|_0^1 - \left[x^2 e^{x^2} \Big|_0^1 - \int_0^1 2x e^{x^2} \, dx \right]$$

$$= \left[e^{x^2} - x^2 e^{x^2} + e^{x^2} \right] \Big|_0^1 = (2 - x^2) e^{x^2} \Big|_0^1 = e - 2$$

10.   Describe R by $1 \leq y \leq 2$ and $y^2 \leq x \leq \frac{8}{y}$. Then,

$$\int_R\int 32xy^3\,dA = \int_0^2\int_{y^2}^{8/y} 32xy^3\,dx\,dy = \int_1^2 16x^2y^3\Big|_{y^2}^{8/y}\,dy$$

$$= 16\int_1^2\left[\frac{64}{y^2}(y^3) - y^4(y^3)\right]\,dy = 16\int_1^2(64y - y^7)\,dy$$

$$= 16\left(32y^2 - \frac{1}{8}y^8\right)\Big|_1^2$$

$$= 16\left[32(4) - \frac{1}{8}(256)\right] - 16\left[32 - \frac{1}{8}\right] = 1{,}536 - 510 = 1{,}026$$

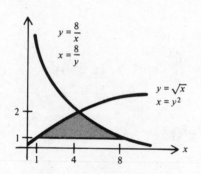

Region for Problem 10.

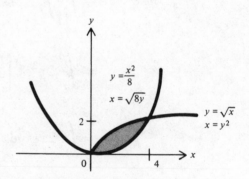

Region for Problem 11.

11. The region for $\int_0^4\int_{x^2/8}^{\sqrt{x}} f(x,y)\,dy\,dx$ is described by

$$0 \le x \le 4 \qquad \text{and} \qquad \frac{1}{8}x^2 \le y \le \sqrt{x}$$

This region can also be described by

$$0 \le y \le 2 \qquad \text{and} \qquad y^2 \le x \le \sqrt{8y}$$

Hence, an equivalent iterated integral is

$$\int_0^2\int_{y^2}^{\sqrt{8y}} f(x,y)\,dx\,dy$$

12. The region for $\int_0^2\int_1^{e^y} f(x,y)\,dx\,dy$ is described by

$$0 \le y \le 2 \qquad \text{and} \qquad 1 \le x \le e^y$$

This region can also be described by

$$1 \leq x \leq e^2 \qquad \text{and} \qquad \ln x \leq y \leq 2$$

Hence, an equivalent iterated integral is

$$\int_1^{e^2} \int_{\ln x}^2 f(x, y)\, dy\, dx$$

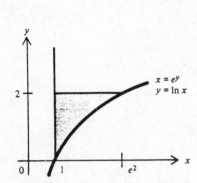

Region for Problem 12.

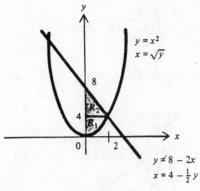

Region for Problem 13.

13.   The region for $\int_0^2 \int_{x^2}^{8-2x} f(x, y)\, dy\, dx$ is described by

$$0 \leq x \leq 2 \qquad \text{and} \qquad x^2 \leq y \leq 8 - 2x$$

This region R can also be written as $R = R_1 + R_2$, where R_1 is described by

$$0 \leq y \leq 4 \qquad \text{and} \qquad 0 \leq x \leq \sqrt{y}$$

and R_2 is described by

$$4 \leq y \leq 8 \qquad \text{and} \qquad 0 \leq x \leq 4 - \frac{y}{2}$$

Hence, an equivalent iterated integral is

$$\int_0^4 \int_0^{\sqrt{y}} f(x, y)\, dx\, dy + \int_4^8 \int_0^{4-y/2} f(x, y)\, dx\, dy$$

14.   The region for $\int_{-3}^1 \int_{x^2}^{6-x} f(x, y)\, dy\, dx$ is described by

$$-3 \leq x \leq 1 \qquad \text{and} \qquad x^2 \leq y \leq 6 - x$$

This region R can also be written as $R = R_1 + R_2 + R_3$, where R_1 is described by

$$0 \leq y \leq 1 \qquad \text{and} \qquad -\sqrt{y} \leq x \leq \sqrt{y}$$

R_2 is described by

$$1 \leq y \leq 5 \qquad \text{and} \qquad -\sqrt{y} \leq x \leq 1$$

and R_3 is described by

$$5 \leq y \leq 9 \qquad \text{and} \qquad -\sqrt{y} \leq x \leq 6 - y$$

Hence, an equivalent iterated integral is

$$\int_1^0 \int_{-\sqrt{y}}^{\sqrt{y}} f(x,y)\,dx\,dy + \int_1^5 \int_{-\sqrt{y}}^{2} f(x,y)\,dx\,dy + \int_5^9 \int_{-\sqrt{y}}^{6-y} f(x,y)\,dx\,dy$$

15. Describe R by $-1 \leq x \leq 2$ and $x^2 - 4 \leq y \leq x - 2$. Then,

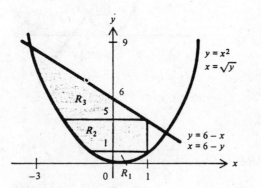

Region for Problem 14.

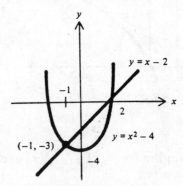

Region for Problem 15.

$$\text{Area of } R = \int_{-1}^{2} \int_{x^2-4}^{x-2} 1\,dy\,dx = \int_{-1}^{2} y \Big|_{x^2-4}^{x-2} dx$$

$$= \int_{1}^{2} (x - 2 - x^2 + 4)\,dx = \int_{-1}^{2} (x + 2 - x^2)\,dx$$

$$= \left[\frac{1}{2}x^2 + 2x - \frac{1}{3}x^3 \right]\Big|_{-1}^{2} = \left[2 + 4 - \frac{8}{3} \right] - \left[\frac{1}{2} - 2 + \frac{1}{3} \right] = \frac{9}{2}$$

16. Describe R by $0 \leq y \leq 1$ and $0 \leq x \leq e^y$. Then,

$$\text{Area of } R = \int_0^1 \int_0^{e^y} 1\,dx\,dy = \int_0^1 x \Big|_0^{e^y} dy$$

$$= \int_0^1 e^y\,dy = e^y \big|_0^1 = e - 1$$

17. Describe R by $0 \leq y \leq 1$ and $\sqrt{y} \leq x \leq 2 - y$. Then,

$$\text{Area of } R = \int_0^1 \int_{\sqrt{y}}^{2-y} 1\, dx\, dy$$

$$= \int_0^1 (2 - y - \sqrt{y})\, dy = \left[2y - \frac{1}{2}y^2 - \frac{2}{3}y^{3/2} \right]\Big|_0^1$$

$$= 2 - \frac{1}{2} - \frac{2}{3} = \frac{5}{6}$$

18. Describe R by $0 \leq x \leq 2$ and $0 \leq y \leq 1 - \frac{1}{2}x$. Then,

$$\text{Volume} = \int_0^2 \int_0^{1-x/2} 2xy\, dy\, dx = \int_0^2 xy^2 \Big|_0^{1-x/2} dx$$

$$= \int_0^2 x \left[1 - \frac{1}{2}x \right]^2 dx = \int_0^2 \left[x - x^2 + \frac{1}{4}x^3 \right] dx$$

$$= \left[\frac{1}{2}x^2 - \frac{1}{3}x^3 + \frac{1}{16}x^4 \right]\Big|_0^2 = 2 - \frac{8}{3} + 1 = \frac{1}{3}$$

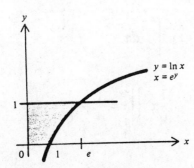

Region for Problem 16.

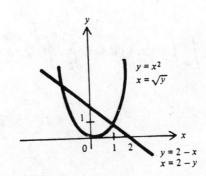

Region for Problem 17.

19. Describe R by $0 \leq x \leq 1$ and $x^2 \leq y \leq x$. Then,

$$\text{Volume} = \int_0^1 \int_{x^2}^x xe^y \, dy \, dx = \int_0^1 xe^y \Big|_{x^2}^x dy$$

$$= \int_0^1 (xe^x - xe^{x^2}) \, dx = \int_0^1 xe^x \, dx - \int_0^1 xe^{x^2} \, dx$$

$$= (xe^x - e^x)\Big|_0^1 - \frac{1}{2}e^{x^2}\Big|_0^1$$

$$= [(e - e) + 1] - \left[\frac{1}{2}e - \frac{1}{2}\right] = \frac{3}{2} - \frac{1}{2}e = \frac{1}{2}(3 - e)$$

20. Describe R by $0 \le x \le 1$ and $x^2 \le y \le x$. Then,

$$\text{Area of } R = \int_0^1 \int_{x^2}^x 1 \, dy \, dx = \int_0^1 y \Big|_{x^2}^x dx$$

$$= \int_0^1 (x - x^2) \, dx = \left[\frac{1}{2}x^2 - \frac{1}{3}x^3\right]\Big|_0^1 = \frac{1}{2} - \frac{1}{3} = \frac{1}{6}$$

and

$$\int_R \int f(x, y) \, dA = \int_0^1 \int_{x^2}^x 2xy \, dy \, dx = \int_0^1 xy^2 \Big|_{x^2}^x dx$$

$$= \int_0^1 (x^3 - x^5) \, dx = \left[\frac{1}{4}x^4 - \frac{1}{6}x^6\right]\Big|_0^1 = \frac{1}{4} - \frac{1}{6} = \frac{1}{12}$$

Hence,

$$\text{Average value} = \frac{1}{\text{area of } R} \int_R \int f(x, y) \, dA = 6 \left[\frac{1}{12}\right] = \frac{1}{2}$$

21. (a) The number of cubic feet of snow is the volume under the surface $f(x, y)$ of the snow. Hence,

$$\text{Volume of snow} = \int_R \int f(x, y)\, dA$$

$$\text{Average depth} = \frac{1}{\text{area of } R} \int_R \int f(x, y)\, dA$$

(b)

$$= \frac{\int_R \int f(x, y)\, dA}{\int_R \int 1\, dA}$$

22. The joint probability density function is $f(x, y) = 6e^{-2x}e^{-3y}$.

(a) $P(0 \leq x \leq 1 \text{ and } 0 \leq y \leq 2) = \int_0^2 \int_0^1 6e^{-2x}e^{-3y}\, dx\, dy$

$$= \int_0^2 -3e^{-2x}e^{-3y}\Big|_0^1 dy$$

$$= \int_0^2 (-3e^{-2}e^{-3y} + 3e^{-3y})\, dy = 3(1 - e^{-2}) \int_0^2 e^{-3y}\, dy$$

$$= -(1 - e^{-2})e^{-3y}\Big|_0^2 = -(1 - e^{-2})(e^{-6} - 1) \simeq 0.8625$$

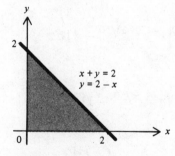

Region for Problem 22b.

(b) The region R defined by $x + y \leq 2$ is described by

$$0 \leq x \leq 2 \quad \text{and} \quad 0 \leq y \leq 2 - x$$

Hence,

$$P(x+y \leq 2) = \int_0^2 \int_0^{2-x} 6e^{-2x}e^{-3y}\, dy\, dx = \int_0^2 -2e^{-2x}e^{-3y}\Big|_0^{2-x}\, dx$$

$$= \int_0^2 [-2e^{-2x}e^{-6+3x} + 2e^{-2x}]\, dx = 2\int_0^2 e^{-2x}\, dx - 2e^{-6}\int_0^2 e^x\, dx$$

$$= -e^{-2x}\Big|_0^2 - 2e^{-6}e^x\Big|_0^2$$

$$= -(e^{-4}-1) - 2e^{-6}(e^2-1) = 1 - 3e^{-4} + 2e^{-6} \simeq 0.9500$$

23. The probability of being late for the meeting is 1 minus the probability of not being late. The probability of not being late is the probability that the total time $x+y$ is less than or equal to 50. The corresponding region is described by

$$0 \leq x \leq 50 \qquad \text{and} \qquad 0 \leq y \leq 50 - x$$

Hence, the probability of being late is

$$1 - \int_0^{50} \int_0^{50-x} \frac{1}{500}e^{-x/10}e^{-y/50}\, dy\, dx$$

$$= 1 - \int_0^{50} -\frac{1}{10}e^{-x/10}e^{-y/50}\Big|_0^{50-x}\, dx$$

$$= 1 - \int_0^{50} -\frac{1}{10}e^{-x/10}[e^{-(50-x)/50} - 1]\, dx$$

$$= 1 - \int_0^{50} -\frac{1}{10}e^{-x/10}[e^{-1}e^{x/50} - 1]\, dx$$

$$= 1 + \frac{1}{10}e^{-1}\int_0^{50} e^{-2x/25}\, dx - \frac{1}{10}\int_0^{50} e^{-x/10}\, dx$$

$$= 1 + \left[-\frac{5}{4}e^{-1}e^{-2x/25}\right]\Big|_0^{50} + e^{-x/10}\Big|_0^{50}$$

$$= 1 - \frac{5}{4}e^{-1}(e^{-4} - 1) + e^{-5} - 1 = \frac{5}{4}e^{-1} - \frac{1}{4}e^{-5} \simeq 0.4582$$

CHAPTER 11

INFINITE SERIES AND TAYLOR APPROXIMATION

1 INFINITE SERIES

2 THE GEOMETRIC SERIES

3 TAYLOR APPROXIMATION

4 NEWTON'S METHOD

REVIEW PROBLEMS

Chapter 11, Section 1

1. $\frac{1}{3} + \frac{1}{9} + \frac{1}{27} + \frac{1}{81} + \cdots = \frac{1}{3^1} + \frac{1}{3^2} + \frac{1}{3^3} + \frac{1}{3^4} + \cdots = \sum_{n=1}^{\infty} \frac{1}{3^n}$

3. $\frac{1}{2} + \frac{2}{3} + \frac{3}{4} + \frac{4}{5} + \cdots = \frac{2-1}{2} + \frac{3-1}{3} + \frac{4-1}{4} + \frac{5-1}{5} + \cdots = \sum_{n=2}^{\infty} \frac{n-1}{n}$

or, equivalently,

$\frac{1}{2} + \frac{2}{3} + \frac{3}{4} + \frac{4}{5} + \cdots = \frac{1}{1+1} + \frac{2}{2+1} + \frac{3}{3+1} + \frac{4}{4+1} + \cdots = \sum_{n=1}^{\infty} \frac{n}{n+1}$

5. $\frac{1}{2} - \frac{4}{3} + \frac{9}{4} - \frac{16}{5} + \cdots = \frac{1^2}{1+1} - \frac{2^2}{2+1} + \frac{3^2}{3+1} - \frac{4^2}{4+1} + \cdots$

$= \sum_{n=1}^{\infty} (-1)^{n+1} \frac{n^2}{n+1}$

7.   For $\sum_{n=1}^{\infty} \frac{1}{2^n}$,

$$S_4 = \frac{1}{2^1} + \frac{1}{2^2} + \frac{1}{2^3} + \frac{1}{2^4} = \frac{1}{2} + \frac{1}{4} + \frac{1}{8} + \frac{1}{16} = \frac{15}{16}$$

9.   For $\sum_{n=1}^{\infty} \frac{(-1)^n}{n}$

$$S_4 = \frac{(-1)^1}{1} + \frac{(-1)^2}{2} + \frac{(-1)^3}{3} + \frac{(-1)^4}{4} = -1 + \frac{1}{2} - \frac{1}{3} + \frac{1}{4} = -\frac{7}{12}$$

11.   For $\sum_{n=1}^{\infty} \left[\frac{1}{n+3} - \frac{1}{n+4} \right]$,

$$S_n = \left[\frac{1}{4} - \frac{1}{5}\right] + \left[\frac{1}{5} - \frac{1}{6}\right] + \left[\frac{1}{6} - \frac{1}{7}\right] + \left[\frac{1}{7} - \frac{1}{8}\right] + \cdots + \left[\frac{1}{n+3} - \frac{1}{n+4}\right]$$

$$= \frac{1}{4} - \frac{1}{n+4}$$

Hence,

$$\lim_{n\to\infty} S_n = \lim_{n\to\infty}\left[\frac{1}{4} - \frac{1}{n+4}\right] = \frac{1}{4}$$

and so

$$\sum_{n=1}^{\infty}\left[\frac{1}{n+3} - \frac{1}{n+4}\right] = \frac{1}{4}$$

13. Since $\sum_{n=1}^{\infty}\frac{1}{(n+1)(n+2)} = \sum_{n=1}^{\infty}\left[\frac{1}{n+1} - \frac{1}{n+2}\right]$,

$$S_n = \left[\frac{1}{2} - \frac{1}{3}\right] + \left[\frac{1}{3} - \frac{1}{4}\right] + \left[\frac{1}{4} - \frac{1}{5}\right] + \left[\frac{1}{5} - \frac{1}{6}\right] + \cdots + \left[\frac{1}{n+1} - \frac{1}{n+2}\right]$$

$$= \frac{1}{2} - \frac{1}{n+2}$$

Hence,

$$\lim_{n\to\infty} S_n = \lim_{n\to\infty}\left[\frac{1}{2} - \frac{1}{n+2}\right] = \frac{1}{2}$$

and so

$$\sum_{n=1}^{\infty}\frac{1}{(n+1)(n+2)} = \frac{1}{2}$$

15. For $\sum_{n=1}^{\infty}\frac{6}{10^n}$,

$$S_n = \frac{6}{10} + \frac{6}{10^2} + \frac{6}{10^3} + \cdots + \frac{6}{10^n}$$

and

$$\frac{1}{10}S_n = \frac{6}{10^2} + \frac{6}{10^3} + \frac{6}{10^4} + \cdots + \frac{6}{10^n} + \frac{6}{10^{n+1}}$$

Hence,

$$S_n - \frac{1}{10}S_n = \frac{9}{10}S_n = \frac{6}{10} - \frac{6}{10^{n+1}} = \frac{6}{10}\left[1 - \frac{1}{10^n}\right]$$

$$S_n = \left[\frac{10}{9}\right]\left[\frac{6}{10}\right]\left[1 - \frac{1}{10^n}\right] = \frac{2}{3}\left[1 - \frac{1}{10^n}\right]$$

Hence,

$$\lim_{n\to\infty} S_n = \lim_{n\to\infty} \frac{2}{3}\left[1 - \frac{1}{10^n}\right] = \frac{2}{3}$$

and so

$$\sum_{n=1}^{\infty} \frac{6}{10^n} = \frac{2}{3}$$

17. For $\sum_{n=1}^{\infty} \frac{4}{3^n}$,

$$S_n = \frac{4}{3} + \frac{4}{3^2} + \frac{4}{3^3} + \cdots + \frac{4}{3^n}$$

and

$$\frac{1}{3}S_n = \frac{4}{3^2} + \frac{4}{3^3} + \frac{4}{3^4} + \cdots + \frac{4}{3^n} + \frac{4}{3^{n+1}}$$

Hence,

$$S_n - \frac{1}{3}S_n = \frac{2}{3}S_n = \frac{4}{3} - \frac{4}{3^{n+1}} = \frac{4}{3}\left[1 - \frac{1}{3^n}\right]$$

or

$$S_n = \left[\frac{3}{2}\right]\left[\frac{4}{3}\right]\left[1 - \frac{1}{3^n}\right] = 2\left[1 - \frac{1}{3^n}\right]$$

Hence,

$$\lim_{n\to\infty} S_n = \lim_{n\to\infty} 2\left[1 - \frac{1}{3^n}\right] = 2$$

and so

$$\sum_{n=1}^{\infty} \frac{4}{3^n} = 2$$

19. The series $\sum_{n=1}^{\infty} \frac{n}{2n+1}$ diverges since

$$\lim_{n\to\infty} a_n = \lim_{n\to\infty} \frac{n}{2n+1} = \lim_{n\to\infty} \frac{1}{2 + 1/n} = \frac{1}{2} \neq 0$$

21. The series $\sum_{n=1}^{\infty} \left[\frac{1}{n+1} - \frac{1}{n}\right]$ converges to -1 since

$$\lim_{n \to \infty} S_n = \lim_{n \to \infty} \left[\left(\frac{1}{2} - 1 \right) + \left(\frac{1}{3} - \frac{1}{2} \right) + \left(\frac{1}{4} - \frac{1}{3} \right) + \cdots + \left(\frac{1}{n+1} - \frac{1}{n} \right) \right]$$

$$= \lim_{n \to \infty} \left[-1 + \frac{1}{n+1} \right] = -1$$

23. The series $\displaystyle\sum_{n=1}^{\infty} \left[-\frac{2}{3} \right]^n$ converges to $-\frac{2}{5}$ since

$$S_n = \left[-\frac{2}{3} \right] + \left[-\frac{2}{3} \right]^2 + \left[-\frac{2}{3} \right]^3 + \cdots + \left[-\frac{2}{3} \right]^n$$

and

$$-\frac{2}{3} S_n = \left[-\frac{2}{3} \right]^2 + \left[-\frac{2}{3} \right]^3 + \left[-\frac{2}{3} \right]^4 + \cdots + \left[-\frac{2}{3} \right]^n + \left[-\frac{2}{3} \right]^{n+1}$$

so that

$$S_n + \frac{2}{3} S_n = -\frac{2}{3} - \left[-\frac{2}{3} \right]^{n+1}$$

$$\frac{5}{3} S_n = -\frac{2}{3} \left[1 - \left(-\frac{2}{3} \right)^n \right] \quad \text{or} \quad S_n = -\frac{2}{5} \left[1 - \left(-\frac{2}{3} \right)^n \right]$$

and so

$$\lim_{n \to \infty} S_n = \lim_{n \to \infty} -\frac{2}{5} \left[1 - \left(-\frac{2}{3} \right)^n \right] = -\frac{2}{5}$$

25. The series $\displaystyle\sum_{n=1}^{\infty} \frac{1}{\sqrt{n}}$ diverges since

$$S_n = \frac{1}{\sqrt{1}} + \frac{1}{\sqrt{2}} + \frac{1}{\sqrt{3}} + \cdots + \frac{1}{\sqrt{n}}$$

$$\geq \frac{1}{\sqrt{n}} + \frac{1}{\sqrt{n}} + \frac{1}{\sqrt{n}} + \cdots + \frac{1}{\sqrt{n}} = \frac{n}{\sqrt{n}} = \sqrt{n}$$

which increases without bound as $n \to \infty$.

Chapter 11, Section 2

1. $\displaystyle\sum_{n=0}^{\infty} \left[\frac{4}{5}\right]^n = \frac{1}{1-4/5} = 5$

3. $\displaystyle\sum_{n=0}^{\infty} \frac{2}{3^n} = 2\sum_{n=0}^{\infty} \left[\frac{1}{3}\right]^n = 2\left[\frac{1}{1-1/3}\right] = 3$

5. For $\displaystyle\sum_{n=1}^{\infty} \left[\frac{3}{2}\right]^n$, the ratio is $r = \frac{3}{2} > 1$, and so the geometric series diverges.

7. $\displaystyle\sum_{n=2}^{\infty} \frac{3}{(-4)^2} = 3\left[\left(-\frac{1}{4}\right)^2 + \left(-\frac{1}{4}\right)^3 + \left(-\frac{1}{4}\right)^4 + \cdots\right]$

    $= 3\left[-\frac{1}{4}\right]^2\left[1 + \left(-\frac{1}{4}\right) + \left(-\frac{1}{4}\right)^2 + \cdots\right] = \frac{3}{16}\left[\frac{1}{1-(-1/4)}\right] = \frac{3}{20}$

9. $\displaystyle\sum_{n=1}^{\infty} 5(0.9)^n = 5[0.9 + (0.9)^2 + (0.9)^3 + \cdots]$

    $= 5(0.9)[1 + 0.9 + (0.9)^2 + \cdots] = 5(0.9)\left[\frac{1}{1-0.9}\right] = 4.5(10) = 45$

11. $\displaystyle\sum_{n=1}^{\infty} \frac{3^n}{4^{n+2}} = \sum_{n=1}^{\infty} \frac{3^n}{4^n 4^2} = \frac{1}{16}\sum_{n=1}^{\infty} \left[\frac{3}{4}\right]^n$

    $= \frac{1}{16}\left[\left(\frac{3}{4}\right) + \left(\frac{3}{4}\right)^2 + \left(\frac{3}{4}\right)^3 + \cdots\right] = \left[\frac{1}{16}\right]\left[\frac{3}{4}\right]\left[1 + \frac{3}{4} + \left(\frac{3}{4}\right)^2 + \cdots\right]$

    $= \left[\frac{1}{16}\right]\left[\frac{3}{4}\right]\left[\frac{1}{1-3/4}\right] = \left[\frac{1}{16}\right]\left[\frac{3}{4}\right][4] = \frac{3}{16}$

13. $\displaystyle\sum_{n=0}^{\infty} \frac{4^{n+1}}{5^{n-1}} = \sum_{n=0}^{\infty} \frac{4^n 4}{5^n 5^{-1}} = \frac{4}{5^{-1}}\sum_{n=0}^{\infty} \left[\frac{4}{5}\right]^n = 20\left[\frac{1}{1-4/5}\right] = 20(5) = 100$

15. Write the decimal as a geometric series as follows:

    $$0.3333\ldots = \frac{3}{10} + \frac{3}{100} + \frac{3}{1,000} + \frac{3}{10,000} + \cdots$$

    $$= \frac{3}{10}\left[1 + \frac{1}{10} + \frac{1}{100} + \frac{1}{1,000} + \cdots\right]$$

    $$= \frac{3}{10}\left[1 + \frac{1}{10} + \left(\frac{1}{10}\right)^2 + \left(\frac{1}{10}\right)^3 + \cdots\right]$$

    $$= \frac{3}{10}\left[\frac{1}{1 - 1/10}\right] = \left[\frac{3}{10}\right]\left[\frac{10}{9}\right] = \frac{1}{3}$$

17. Write the decimal as a geometric series as follows:

$$0.252525\ldots = \frac{25}{100} + \frac{25}{10,000} + \frac{25}{100,000} + \cdots$$

$$= \frac{25}{100}\left[1 + \frac{1}{100} + \frac{1}{10,000} + \cdots\right] = \frac{25}{100}\left[1 + \frac{1}{100} + \left(\frac{1}{100}\right)^2 + \cdots\right]$$

$$= \frac{25}{100}\left[\frac{1}{1 - 1/100}\right] = \left[\frac{25}{100}\right]\left[\frac{100}{99}\right] = \frac{25}{99}$$

19. Since 92 percent of all income is spent, the amount (in billions) spent by beneficiaries of the 50 billion dollar tax cut is

$$0.92(50)$$

This becomes new income, of which 92 percent or

$$0.92[0.92(50)] = (0.92)^2(50)$$

is spent. this, in turn, generates additional spending of

$$0.92[(0.92)^2(50)] = (0.92)^3(50)$$

and so on. The total amount spent if this process continues indefinitely is

$$(0.92)(50) + (0.92)^2(50) + (0.92)^3(50) + \cdots$$

$$= (0.92)(50)[1 + (0.92) + (0.92)^2 + \cdots]$$

$$= (0.92)(50)\left[\frac{1}{1 - 0.92}\right] = 46\left[\frac{1}{0.08}\right] = 575 \text{ billion dollars}$$

21. The present value of the investment is the sum of present values of the individual payments. Thus,

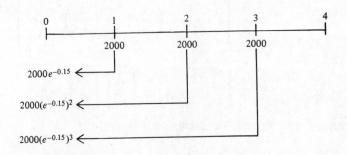

Present values of payments for Problem 21.

$$\text{Present value} = 2,000e^{-0.15} + 2,000(e^{-0.15})^2 + 2,000(e^{-0.15})^3 + \cdots$$

$$= 2,000e^{-0.15}[1 + e^{-0.15} + (e^{-0.15})^2 + \cdots]$$

$$= 2,000e^{-0.15}\left[\frac{1}{1 - e^{-0.15}}\right] \simeq \$12,358.32$$

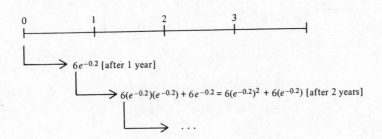

Group membership for Problem 23.

23. As suggested by the figure, the number of active trustees in the long run is

$$6(e^{-0.2}) + 6(e^{-0.2})^2 + 6(e^{-0.2})^3 + 6(e^{-0.2})^4 + \cdots$$

$$= 6e^{-0.2}[1 + e^{-0.2} + (e^{-0.2})^2 + (e^{-0.2})^3 + \cdots]$$

$$= 6e^{-0.2}\left[\frac{1}{1 - e^{-0.2}}\right] \simeq 27$$

25. Let x denote the number of the trial on which the first old penny is found. Then,

$$P(x \geq 100) = \sum_{n=100}^{\infty} P(x = n) = \sum_{n=100}^{\infty} (1 - p)^{n-1}p$$

where $p = 0.002$ is the probability of finding an old penny and $1 - p = 0.998$. Hence,

$$P(x \geq 100) = \sum_{n=100}^{\infty} (0.998)^{n-1}(0.002)$$

$$= (0.998)^{99}(0.002) + (0.998)^{100}(0.002) + \cdots$$

$$= (0.002)(0.998)^{99}[1 + 0.998 + (0.998)^2 + \cdots]$$

$$= (0.002)(0.998)^{99} \left[\frac{1}{1 - 0.998} \right]$$

$$= (0.002)(0.998)^{99} \left[\frac{1}{0.002} \right]$$

$$= (0.998)^{99} \simeq 0.8202$$

Chapter 11, Section 3

1. For $f(x) = e^{3x}$ about $x = 0$, compute the Taylor coefficients as follows:

$$f(x) = e^{3x}, \ f'(x) = 3e^{3x}, \ f''(x) = 3^2 e^{3x}, \ f^{(3)}(x) = 3^3 e^{3x}, \dots, \ f^{(n)}(x) = 3^n e^{3x}$$

and so,

$$f(0) = 1, \ f'(0) = 3, \ f''(0) = 3^2, \ f^{(3)}(0) = 3^3, \dots, \ f^{(n)}(0) = 3^n$$

Hence, for $n = 0, 1, 2, \dots,$

$$a_n = \frac{f^{(n)}(0)}{n!} = \frac{3^n}{n!}$$

and the corresponding Taylor series is $\sum_{n=0}^{\infty} \frac{3^n}{n!} x^n$.

3. For $f(x) = \ln(1 + x)$ about $x = 0$, compute the Taylor coefficients as follows:

$$f(x) = \ln(1 + x), \quad f'(x) = \frac{1}{1 + x}, \quad f''(x) = \frac{-1}{(1 + x)^2}, \quad f^{(3)}(x) = \frac{-2(-1)}{(1 + x)^3},$$

$$f^{(4)}(x) = \frac{(-3)(-2)(-1)}{(1 + x)^4}, \dots, \ f^{(n)}(x) = (-1)^{n+1}\frac{(n - 1)!}{(1 + x)^n}$$

and so

$$f(0) = 0, \quad f'(0) = 1, \quad f''(0) = -1, \quad f^{(3)}(0) = 2!,$$

$$f^{(4)}(0) = -3!, \dots, \ f^{(n)}(0) = (-1)^{n+1}(n - 1)!$$

Hence, for $n = 1, 2, 3, \dots$

$$a_n = \frac{f^{(n)}(0)}{n!} = \frac{(-1)^{n+1}(n - 1)!}{n!} = \frac{(-1)^{n+1}}{n}$$

and the corresponding Taylor series is $\sum_{n=1}^{\infty} \frac{(-1)^{n+1}}{n} x^n$.

5. For $f(x) = \frac{e^x + e^{-x}}{2}$ about $x = 0$, compute the Taylor coefficients as follows:

$$f(x) = \frac{e^x + e^{-x}}{2}, \quad f'(x) = \frac{e^x - e^{-x}}{2}, \quad f''(x) = \frac{e^x + e^{-x}}{2}, \dots$$

and so

$$f(0) = 1, \quad f'(0) = 0, \quad f''(0) = 1, \dots$$

Hence,

$$a_n = \frac{f^{(n)}(0)}{n!} = \begin{cases} \frac{1}{n!} & \text{if } n \text{ is even} \\ 0 & \text{if } n \text{ is odd} \end{cases}$$

or

$$a_{2n} = \frac{1}{(2n)!} \quad \text{for} \quad n = 0, 1, 2, 3, \dots$$

The corresponding Taylor series is $\sum_{n=0}^{\infty} \frac{1}{(2n)!} x^{2n}$.

7. For $f(x) = (1+x)e^x$ about $x = 0$, compute the Taylor coefficients (using the product rule) as follows:

$$f(x) = (1+x)e^x, \quad f'(x) = (2+x)e^x, \quad f''(x) = (3+x)e^x$$

$$f^{(3)}(x) = (4+x)e^x, \dots, f^{(n)}(x) = (n+1+x)e^x$$

and so

$$f(0) = 1, \quad f'(0) = 2, \quad f''(0) = 3, \quad f^{(3)}(0) = 4, \dots, f^{(n)}(0) = n+1$$

Hence, for $n = 0, 1, 2, \dots$

$$a_n = \frac{f^{(n)}(0)}{n!} = \frac{n+1}{n!}$$

and the corresponding Taylor series is $\sum_{n=0}^{\infty} \frac{n+1}{n!} x^n$.

9. For $f(x) = e^{2x}$ about $x = 1$, compute the Taylor coefficients as follows:

$$f(x) = e^{2x}, \ f'(x) = 2e^{2x}, \ f''(x) = 2^2 e^{2x}, \ f^{(3)}(x) = 2^3 e^{2x}, \dots, f^{(n)}(x) = 2^n e^{2x}$$

and so

$$f(1) = e^2, \ f'(1) = 2e^2, \ f''(1) = 2^2 e^2, \ f^{(3)}(1) = 2^3 e^2, \dots, f^{(n)}(1) = 2^n e^2$$

Hence, for $n = 0, 1, 2, \dots$

$$a_n = \frac{f^{(n)}(1)}{n!} = \frac{2^n e^2}{n!}$$

and the corresponding Taylor series is $\sum_{n=0}^{\infty} \frac{2^n e^2}{n!} (x-1)^n$.

11. For $f(x) = \frac{1}{x}$ about $x = 1$, compute the Taylor coefficients as follows:

$$f(x) = \frac{1}{x}, \quad f'(x) = \frac{-1}{x^2}, \quad f''(x) = \frac{(-2)(-1)}{x^3},$$

$$f^{(3)}(x) = \frac{(-3)(-2)(-1)}{n^4}, \dots, f^{(n)}(x) = \frac{(-1)^n n!}{x^{n+1}}$$

and so

$$f(1) = 1, \quad f'(1) = -1, \quad f''(1) = 2!, \quad f^{(3)}(1) = -3!, \ldots, f^{(n)}(1) = (-1)^n n!$$

Hence, for $n = 0, 1, 2, \ldots$

$$a_n = \frac{f^{(n)}(1)}{n!} = \frac{(-1)^n n!}{n!} = (-1)^n$$

and the corresponding Taylor series is $\displaystyle\sum_{n=0}^{\infty} (-1)^n (x-1)^n$.

13. For $f(x) = \frac{1}{2-x}$ about $x = 1$, compute the Taylor coefficients as follows:

$$f(x) = \frac{1}{2-x}, \quad f'(x) = \frac{1}{(2-x)^2}, \quad f''(x) = \frac{2}{(2-x)^3},$$

$$f^{(3)}(x) = \frac{3(2)}{(2-x)^4}, \ldots, f^{(n)}(x) = \frac{n!}{(2-x)^{n+1}}$$

and so,

$$f(1) = 1, \quad f'(1) = 1, \quad f''(1) = 2, \quad f^{(3)}(1) = 3!, \ldots, f^{(n)}(1) = n!$$

Hence, for $n = 0, 1, 2, \ldots$

$$a_n = \frac{f^{(n)}(1)}{n!} = \frac{n!}{n!} = 1$$

and the corresponding Taylor series is $\displaystyle\sum_{n=0}^{\infty} (x-1)^n$.

15. Start with the fact that

$$\frac{1}{1-x} = \sum_{n=0}^{\infty} x^n \qquad \text{for} \quad |x| < 1$$

replace x by $5x$ to get

$$\frac{1}{1-5x} = \sum_{n=0}^{\infty} (5x)^n \qquad \text{for} \quad |5x| < 1 \text{ or } |x| < \frac{1}{5}$$

and multiply by x to get

$$\frac{x}{1-5x} = \sum_{n=0}^{\infty} 5^n x^{n+1} \qquad \text{for} \quad |x| < \frac{1}{5}$$

17. Start with the fact that

$$e^x = \sum_{n=0}^{\infty} \frac{x^n}{n!} \qquad \text{for all } x$$

and replace x by $-\frac{x}{2}$ to get

$$e^{-x/2} = \sum_{n=0}^{\infty} \frac{1}{n!}\left[-\frac{x}{2}\right]^n = \sum_{n=0}^{\infty} \frac{x^n}{n!(-2)^n} \qquad \text{for all } x$$

19. From Example 3.6, the Taylor polynomial of degree 3 for $f(x) = \sqrt{x}$ about $x = 4$ is

$$P_3(x) = 2 + \frac{1}{4}(x-4) - \frac{1}{64}(x-4)^2 + \frac{1}{512}(x-4)^3$$

and so

$$\sqrt{3.8} \simeq P_3(3.8) = 2 + \frac{1}{4}(-0.2) - \frac{1}{64}(-0.2)^2 + \frac{1}{512}(-0.2)^3 \simeq 1.94936$$

21. From Example 3.4, the Taylor polynomial of degree 5 for $f(x) = \ln x$ about $x = 1$ is

$$P_5(x) = (x-1) - \frac{1}{2}(x-1)^2 + \frac{1}{3}(x-1)^3 - \frac{1}{4}(x-1)^4 + \frac{1}{5}(x-1)^5$$

and so

$$\ln(1.1) \simeq P_5(1.1) = (0.1) - \frac{1}{2}(0.1)^2 + \frac{1}{3}(0.1)^3 - \frac{1}{4}(0.1)^4 + \frac{1}{5}(0.1)^5$$

$$\simeq 0.09531$$

23. From Example 3.5, the Taylor polynomial of degree 4 for $f(x) = e^x$ about $x = 0$ is

$$P_4(x) = 1 + x + \frac{x^2}{2!} + \frac{x^3}{3!} + \frac{x^4}{4!}$$

and so

$$e^{0.3} \simeq P_4(0.3) = 1 + 0.3 + \frac{(0.3)^2}{2!} + \frac{(0.3)^3}{3!} + \frac{(0.3)^4}{4!} \simeq 1.34984$$

25. Since $e^x = \sum_{n=0}^{\infty} \frac{x^n}{n!}$, it follows that

$$e^{-x^2} = \sum_{n=0}^{\infty} \frac{(-x^2)^n}{n!} = \sum_{n=0}^{\infty} \frac{(-1)^n}{n!} x^{2n}$$

so that

$$P_6(x) = 1 - x^2 + \frac{1}{2}x^4 - \frac{1}{6}x^6$$

Hence,

$$\int_0^{1/2} e^{-x^2}\, dx = \int_0^{1/2} P_6(x)\, dx = \int_0^{1/2} \left(1 - x^2 + \frac{1}{2}x^4 - \frac{1}{6}x^6\right) dx$$

$$= \left[x - \frac{1}{3}x^3 + \frac{1}{10}x^5 - \frac{1}{42}x^7\right]\Big|_0^{1/2}$$

$$= \frac{1}{2} - \frac{1}{3}\left[\frac{1}{2}\right]^3 + \frac{1}{10}\left[\frac{1}{2}\right]^5 - \frac{1}{42}\left[\frac{1}{2}\right]^7 \simeq 0.46127$$

27. Start with the fact that

$$\frac{1}{1-x} = \sum_{n=0}^{\infty} x^n \qquad \text{for} \qquad |x| < 1$$

and replace x by $-x^2$ to get

$$\frac{1}{1+x^2} = \sum_{n=0}^{\infty}(-x^2)^n = \sum_{n=0}^{\infty}(-1)^n x^{2n} \quad \text{for } |x| < 1$$

so that

$$P_2(x) = 1 - x^2 + x^4$$

Hence,

$$\int_0^{0.1} \frac{1}{1+x^2}\, dx \simeq \int_0^{0.1} P_2(x)\, dx = \int_0^{0.1}(1 - x^2 + x^4)\, dx$$

$$= \left[x - \frac{1}{3}x^3 + \frac{1}{5}x^5\right]\Big|_0^{0.1} = 0.1 - \frac{1}{3}(0.1)^3 + \frac{1}{5}(0.1)^5$$

$$\simeq 0.9967$$

Chapter 11, Section 4

1.   Let $f(x) = x^2 - 12$. Then $f'(x) = 2x$ and so

$$x - \frac{f(x)}{f'(x)} = x - \frac{x^2 - 12}{2x} = \frac{x^2 + 12}{2x}$$

Hence, for $n = 1, 2, 3, \ldots$, Newton's method gives

$$x_n = \frac{x_{n-1}^2 + 12}{2x_{n-1}}$$

Since $3^2 = 9$, a convenient first estimate is $x_0 = 3$. Then,

$$x_1 = \frac{3^2 + 12}{2(3)} = 3.5$$

$$x_2 = \frac{(3.5)^2 + 12}{2(3.5)} = 3.4643$$

$$x_3 = \frac{(3.4643)^2 + 12}{2(3.4643)} \approx 3.4641$$

Hence, 3.464 is the desired approximation to the root of $x^2 - 12 = 0$.

3.   Let $f(x) = x^3 + x^2 - 1$. Then, $f'(x) = 3x^2 + 2x$, and so

$$x - \frac{f(x)}{f'(x)} = x - \frac{x^3 + x^2 - 1}{3x^2 + 2x} = \frac{2x^3 + x^2 + 1}{3x^2 + 2x}$$

Hence, for $n = 1, 2, 3, \ldots$, Newton's method gives

$$x_n = \frac{2x_{n-1}^3 + x_{n-1}^2 + 1}{3x_{n-1}^2 + 2x_{n-1}}$$

Since $f(1) = 1$, a convenient first estimate is $x_0 = 1$. Then,

$$x_1 = \frac{2(1)^3 + (1)^2 + 1}{3(1)^2 + 2(1)} = 0.8$$

$$x_2 = \frac{2(0.8)^3 + (0.8)^2 + 1}{3(0.8)^2 + 2(0.8)} \approx 0.7568$$

$$x_3 = \frac{2(0.7568)^3 + (0.7568)^2 + 1}{3(0.7568)^2 + 2(0.7568)} \approx 0.7549$$

$$x_4 = \frac{2(0.7549)^3 + (0.7549)^2 + 1}{3(0.7549)^2 + 2(0.7549)} \approx 0.7549$$

Hence, 0.755 is the desired approximation to the root of $x^3 + x^2 - 1 = 0$.

5.   Write $e^x = -2x$ as $e^x + 2x = 0$ and let $f(x) = e^x + 2x$. Then, $f'(x) = e^x + 2$, and so

$$x - \frac{f(x)}{f'(x)} = x - \frac{e^x + 2x}{e^x + 2} = \frac{e^x(x - 1)}{e^x + 2}$$

Hence, for $n = 1, 2, 3, \ldots$, Newton's method gives

$$x_n = \frac{e^{x_{n-1}}(x_{n-1} - 1)}{e^{x_{n-1}} + 2}$$

Take $x_0 = 0$ as the first estimate. Then,

$$x_1 = \frac{e^0(0 - 1)}{e^0 + 2} = -0.5$$

$$x_2 = \frac{e^{-0.5}(-0.5 - 1)}{e^{-0.5} + 2} \approx -0.3490$$

$$x_3 = \frac{e^{-0.3490}(-0.3490 - 1)}{e^{-0.3490} + 2} \approx -0.3517$$

$$x_4 = \frac{e^{-0.3517}(-0.3517 - 1)}{e^{-0.3517} + 2} \approx -0.3517$$

Hence, -0.352 is the desired approximation to the root of $e^x = -2x$.

7.   Let $f(x) = x^2 - 5x + 1$. Then, $f'(x) = 2x - 5$, and so

$$x - \frac{f(x)}{f'(x)} = x - \frac{x^2 - 5x + 1}{2x - 5} = \frac{x^2 - 1}{2x - 5}$$

Hence, for $n = 1, 2, 3, \ldots$, Newton's method gives

$$x_n = \frac{x_{n-1}^2 - 1}{2x_{n-1} - 5}$$

Since $f(0) = 1$ and $f(1) = -3$, there is a root between 0 and 1. A convenient first estimate is $x_0 = 0$. Then,

$$x_1 = \frac{0^2 - 1}{2(0) - 5} = 0.2$$

$$x_2 = \frac{(0.2)^2 - 1}{2(0.2) - 5} \approx 0.2087$$

$$x_3 = \frac{(0.2087)^2 - 1}{2(0.2087) - 5} \approx 0.2087$$

Hence, an approximation to one of the roots is 0.209..

Since $f(4) = -3$ and $f(5) = 1$, there is a root between 4 and 5. A convenient first estimate is $x_0 = 5$. Then,

$$x_1 = \frac{(5)^2 - 1}{2(5) - 5} = 4.80$$

$$x_2 = \frac{(4.80)^2 - 1}{2(4.80) - 5} \approx 4.7913$$

$$x_3 = \frac{(4.7913)^2 - 1}{2(4.7913) - 5} \approx 4.7913$$

Hence, an approximation to the other root is 4.791.

9.   Observe that $\sqrt{2}$ is a root of the equation $f(x) = 0$, where $f(x) = x^2 - 2$. Since $f'(x) = 2x$,

$$x - \frac{f(x)}{f'(x)} = x - \frac{x^2 - 2}{2x} = \frac{x^2 + 2}{2x}$$

Hence, for $n = 1, 2, 3, \ldots$, Newton's method gives

$$x_n = \frac{x_{n-1}^2 + 2}{2x_{n-1}}$$

A convenient first estimate is $x_0 = 1$. Then,

$$x_1 = \frac{(1)^2 + 2}{2(1)} = 1.5$$

$$x_2 = \frac{(1.5)^2 + 2}{2(1.5)} \approx 1.41667$$

$$x_3 = \frac{(1.41667)^2 + 2}{2(1.41667)} \approx 1.414216$$

$$x_4 = \frac{(1.414216)^2 + 2}{2(1.414216)} \approx 1.414214$$

$$x_5 = \frac{(1.414214)^2 + 2}{2(1.414214)} \approx 1.414214$$

Hence, the desired approximation to $\sqrt{2}$ is 1.41421.

11.   Observe that $\sqrt[3]{9}$ is the root of the equation $f(x) = 0$, where $f(x) = x^3 - 9$. Since $f'(x) = 3x^2$,

$$x - \frac{f(x)}{f'(x)} = x - \frac{x^3 - 9}{3x^2} = \frac{2x^3 + 9}{3x^2}$$

Hence, for $n = 1, 2, 3, \ldots$, Newton's method gives

$$x_n = \frac{2x_{n-1}^3 + 9}{3x_{n-1}^2}$$

A convenient first estimate is $x_0 = 2$, since $2^3 = 8$. Then,

$$x_1 = \frac{2(2)^3 + 9}{3(2)^2} \approx 2.083333$$

$$x_2 = \frac{2(2.083333)^3 + 9}{3(2.083333)^2} \approx 2.080089$$

$$x_3 = \frac{2(2.080089)^3 + 9}{3(2.080089)^2} \approx 2.080084$$

$$x_4 = \frac{2(2.080084)^3 + 9}{3(2.080084)^2} \approx 2.080084$$

Hence, the desired approximation to $\sqrt[3]{9}$ is 2.08008.

Chapter 11, Review Problems

1. From the series $\sum_{n=1}^{\infty} \left[\frac{1}{n+1} - \frac{1}{n+3}\right]$, the nth partial sum is

$$S_n = \left[\frac{1}{2} - \frac{1}{4}\right] + \left[\frac{1}{3} + \frac{1}{5}\right] + \left[\frac{1}{4} - \frac{1}{6}\right] + \left[\frac{1}{5} - \frac{1}{7}\right] + \left[\frac{1}{6} - \frac{1}{8}\right]$$

$$+ \cdots + \left[\frac{1}{n-1} - \frac{1}{n+1}\right] + \left[\frac{1}{n} - \frac{1}{n+2}\right] + \left[\frac{1}{n+1} - \frac{1}{n+3}\right]$$

$$= \frac{1}{2} + \frac{1}{3} - \frac{1}{n+2} - \frac{1}{n+3} = \frac{5}{6} - \frac{1}{n+2} - \frac{1}{n+3}$$

and so the sum of the series is

$$\lim_{n\to\infty} S_n = \lim_{n\to\infty}\left[\frac{5}{6} - \frac{1}{n+2} - \frac{1}{n+3}\right] = \frac{5}{6}$$

2. The series $\sum_{n=2}^{\infty} \frac{1}{n(n-1)}$ can be rewritten as $\sum_{n=2}^{\infty} \left[\frac{1}{n-1} - \frac{1}{n}\right]$. For this series, the nth partial sum is

$$S_n = \left[1 - \frac{1}{2}\right] + \left[\frac{1}{2} - \frac{1}{3}\right] + \left[\frac{1}{3} - \frac{1}{4}\right] + \cdots + \left[\frac{1}{n-2} - \frac{1}{n-1}\right] + \left[\frac{1}{n-1} - \frac{1}{n}\right]$$

$$= 1 - \frac{1}{n}$$

and so the sum of the series is

$$\lim_{n\to\infty} S_n = \lim_{n\to\infty}\left[1 - \frac{1}{n}\right] = 1$$

3. The series $\sum_{n=1}^{\infty} \frac{2}{(-3)^n} = 2\sum_{n=1}^{\infty} \left[-\frac{1}{3}\right]^n$ is a geometric series with nth partial sum

$$S_n = 2\left[-\frac{1}{3} + (-\frac{1}{3})^2 + (-\frac{1}{3})^3 + \cdots + (-\frac{1}{3})^n\right]$$

Multiplying S_n by $-\frac{1}{3}$ gives

$$-\frac{1}{3}S_n = 2\left[(-\frac{1}{3})^2 + (-\frac{1}{3})^3 + (-\frac{1}{3})^4 + \cdots + (-\frac{1}{3})^{n+1}\right]$$

and so

$$S_n + \frac{1}{3}S_n = \frac{4}{3}S_n = 2\left[-\frac{1}{3} - (-\frac{1}{3})^{n+1}\right] = -\frac{2}{3}\left[1 - (-\frac{1}{3})^n\right]$$

Hence,

$$S_n = \frac{3}{4}\left[-\frac{2}{3}\right]\left[1-(-\frac{1}{3})^n\right] = -\frac{1}{2}\left[1-(-\frac{1}{3})^n\right]$$

and the sum of the series is

$$\lim_{n\to\infty} S_n = \lim_{n\to\infty} -\frac{1}{2}\left[1-(-\frac{1}{3})^n\right] = -\frac{1}{2}$$

4. $$\sum_{n=1}^{\infty}\frac{3}{(-5)^n} = 3\sum_{n=1}^{\infty}\left[-\frac{1}{5}\right]^n = -\frac{3}{5}\sum_{n=1}^{\infty}\left[-\frac{1}{5}\right]^n$$

$$= -\frac{3}{5}\left[\frac{1}{1-(-1/5)}\right] = -\frac{3}{5}\left[\frac{5}{6}\right] = -\frac{1}{2}$$

5. The series $\sum_{n=1}^{\infty}\left[-\frac{3}{2}\right]^n$ is a geometric series with ratio $-\frac{3}{2}$ and hence diverges since $\left|-\frac{3}{2}\right| > 1$.

6. $$\sum_{n=1}^{\infty} e^{-0.5n} = \sum_{n=1}^{\infty}(e^{-0.5})^n = e^{-0.5}\sum_{n=0}^{\infty}(e^{-0.5})^n$$

$$= e^{-0.5}\left[\frac{1}{1-e^{-0.5}}\right] \simeq 1.5415$$

7. $$\sum_{n=2}^{\infty}\frac{2^{n+1}}{3^{n-3}} = \sum_{n=2}^{\infty}\frac{2^n(2)}{3^n(3^{-3})} = \frac{2}{3^{-3}}\sum_{n=2}^{\infty}\frac{2^n}{3^n} = 54\left[(\frac{2}{3})^2 + (\frac{2}{3})^3 + (\frac{2}{3})^4 + \cdots\right]$$

$$= 54\left[\frac{2}{3}\right]^2\left[1+\frac{2}{3}+(\frac{2}{3})^2+\cdots\right] = 54\left[\frac{2}{3}\right]^2\left[\frac{1}{1-2/3}\right] = 54\left[\frac{4}{9}\right][3] = 72$$

8. Write the decimal as a geometric series as follows:

$$1.545454\ldots = 1 + \frac{54}{100} + \frac{54}{10,000} + \frac{54}{1,000,000} + \cdots$$

$$= 1 + \frac{54}{100}\left[1+\frac{1}{100}+(\frac{1}{100})^2+\cdots\right] = 1 + \frac{54}{100}\left[\frac{1}{1-1/100}\right]$$

$$= 1 + \frac{54}{100}\left[\frac{100}{99}\right] = 1 + \frac{54}{99} = \frac{17}{11}$$

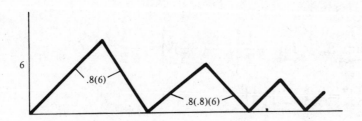

Figure for Problem 9.

9. As indicated in the figure, the distance the ball travels is

$$6 + 2(0.8)(6) + 2(0.8)^2(6) + 2(0.8)^3(6) + \cdots$$

$$= 6 + 12(0.8)[1 + 0.8 + (0.8)^2 + \cdots] = 6 + 12(0.8)\left[\frac{1}{1 - 0.8}\right] = 54 \text{ feet}$$

10. At 12 percent annual interest compounded continuously, the present value of the first withdrawal of \$500 is

$$P_1 = 500e^{(-0.12)(1)}$$

the present value of the second withdrawal of \$500 is

$$P_2 = 500e^{(-0.12)(2)}$$

and, in general, the present value of the n th withdrawal of \$500 is

$$P_n = 500e^{(-0.12)(n)}$$

Hence, the amount to be invested is

$$\sum_{n=1}^{\infty} P_n = \sum_{n=1}^{\infty} 500e^{(-0.12)(n)} = 500 \sum_{n=1}^{\infty} (e^{-0.12})^n$$

$$= 500e^{-0.12} \sum_{n=0}^{\infty} (e^{-0.12})^n = 500e^{-0.12}\left[\frac{1}{1 - e^{-0.12}}\right] \simeq \$3,921.67$$

11. Of the original dose of 10 units, only $10e^{-0.8}$ units are left in the patient's body just prior to the second injection one day later. Hence, the number of units in the patient's body after 1 day (just after the next injection) is

$$S_1 = 10 + 10e^{-0.8}$$

The medication in the patient's body after 2 days (immediately following the injection) consists of the 10 units just injected plus what remains from the first two injections. Of the original dose, only $10e^{(-0.8)(2)}$ units are left, and of the second dose, only $10e^{-0.8}$ units are left. Hence, the number of units in the patient's body after 2 days is

$$S_2 = 10 + 10e^{-0.8} + 10e^{(-0.8)(2)}$$

In general, the number of units in the patient's body after n days is

$$S_n = 10 + 10e^{-0.8} + 10e^{(-0.8)(2)} + \cdots + 10e^{(-0.8)(n)}$$

The number S of units in the patient's body in the long run is the limit of S_n as n increases without bound. That is,

$$S = \lim_{n \to \infty} S_n = \sum_{n=0}^{\infty} 10e^{(-0.8)(n)} = 10 \sum_{n=0}^{\infty} (e^{-0.8})^n$$

$$= 10 \left[\frac{1}{1 - e^{-0.8}} \right] \simeq 18.16 \text{ units}$$

12. Let x denote the number of the roll on which the first six comes up. If you roll second, you will get the first 6 if $x = 2, 5, 8, \ldots$. Since the probability of rolling a 6 is $\frac{1}{6}$ and the probability of not rolling a 6 is $\frac{5}{6}$, the probability that you win is

$$P(x = 2) + P(x = 5) + P(x = 8) + \cdots = \frac{1}{6}\left[\frac{5}{6}\right] + \frac{1}{6}\left[\frac{5}{6}\right]^4 + \frac{1}{6}\left[\frac{5}{6}\right]^7 + \cdots$$

$$= \frac{1}{6}\left[\frac{5}{6}\right]\left[1 + \left(\frac{5}{6}\right)^3 + \left(\frac{5}{6}\right)^6 + \cdots\right]$$

$$= \frac{5}{36}\left[1 + \left(\frac{5}{6}\right)^3 + \left[\left(\frac{5}{6}\right)^3\right]^2 + \cdots\right] = \frac{5}{36}\left[1 + \frac{125}{216} + (\frac{125}{216})^2 + \cdots\right]$$

$$= \frac{5}{36}\left[\frac{1}{1 - (125/216)}\right] = \frac{5}{36}\left[\frac{216}{91}\right] = 0.3297$$

13. For $f(x) = \frac{1}{x+3}$ about $x = 0$, compute the Taylor coefficients as follows:

$$f(x) = \frac{1}{x+3}, \quad f'(x) = \frac{-1}{(x+3)^2}, \quad f''(x) = \frac{2(1)}{(x+3)^3},$$

$$f^{(3)}(x) = \frac{-3(2)(1)}{(x+3)^4}, \dots, f^{(n)}(x) = \frac{(-1)^n n!}{(x+3)^{n+1}}$$

and so

$$f(0) = \frac{1}{3}, \quad f'(0) = \frac{-1}{3^2}, \quad f''(0) = \frac{2!}{3^3},$$

$$f^{(3)}(0) = \frac{-(3!)}{3^4}, \dots, f^{(n)}(0) = \frac{(-1)^n n!}{3^{n+1}}$$

Hence, for $n = 0, 1, 2, \dots$

$$a_n = \frac{f^{(n)}(0)}{n!} = \frac{(-1)^n n!/3^{n+1}}{n!} = \frac{(-1)^n}{3^{n+1}}$$

and the corresponding Taylor series is $\sum_{n=0}^{\infty} \frac{(-1)^n}{3^{n+1}} x^n$.

14. For $f(x) = e^{-3x}$ about $x = 0$, compute the Taylor coefficients as follows:

$$f(x) = e^{-3x}, \quad f'(x) = -3e^{-3x}, \quad f''(x) = 3^2 e^{-3x}, \dots, f^{(n)}(x) = (-1)^n 3^n e^{-3x}$$

and so

$$f(0) = 1, \quad f'(0) = -3, \quad f''(0) = 3^2, \dots, f^{(n)}(0) = (-1)^n 3^n$$

Hence, for $n = 0, 1, 2, \dots$

$$a_n = \frac{f^{(n)}(0)}{n!} = \frac{(-1)^n 3^n}{n!} = \frac{(-3)^n}{n!}$$

and the corresponding Taylor series is $\sum_{n=0}^{\infty} \frac{(-3)^n}{n!} x^n$.

15. For $f(x) = (1+2x)e^x$ about $x = 0$, compute the Taylor coefficients (using the product rule) as follows:

$$f(x) = (1+2x)e^x, \quad f'(x) = (3+2x)e^x, \quad f''(x) = (5+2x)e^x,$$

$$\dots, f^{(n)}(x) = [(2n+1) + 2x]e^x$$

and so

$$f(0) = 1, \quad f'(0) = 3, \quad f''(0) = 5, \dots, f^{(n)}(0) = 2n+1$$

Hence, for $n = 0, 1, 2, \dots$

$$a_n = \frac{f^{(n)}(0)}{n!} = \frac{2n+1}{n!}$$

and the corresponding Taylor series is $\sum\limits_{n=0}^{\infty} \frac{2n+1}{n!} x^n$.

16. For $f(x) = \frac{1}{(1+x)^2}$ about $x = -2$, compute the Taylor coefficients as follows:

$$f(x) = \frac{1}{(1+x)^2}, \quad f'(x) = \frac{-2}{(1+x)^3}, \quad f''(x) = \frac{3!}{(1+x)^4},$$

$$\dots, f^{(n)}(x) = \frac{(-1)^n (n+1)!}{(1+x)^{n+2}}$$

and so

$$f(-2) = 1, \quad f'(-2) = 2, \quad f''(-2) = 3!,$$

$$\dots, f^{(n)}(-2) = \frac{(-1)^n (n+1)!}{(-1)^{n+2}} = (n+1)!$$

Hence, for $n = 0, 1, 2, \dots$

$$a_n = \frac{f^{(n)}(-2)}{n!} = \frac{(n+1)!}{n!} = n+1$$

and the corresponding Taylor series is $\sum\limits_{n=0}^{\infty} (n+1)(x+2)^n$.

17. Start with the fact that

$$\frac{1}{1-x} = \sum_{n=0}^{\infty} x^n \qquad \text{for} \quad |x| < 1$$

and replace x by $-4x^2$ to get

$$\frac{1}{1+4x^2} = \sum_{n=0}^{\infty} (-4x^2)^n = \sum_{n=0}^{\infty} (-1)^n 4^n x^{2n} \qquad \text{for } |-4x^2| < 1 \text{ or } |x| < \frac{1}{2}$$

Finally, multiply by $4x$ to get

$$\frac{4x}{1+4x^2} = \sum_{n=0}^{\infty} (-1)^n 4^{n+1} x^{2n+1} \qquad \text{for } |x| < \frac{1}{2}$$

18. For $f(x) = \sqrt{x} = x^{1/2}$ about $x = 1$, compute the Taylor coefficients as follows:

$$f(x) = x^{1/2}, \quad f'(x) = \frac{1}{2} x^{-1/2}, \quad f''(x) = -\frac{1}{4} x^{-3/2}, \quad f^{(3)}(x) = \frac{3}{8} x^{-5/2}$$

and so

$$f(1) = 1, \quad f'(1) = \frac{1}{2}, \quad f''(1) = -\frac{1}{4}, \quad f^{(3)}(1) = \frac{3}{8}$$

Hence, the Taylor polynomial of degree 3 is

$$P_3(x) = 1 + \frac{1}{2}(x - 1) - \frac{1}{8}(x - 1)^2 + \frac{1}{16}(x - 1)^3$$

and

$$\sqrt{0.9} \simeq P_3(0.9) = 1 + \frac{1}{2}(-0.1) - \frac{1}{8}(-0.1)^2 + \frac{1}{16}(-0.1)^3 \simeq 0.94869$$

19. For $f(x) = \frac{x}{1+x^3}$, it would be time-consuming to calculate the Taylor coefficients directly. Instead, start with

$$\frac{1}{1 - x} = \sum_{n=0}^{\infty} x^n \qquad \text{for } |x| < 1$$

and replace x by $-x^3$ to get

$$\frac{1}{1 + x^3} = \sum_{n=0}^{\infty} (-x^3)^n = \sum_{n=0}^{\infty} (-1)^n x^{3n} \qquad \text{for } |-x^3| < 1 \text{ or } |x| < 1$$

and multiply by x to get

$$\frac{x}{1 + x^3} = \sum_{n=0}^{\infty} (-1)^n x^{3n+1} \qquad \text{for } |x| < 1$$

Hence, the Taylor polynomial of degree 10 is

$$P_{10}(x) = x - x^4 + x^7 - x^{10}$$

and

$$\int_0^{1/2} \frac{x}{1 + x^3} \, dx = \int_0^{1/2} P_{10}(x) \, dx = \int_0^{1/2} (x - x^4 + x^7 - x^{10}) \, dx$$

$$= \left[\frac{1}{2}x^2 - \frac{1}{5}x^5 + \frac{1}{8}x^8 - \frac{1}{11}x^{11} \right]\Big|_0^{1/2}$$

$$= \frac{1}{2}\left[\frac{1}{2}\right]^2 - \frac{1}{5}\left[\frac{1}{2}\right]^5 + \frac{1}{8}\left[\frac{1}{2}\right]^8 - \frac{1}{11}\left[\frac{1}{2}\right]^{11} \simeq 0.11919$$

20. Let $f(x) = x^2 - 55$. Then $f'(x) = 2x$ and

$$x - \frac{f(x)}{f'(x)} = x - \frac{x^2 - 55}{2x} = \frac{x^2 + 55}{2x}$$

Hence, for $n = 1, 2, 3, \ldots$, Newton's method gives

$$x_n = \frac{x_{n-1}^2 + 55}{2x_{n-1}}$$

Since $f(7) < 0$ and $f(8) > 0$, there is a root between 7 and 8. If $x_0 = 7$,

$$x_1 = \frac{7^2 + 55}{2(7)} \approx 7.428571$$

$$x_2 = \frac{(7.428571)^2 + 55}{2(7.428571)} \approx 7.416209$$

$$x_3 = \frac{(7.416209)^2 + 55}{2(7.416209)} \approx 7.416199$$

$$x_4 = \frac{(7.416199)^2 + 55}{2(7.416199)} \approx 7.416199$$

Hence, rounded to five decimal places, the root is 7.41620.

21.   Let $f(x) = x^3 + 3x^2 + 1$. Then $f'(x) = 3x^2 + 6x$, and

$$x - \frac{f(x)}{f'(x)} = x - \frac{x^3 + 3x^2 + 1}{3x^2 + 6x} = \frac{2x^3 + 3x^2 - 1}{3x(x + 2)}$$

Hence, for $n = 1, 2, 3, \ldots$, Newton's method gives

$$x_n = \frac{2x_{n-1}^3 + 3x_{n-1}^2 - 1}{3x_{n-1}(x_{n-1} + 2)}$$

Since $f(-3) = 1 > 0$ and $f(-4) = -15 > 0$, there is a root between -4 and -3. A convenient first estimates is $x_0 = -3$. Then

$$x_1 = \frac{2(-3)^3 + 3(-3)^2 - 1}{3(-3)(-3 + 2)} \approx -3.1111$$

$$x_2 = \frac{2(-3.1111)^3 + 3(-3.1111)^2 - 1}{3(-3.1111)(-3.1111 + 2)} \approx -3.1038$$

$$x_3 = \frac{2(-3.1038)^3 + 3(-3.1038)^2 - 1}{3(-3.1038)(-3.1038 + 2)} \approx -3.1038$$

Hence, rounded to 3 decimal places, the root is -3.104.

22.   Write $e^{-x} = 3x$ as $e^{-x} - 3x = 0$ and let $f(x) = e^{-x} - 3x$. Then $f'(x) = -e^{-x} - 3$, and

$$x - \frac{f(x)}{f'(x)} = x - \frac{e^{-x} - 3x}{-e^{-x} - 3} = \frac{-xe^{-x} - e^{-x}}{-e^{-x} - 3} = \frac{-(xe^{-x} + e^{-x})}{-(e^{-x} + 3)} = \frac{e^{-x}(1 + x)}{3 + e^{-x}}$$

Hence, for $n = 1, 2, 3, \ldots$, Newton's method gives

$$x_n = \frac{e^{-x_{n-1}}(1 + x_{n-1})}{3 + e^{-x_{n-1}}}$$

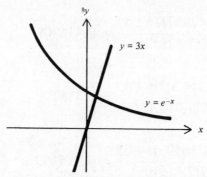

Graph for Problem 22.

From the graph, the curves intersect at a point in the first quadrant near the origin. A convenient first estimate is $x_0 = 0.3$. Then

$$x_1 = \frac{e^{-0.3}(1 + 0.3)}{3 + e^{-0.3}} \approx 0.2040$$

$$x_2 = \frac{e^{-0.2040}(1 + 0.2040)}{3 + e^{-0.2040}} \approx 0.2573$$

$$x_3 = \frac{e^{-0.2573}(1 + 0.2573)}{3 + e^{-0.2573}} \approx 0.2576$$

$$x_4 = \frac{e^{-0.2576}(1 + 0.2576)}{3 + e^{-0.2576}} \approx 0.2576$$

Hence, rounded to 3 decimal places, the root is 0.258.

CHAPTER 12
TRIGONOMETRIC FUNCTIONS
1 THE TRIGONOMETRIC FUNCTIONS 2 DIFFERENTIATION OF TRIGONOMETRIC FUNCTIONS 3 APPLICATIONS OF TRIGONOMETRIC FUNCTIONS REVIEW PROBLEMS

Chapter 12, Section 1

1.　The angle is $\frac{1}{3}$ of 90°, or 30°.

3.　The angle is 90° plus $\frac{1}{3}$ of 90°, or 120°.

5.　The angle is 90° plus $\frac{1}{3}$ of 90° in a clockwise direction, or −120°.

Problem 7.

Problem 9.

Problem 11.

13.　From the proportion $\frac{15}{180} = \frac{\text{radians}}{\pi}$, it follows that

$$\text{Radians} = \frac{15\pi}{180} = \frac{\pi}{12}$$

15.　From the proportion $\frac{-150}{180} = \frac{\text{radians}}{\pi}$, it follows that

$$\text{Radians} = \frac{-150\pi}{180} = -\frac{5\pi}{6}$$

17.　From the proportion $\frac{540}{180} = \frac{\text{radians}}{\pi}$, it follows that

$$\text{Radians} = \frac{540\pi}{180} = 3\pi$$

19. From the proportion $\frac{\text{degrees}}{180} = \frac{5\pi/6}{\pi}$, it follows that

$$\text{Degrees} = \frac{180(5\pi/6)}{\pi} = 150°$$

21. From the proportion $\frac{\text{degrees}}{180} = \frac{3\pi/6}{\pi}$, it follows that

$$\text{Degrees} = \frac{180(3\pi/2)}{\pi} = 270°$$

23. From the proportion $\frac{\text{degrees}}{180} = \frac{-3\pi/4}{\pi}$, it follows that

$$\text{Degrees} = \frac{180(-3\pi/4)}{\pi} = -135°$$

25. The angle is $\frac{\pi}{2}$ radians plus $\frac{2}{3}$ of $\frac{\pi}{2}$ radians, or $\frac{\pi}{2} + \frac{2}{3}\left[\frac{\pi}{2}\right] = \frac{5\pi}{6}$ radians.

27. The angle is $\frac{\pi}{2}$ radians plus $\frac{1}{2}$ of $\frac{\pi}{2}$ radians, or $\frac{\pi}{2} + \frac{1}{2}\left[\frac{\pi}{2}\right] = \frac{3\pi}{4}$ radians.

29. The angle is $-\frac{\pi}{2}$ radians plus $\frac{1}{3}$ of $-\frac{\pi}{2}$ radians, or $-\frac{\pi}{2} + \frac{1}{3}\left[-\frac{\pi}{2}\right] = -\frac{2\pi}{3}$ radians.

Problem 31.

Problem 33.

Problem 35.

37. $\cos\frac{7\pi}{2} = \cos\left[2\pi - \frac{\pi}{2}\right] = \cos\left[-\frac{\pi}{2}\right] = \cos\frac{\pi}{2} = 0$

39. $\cos\left[-\frac{5\pi}{2}\right] = \cos\frac{5\pi}{2} = \cos\left[2\pi + \frac{\pi}{2}\right] = \cos\frac{\pi}{2} = 0$

41. $\cot\frac{5\pi}{2} = \frac{\cos 5\pi/2}{\sin 5\pi/2} = \frac{\cos(2\pi+\pi/2)}{\sin(2\pi+\pi/2)} = \frac{\cos \pi/2}{\sin \pi/2} = \frac{0}{1} = 0$

43. $\tan(-\pi) = \frac{\sin(-\pi)}{\cos(-\pi)} = \frac{-\sin\pi}{\cos\pi} = \frac{0}{-1} = 0$

45. Since $\frac{\pi}{3}$ radians is $60°$, it follows from the well-known $30° - 60° - 90°$ right triangle that

$$\cos\frac{\pi}{3} = \cos 60° = \frac{1}{2} \qquad \text{and} \qquad \sin\frac{\pi}{3} = \sin 60° = \frac{\sqrt{3}}{2}$$

47. $\sin \frac{7\pi}{6} = \sin\left[\pi + \frac{\pi}{6}\right] = \sin\pi\cos\frac{\pi}{6} + \cos\pi\sin\frac{\pi}{6} = 0 - 1\left[\frac{1}{2}\right] = -\frac{1}{2}$

$\cos \frac{7\pi}{6} = \cos\left[\pi + \frac{\pi}{6}\right] = \cos\pi\cos\frac{\pi}{6} - \sin\pi\sin\frac{\pi}{6} = -1\left[\frac{\sqrt{3}}{2}\right] - 0 = -\frac{\sqrt{3}}{2}$

$\sin \frac{5\pi}{4} = \sin\left[\pi + \frac{\pi}{4}\right] = \sin\pi\cos\frac{\pi}{4} + \cos\pi\sin\frac{\pi}{4} = 0 - 1\left[\frac{\sqrt{2}}{2}\right] = -\frac{\sqrt{2}}{2}$

$\cos \frac{5\pi}{4} = \cos\left[\pi + \frac{\pi}{4}\right] = \cos\pi\cos\frac{\pi}{4} - \sin\pi\sin\frac{\pi}{4} = -1\left[\frac{\sqrt{2}}{2}\right] - 0 = -\frac{\sqrt{2}}{2}$

$\sin \frac{4\pi}{3} = \sin\left[\pi + \frac{\pi}{3}\right] = \sin\pi\cos\frac{\pi}{3} + \cos\pi\sin\frac{\pi}{3} = 0 - 1\left[\frac{\sqrt{3}}{2}\right] = -\frac{\sqrt{3}}{2}$

$\cos \frac{4\pi}{3} = \cos\left[\pi + \frac{\pi}{3}\right] = \cos\pi\cos\frac{\pi}{3} - \sin\pi\sin\frac{\pi}{3} = -1\left[\frac{1}{2}\right] - 0 = -\frac{1}{2}$

$\sin \frac{3\pi}{2} = -1$ and $\cos\frac{3\pi}{2} = 0$ (from the graphs on pages 618 and 619)

$\sin \frac{5\pi}{3} = \sin\left[\pi + \frac{2\pi}{3}\right] = \sin\pi\cos\frac{2\pi}{3} + \cos\pi\sin\frac{2\pi}{3} = 0 - 1\left[\frac{\sqrt{3}}{2}\right] = -\frac{\sqrt{3}}{2}$

$\cos \frac{5\pi}{3} = \cos\left[\pi + \frac{2\pi}{3}\right] = \cos\pi\cos\frac{2\pi}{3} - \sin\pi\sin\frac{2\pi}{3} = -1\left[-\frac{1}{2}\right] - 0 = \frac{1}{2}$

$\sin \frac{7\pi}{4} = \sin\left[\pi + \frac{3\pi}{4}\right] = \sin\pi\cos\frac{3\pi}{4} + \cos\pi\sin\frac{3\pi}{4} = 0 - 1\left[\frac{\sqrt{2}}{2}\right] = -\frac{\sqrt{2}}{2}$

$\cos \frac{7\pi}{4} = \cos\left[\pi + \frac{3\pi}{4}\right] = \cos\pi\cos\frac{3\pi}{4} - \sin\pi\sin\frac{3\pi}{4} = -1\left[-\frac{\sqrt{2}}{2}\right] - 0 = \frac{\sqrt{2}}{2}$

$\sin \frac{11\pi}{6} = \sin\left[\pi + \frac{5\pi}{6}\right] = \sin\pi\cos\frac{5\pi}{6} + \cos\pi\sin\frac{5\pi}{6} = 0 - 1\left[\frac{1}{2}\right] = -\frac{1}{2}$

$\cos \frac{11\pi}{6} = \cos\left[\pi + \frac{5\pi}{6}\right] = \cos\pi\cos\frac{5\pi}{6} - \sin\pi\sin\frac{5\pi}{6} = -\left[-\frac{\sqrt{3}}{2}\right] - 0 = \frac{\sqrt{3}}{2}$

$\sin 2\pi = 0$ and $\cos 2\pi = 1$ (from the graphs on pages 618 and 619).

49. $\sin\left[-\frac{\pi}{6}\right] = -\sin\frac{\pi}{6} = -\frac{1}{2}$

51. $\cos\frac{7\pi}{3} = \cos\left[2\pi + \frac{\pi}{3}\right] = \cos\frac{\pi}{3} = \frac{1}{2}$

53. $\tan\frac{\pi}{6} = \frac{\sin\pi/6}{\cos\pi/6} = \frac{1/2}{\sqrt{3}/2} = \frac{1}{\sqrt{3}} = \frac{\sqrt{3}}{3}$

55. $\sec\frac{\pi}{3} = \frac{1}{\cos\pi/3} = \frac{1}{1/2} = 2$

57. $\tan\left[-\frac{\pi}{4}\right] = \frac{\sin(-\pi/4)}{\cos(-\pi/4)} = \frac{-\sin\pi/4}{\cos\pi/4} = \frac{-\sqrt{2}/2}{\sqrt{2}} = -1$

59.   Since $\sec\theta = \frac{5}{4}$, the corresponding right triangle shows that $\tan\theta = \frac{3}{4}$.

61. Since $\csc\theta = \frac{1}{\sin\theta} = \frac{5}{3}$, the corresponding right triangle shows that $\tan\theta = \frac{3}{4}$.

63. Since $\tan\theta = \frac{3}{4}$, the corresponding right triangle shows that $\cos\theta = \frac{4}{5}$.

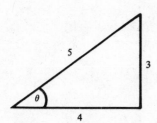

Triangle for Problem 59.

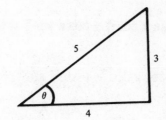

Triangle for Problem 61.

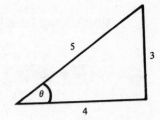

Triangle for Problem 63.

65. Since $\sin 2\theta = 2\sin\theta\cos\theta$, the equation can be rewritten as

$$0 = \cos\theta - \sin 2\theta = \cos\theta - 2\sin\theta\cos\theta = \cos\theta(1 - 2\sin\theta)$$

from which it follows that

$$\cos\theta = 0 \quad\text{or}\quad \theta = \frac{\pi}{2} \quad\text{and}\quad \theta = \frac{3\pi}{2}$$

or

$$1 - 2\sin\theta = 0 \quad 2\sin\theta = 1 \quad \sin\theta = \frac{1}{2} \quad\text{or}\quad \theta = \frac{\pi}{6} \text{ and } \theta = \frac{5\pi}{6}$$

Hence, the solutions are $\theta = \frac{\pi}{6}$, $\theta = \frac{\pi}{2}$, $\theta = \frac{5\pi}{6}$, and $\theta = \frac{3\pi}{2}$.

67. Since $\sin 2\theta = 2\sin\theta\cos\theta$, the equation can be rewritten as

$$0 = \sin 2\theta - \sqrt{3}\cos\theta = 2\sin\theta\cos\theta - \sqrt{3}\cos\theta = \cos\theta(2\sin\theta - \sqrt{3})$$

from which it follows that

$$\cos\theta = 0 \qquad\text{or}\qquad \theta = \frac{\pi}{2}$$

or

$$2\sin\theta - \sqrt{3} = 0 \quad \sin\theta = \frac{\sqrt{3}}{2} \quad\text{or}\quad \theta = \frac{\pi}{3} \quad\text{and}\quad \theta = \frac{2\pi}{3}$$

Hence, the solutions are $\theta = \frac{\pi}{3}$, $\theta = \frac{\pi}{2}$, and $\theta = \frac{2\pi}{3}$.

69. Since $\sin 2\theta = 2\sin\theta\cos\theta$, the equation can be rewritten as

$$0 = 2\cos^2\theta - \sin 2\theta = 2\cos^2\theta - 2\sin\theta\cos\theta = 2\cos\theta(\cos\theta - \sin\theta)$$

from which it follows that

$$2\cos\theta = 0 \qquad\text{or}\qquad \theta = \frac{\pi}{2}$$

or

$$\cos\theta - \sin\theta = 0 \quad \cos\theta = \sin\theta \quad \text{or} \quad \theta = \frac{\pi}{4}$$

Hence, the solutions are $\theta = \frac{\pi}{4}$ and $\theta = \frac{\pi}{2}$.

71.   Since $\cos^2\theta = 1 - \sin^2\theta$, the equation can be rewritten as

$$1 = 2\cos^2\theta + \sin\theta = 2(1 - \sin^2\theta) + \sin\theta = 2 - 2\sin^2\theta + \sin\theta$$

$$2\sin^2\theta - \sin\theta - 1 = 0 \quad \text{or} \quad (2\sin\theta + 1)(\sin\theta - 1) = 0$$

from which it follows that

$$2\sin\theta + 1 = 0 \quad \text{or} \quad \sin\theta = -\frac{1}{2}$$

which has no solution in the interval $0 \le \theta \le \pi$, or

$$\sin\theta - 1 = 0 \quad \sin\theta = 1 \quad \text{or} \quad \theta = \frac{\pi}{2}$$

Hence, the only solution is $\theta = \frac{\pi}{2}$.

73.   Since $\cos^2\theta = 1 - \sin^2\theta$, the equation can be rewritten as

$$1 = \sin^2\theta - \cos^2\theta + 3\sin\theta = \sin^2\theta - (1 - \sin^2\theta) + 3\sin\theta$$

$$2\sin^2\theta + 3\sin\theta - 2 = 0 \quad \text{or} \quad (2\sin\theta - 1)(\sin\theta + 2) = 0$$

from which it follows that

$$2\sin\theta - 1 = 0 \quad \sin\theta = \frac{1}{2} \quad \text{or} \quad \theta = \frac{\pi}{6} \quad \text{and} \quad \theta = \frac{5\pi}{6}$$

or

$$\sin\theta + 2 = 0$$

which has no solutions. Hence, the solutions are $\theta = \frac{\pi}{6}$ and $\theta = \frac{5\pi}{6}$.

75.   Replace b with $-b$ in the identity

$$\sin(a + b) = \sin a \cos b + \cos a \sin b$$

to get

$$\sin(a - b) = \sin a \cos(-b) + \cos a \sin(-b) = \sin a \cos b - \cos a \sin b$$

since $\cos(-b) = \cos b$ and $\sin(-b) = -\sin b$.

77.   Since

$$\sin(a + b) = \sin a \cos b + \cos a \sin b$$

and

$$\cos(a + b) = \cos a \cos b - \sin a \sin b$$

it follows that

$$\tan(a+b) = \frac{\sin(a+b)}{\cos(a+b)} = \frac{\sin a \cos b + \cos a \sin b}{\cos a \cos b - \sin a \sin b}$$

Now divide each term in the quotient by $\cos a \cos b$ to get

$$\tan(a+b) = \frac{\left[\frac{\sin a \cos b}{\cos a \cos b}\right] + \left[\frac{\cos a \sin b}{\cos a \cos b}\right]}{\left[\frac{\cos a \cos b}{\cos a \cos b}\right] - \left[\frac{\sin a \sin b}{\cos a \cos b}\right]}$$

$$= \frac{\left[\frac{\sin a}{\cos a}\right] - \left[\frac{\sin b}{\cos b}\right]}{1 - \left[\frac{\sin a}{\cos a}\right]\left[\frac{\sin b}{\cos b}\right]}$$

$$= \frac{\tan a - \tan b}{1 - \tan a \tan b}$$

79. Since $\cos(a+b) = \cos a \cos b - \sin a \sin b$,

$$\cos\left[\frac{\pi}{2} - \theta\right] = \cos\frac{\pi}{2}\cos(-\theta) - \sin\frac{\pi}{2}\sin(-\theta)$$

$$= \cos\frac{\pi}{2}\cos\theta + \sin\frac{\pi}{2}\sin\theta = 0\cos\theta + 1\sin\theta = \sin\theta$$

Chapter 12, Section 2

1.   If $f(\theta) = \sin 3\theta$, then

$$f'(\theta) = \cos 3\theta \frac{d}{d\theta}(3\theta) = 3\cos 3\theta$$

3.   If $f(\theta) = \sin(1 - 2\theta)$, then

$$f'(\theta) = \cos(1 - 2\theta)\frac{d}{d\theta}(1 - 2\theta) = -2\cos(1 - 2\theta)$$

5.   If $f(\theta) = \cos(\theta^3 + 1)$, then

$$f'(\theta) = -\sin(\theta^3 + 1)\frac{d}{d\theta}(\theta^3 + 1) = -3\theta^2 \sin(\theta^3 + 1)$$

7.   If $f(\theta) = \cos^2(\frac{\pi}{2} - \theta) = \left[\cos(\frac{\pi}{2} - \theta)\right]^2$, then

$$f'(\theta) = 2\cos(\frac{\pi}{2} - \theta)\frac{d}{d\theta}\left[\cos(\frac{\pi}{2} - \theta)\right]$$

$$= 2\cos(\frac{\pi}{2} - \theta)\left[-\sin(\frac{\pi}{2} - \theta)\right]\frac{d}{d\theta}(\frac{\pi}{2} - \theta)$$

$$= 2\cos(\frac{\pi}{2} - \theta)\left[-\sin(\frac{\pi}{2} - \theta)\right](-1)$$

$$= 2\cos(\frac{\pi}{2} - \theta)\sin(\frac{\pi}{2} - \theta)$$

$$= \sin 2(\frac{\pi}{2} - \theta) = \sin(\pi - 2\theta) \qquad (\text{since } \sin 2a = 2\sin a \cos a)$$

9.   If $f(\theta) = \cos(1 + 3\theta)^2$, then

$$f'(\theta) = -\sin(1 + 3\theta)^2 \frac{d}{d\theta}(1 + 3\theta)^2$$

$$= -\sin(1 + 3\theta)^2 [2(1 + 3\theta)(3)]$$

$$= -6(1 + 3\theta)\sin(1 + 3\theta)^2$$

11.   If $f(\theta) = e^{-\theta/2}\cos 2\pi\theta$, then, the by product rule,

$$f'(\theta) = e^{-\theta/2}\frac{d}{d\theta}\cos 2\pi\theta + \cos 2\pi\theta\frac{d}{d\theta}e^{-\theta/2}$$

$$= e^{-\theta/2}\left[-\sin 2\pi\theta\frac{d}{d\theta}(2\pi\theta)\right] + \cos 2\pi\theta\left[e^{-\theta/2}\frac{d}{d\theta}(-\frac{\theta}{2})\right]$$

$$= e^{-\theta/2}(-2\pi\sin 2\pi\theta) + \cos 2\pi\theta\left[-\frac{1}{2}e^{-\theta/2}\right]$$

$$= -e^{-\theta/2}\left[2\pi\sin 2\pi\theta + \frac{1}{2}\cos 2\pi\theta\right]$$

13. If $f(\theta) = \frac{\sin\theta}{1+\sin\theta}$, then, by the quotient rule,

$$f'(\theta) = \frac{(1+\sin\theta)\frac{d}{d\theta}(\sin\theta) - (\sin\theta)\frac{d}{d\theta}(1+\sin\theta)}{(1+\sin\theta)^2}$$

$$= \frac{(1+\sin\theta)(\cos\theta) - (\sin\theta)(\cos\theta)}{(1+\sin\theta)^2}$$

$$= \frac{\cos\theta}{(1+\sin\theta)^2}$$

15. If $f(\theta) = \tan(1-\theta^5)$, then

$$f'(\theta) = \sec^2(1-\theta^5)\frac{d}{d\theta}(1-\theta^5) = -5\theta^4\sec^2(1-\theta^5)$$

17. If $f(\theta) = \tan^2\left[\frac{\pi}{2} - 2\pi\theta\right] = \left[\tan\left[\frac{\pi}{2} - 2\pi\theta\right]\right]^2$, then

$$f'(\theta) = 2\tan\left[\frac{\pi}{2} - 2\pi\theta\right]\frac{d}{d\theta}\tan\left[\frac{\pi}{2} - 2\pi\theta\right]$$

$$= 2\tan\left[\frac{\pi}{2} - 2\pi\theta\right]\sec^2\left[\frac{\pi}{2} - 2\pi\theta\right]\frac{d}{d\theta}\left[\frac{\pi}{2} - 2\pi\theta\right]$$

$$= 2\tan\left[\frac{\pi}{2} - 2\pi\theta\right]\sec^2\left[\frac{\pi}{2} - 2\pi\theta\right](-2\pi)$$

$$= -4\pi\tan\left[\frac{\pi}{2} - 2\pi\theta\right]\sec^2\left[\frac{\pi}{2} - 2\pi\theta\right]$$

19. If $f(\theta) = \ln\sin^2\theta = \ln(\sin\theta)^2 = 2\ln\sin\theta$, then

$$f'(\theta) = \frac{2}{\sin\theta} \frac{d}{d\theta}(\sin\theta) = \frac{2\cos\theta}{\sin\theta} = 2\cot\theta$$

21. $\dfrac{d}{d\theta}\sec\theta = \dfrac{d}{d\theta}\left[\dfrac{1}{\cos\theta}\right] = \dfrac{d}{d\theta}(\cos\theta)^{-1} = -(\cos\theta)^{-2}\dfrac{d}{d\theta}\cos\theta$

$$= -(\cos\theta)^{-2}(-\sin\theta) = \frac{\sin\theta}{\cos^2\theta}$$

$$= \left[\frac{\sin\theta}{\cos\theta}\right]\left[\frac{1}{\cos\theta}\right] = \tan\theta\sec\theta$$

23. $\dfrac{d}{d\theta}\cot\theta = \dfrac{d}{d\theta}\left[\dfrac{\cos\theta}{\sin\theta}\right] = \dfrac{\sin\theta\frac{d}{d\theta}(\cos\theta)-\cos\theta\frac{d}{d\theta}(\sin\theta)}{\sin^2\theta}$

$$= \frac{\sin\theta(-\sin\theta)-\cos\theta(\cos\theta)}{\sin^2\theta}$$

$$= \frac{-\sin^2\theta-\cos^2\theta}{\sin^2\theta}$$

$$= -\frac{\sin^2\theta+\cos^2\theta}{\sin^2\theta} = -\frac{1}{\sin^2\theta} = -\csc^2\theta$$

25. $\dfrac{d}{d\theta}\cos\theta = \lim\limits_{\Delta\theta\to 0}\dfrac{\cos(\theta+\Delta\theta)-\cos\theta}{\Delta\theta}$ (definition of derivative)

$$= \lim_{\Delta\theta\to 0}\frac{\cos\theta\cos\Delta\theta-\sin\theta\sin\Delta\theta-\cos\theta}{\Delta\theta} \qquad \text{(addition formula)}$$

$$= \lim_{\Delta\theta\to 0}\left[\frac{\cos\theta\cos\Delta\theta-\cos\theta}{\Delta\theta} - \frac{\sin\theta\sin\Delta\theta}{\Delta\theta}\right]$$

$$= \lim_{\Delta\theta\to 0}\cos\theta\left[\frac{\cos\Delta\theta-1}{\Delta\theta}\right] - \lim_{\Delta\theta\to 0}\sin\theta\left[\frac{\sin\Delta\theta}{\Delta\theta}\right]$$

$$= \cos\theta\lim_{\Delta\theta\to 0}\frac{\cos\Delta\theta-1}{\Delta\theta} - \sin\theta\lim_{\Delta\theta\to 0}\frac{\sin\Delta\theta}{\Delta\theta}$$

$$= (\cos\theta)(0) - (\sin\theta)(1) = -\sin\theta \qquad \text{(by the limits on page 634)}$$

Chapter 12, Section 3

1. Let x denote the horizontal distance (in miles) between the plane and the observer, let t denote time (in hours), and draw a diagram representing the situation.

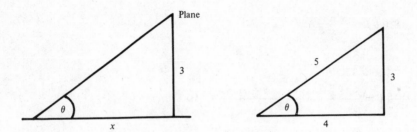

Figure and 3-4-5 right triangle for Problem 1.

It is given that $\frac{dx}{dt} = -500$ miles per hour. From the right triangle in the first figure,

$$\tan \theta = \frac{3}{x}$$

Differentiating both sides of this equation with repsect to t yields

$$\sec^2 \theta \frac{d\theta}{dt} = -\frac{3}{x^2} \frac{dx}{dt}$$

or

$$\frac{d\theta}{dt} = -\frac{3}{x^2} \frac{1}{\sec^2 \theta} \frac{dx}{dt} = \frac{-3 \cos^2 \theta}{x^2} \frac{dx}{dt}$$

When $x = 4$, the corresponding $3 - 4 - 5$ right triangle gives $\cos \theta = \frac{4}{5}$. Hence,

$$\frac{d\theta}{dt} = -\frac{3}{(4)^2} \left[\frac{4}{5}\right]^2 (-500) = 60 \text{ radians per hour}$$

3. Let x denote the length of the rope (from the rowboat to the end of the pier), let t denote time (in minutes), and draw a diagram representing the situation.
    It is given that $\frac{dx}{dt} = -4$ feet per minute. From the right triangle in the first figure,

$$\sin \theta = \frac{12}{x}$$

Differentiating both sides of this equation with respect to t yields

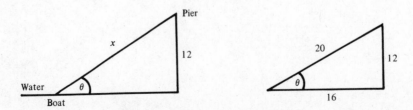

Figure and 12-16-20 right triangle for Problem 3.

$$\cos\theta \frac{d\theta}{dt} = -\frac{12}{x^2}\frac{dx}{dt}$$

or

$$\frac{d\theta}{dt} = -\frac{12}{x^2}\frac{1}{\cos\theta}\frac{dx}{dt} = -\frac{12\sec\theta}{x^2}\frac{dx}{dt}$$

When the distance from the boat to the pier is 16 feet, the corresponding $12-16-20$ right triangle gives $\sec\theta = \frac{20}{16} = \frac{5}{4}$. Hence,

$$\frac{d\theta}{dt} = -\frac{12}{(20)^2}\left[\frac{5}{4}\right](-4) = 0.15 \text{ radian per minute}$$

5.   Let V denote the volume of the trough. Then,

$$V = (\text{area of triangular cross section})(\text{length of trough})$$

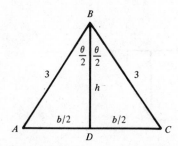

Cross section for Problem 5.

The area A of the triangular cross section is $\frac{1}{2}(\text{base})(\text{height}) = \frac{1}{2}bh$. From the right triangle ABD.

$$\cos\frac{\theta}{2} = \frac{h}{3} \quad \text{or} \quad h = 3\cos\frac{\theta}{2} \quad \text{and} \quad \sin\frac{\theta}{2} = \frac{b/2}{3} \quad \text{or} \quad b = 6\sin\frac{\theta}{2}$$

Hence,

$$A = \frac{1}{2}bh = \frac{1}{2}\left[6\sin\frac{\theta}{2}\right]\left[3\cos\frac{\theta}{2}\right] = 9\sin\frac{\theta}{2}\cos\frac{\theta}{2}$$

By the double-angle formula $\sin 2\alpha = 2\sin\alpha\cos\alpha$ with $\alpha = \frac{\theta}{2}$,

$$\sin\frac{\theta}{2}\cos\frac{\theta}{2} = \frac{1}{2}\sin\theta$$

and so

$$A = \frac{9}{2}\sin\theta$$

Hence,

$$V(\theta) = \left[\frac{9}{2}\sin\theta\right](20) = 90\sin\theta$$

The goal is to maximize $V(\theta)$ for $0 \le \theta \le \pi$. The derivative is

$$V'(\theta) = 90\cos\theta$$

which is zero when $\theta = \frac{\pi}{2}$. Since

$$V(0) = 0 \qquad V(\pi) = 0 \quad \text{and} \quad V(\frac{\pi}{2}) > 0$$

it follows that the volume is greatest when $\theta = \frac{\pi}{2}$ radians.

7.   Let A denote the area of the trapezoid. Then,

$$A = 2(\text{area of triangle}) + (\text{area of rectangle})$$

as indicated in the figure.

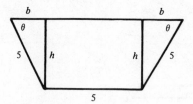

Trapezoid for Problem 7.

The area of the rectangle is $5h$, and the area of each triangles is $\frac{1}{2}bh$. Hence,

$$A = 2\left[\frac{1}{2}bh\right] + 5h = bh + 5h$$

From the right triangle,

$$\sin\theta = \frac{h}{5} \quad \text{or} \quad h = 5\sin\theta \quad \text{and} \quad \cos\theta = \frac{b}{5} \quad \text{or} \quad b = 5\cos\theta$$

Hence,

$$A(\theta) = 25\sin\theta\cos\theta + 25\sin\theta$$

which is to be maximized for $0 \le \theta \le \pi$. By the product rule, the derivative is

$$A'(\theta) = 25 \sin \theta(-\sin \theta) + 25 \cos \theta \cos \theta + 25 \cos \theta$$

$$= -25 \sin^2 \theta + 25 \cos^2 \theta + 25 \cos \theta$$

$$= 25[\cos^2 \theta - 1 + \cos^2 \theta + \cos \theta] \qquad (\text{since } \sin^2 \theta = 1 - \cos^2 \theta)$$

$$= 25(2 \cos^2 \theta + \cos \theta - 1) = 25(2 \cos \theta - 1)(\cos \theta + 1)$$

which is zero when

$$2 \cos \theta - 1 = 0 \qquad \cos \theta = \frac{1}{2} \quad \text{or} \quad \theta = \frac{\pi}{3}$$

and when

$$\cos \theta + 1 = 0 \qquad \cos \theta = -1 \quad \text{or} \quad \theta = \pi$$

Since

$$A(0) = 0 \qquad A(\pi) = 0 \quad \text{and} \quad A(\frac{\pi}{3}) > 0$$

it follows that the area is greatest when $\theta = \frac{\pi}{3}$ radians.

9. Let C denote the horizontal clearance at the corner, as shown in the figure. Then,

$$C = x + y$$

From the right triangle ABC,

$$\sin \theta = \frac{2\sqrt{2}}{x} \quad \text{or} \quad x = \frac{2\sqrt{2}}{\sin \theta}$$

and from the right triangle CED,

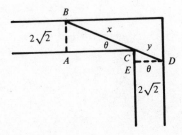

Figure for Problem 9.

$$\cos \theta = \frac{2\sqrt{2}}{y} \quad \text{or} \quad y = \frac{2\sqrt{2}}{\cos \theta}$$

Hence,

$$C(\theta) = \frac{2\sqrt{2}}{\sin \theta} + \frac{2\sqrt{2}}{\cos \theta} = 2\sqrt{2}(\sin \theta)^{-1} + 2\sqrt{2}(\cos \theta)^{-1}$$

where θ is restricted to be in the interval $0 \leq \theta \leq \frac{\pi}{2}$. Since $C(\theta)$ is undefined at the endpoints $\theta = 0$ and $\theta = \frac{\pi}{2}$, the minimum value of C occurs when the derivative is zero. By the chain rule,

$$C'(\theta) = -2\sqrt{2}(\sin \theta)^{-2}(\cos \theta) - 2\sqrt{2}(\cos \theta)^{-2}(-\sin \theta)$$

$$= 2\sqrt{2}\left[\frac{\sin \theta}{\cos^2 \theta} - \frac{\cos \theta}{\sin^2 \theta}\right] = 2\sqrt{2}\left[\frac{\sin^3 \theta - \cos^3 \theta}{\cos^2 \theta \sin^2 \theta}\right]$$

which is zero when

$$\sin^3 \theta = \cos^3 \theta \qquad \sin \theta = \cos \theta \quad \text{or} \quad \theta = \frac{\pi}{4} \text{ radians}$$

Hence, the minimal clearance is

$$C(\frac{\pi}{4}) = \frac{2\sqrt{2}}{\sin \pi/4} + \frac{2\sqrt{2}}{\cos \pi/4}$$

$$= \frac{2\sqrt{2}}{\sqrt{2}/2} + \frac{2\sqrt{2}}{\sqrt{2}/2} = 4 + 4 = 8 \text{ feet}$$

Hence, the longest pipe that can clear the corner is 8 feet long.

Chapter 12, Review Problems

1.   (a)   The angle is

$$90° + \frac{1}{3}(90°) = 120° \quad \text{or} \quad \frac{\pi}{2} + \frac{1}{3}\left[\frac{\pi}{2}\right] = \frac{2\pi}{3} \text{ radians}$$

    (b)   The angle is

$$-180° + \frac{1}{2}(-90°) = -225° \quad \text{or} \quad -\pi + \frac{1}{2}\left[-\frac{\pi}{2}\right] = -\frac{5\pi}{4} \text{ radians}$$

2.   From the proportion $\frac{50}{180} = \frac{\text{radians}}{\pi}$, it follows that

$$\text{Radians} = \frac{50\pi}{180} = \frac{5\pi}{18} \simeq 0.8727$$

3.   From the proportion $\frac{\text{degrees}}{180} = \frac{0.25}{\pi}$, it follows that

$$\text{Degrees} = \frac{180(0.25)}{\pi} \simeq 14.3239°$$

4.   (a) $\sin\left[-\frac{5\pi}{3}\right] = -\sin\frac{5\pi}{3} = -\sin\left[2\pi - \frac{\pi}{3}\right]$

$$= -\sin\left[-\frac{\pi}{3}\right] = \sin\frac{\pi}{3} = \frac{\sqrt{3}}{2}$$

    (b) $\cos\frac{15\pi}{4} = \cos\left[4\pi - \frac{\pi}{4}\right] = \cos\left[-\frac{\pi}{4}\right] = \cos\frac{\pi}{4} = \frac{\sqrt{2}}{2}$

    (c) $\sec\frac{7\pi}{3} = \sec\left[2\pi + \frac{\pi}{3}\right] = \sec\frac{\pi}{3} = \frac{1}{\cos\pi/3} = 2$

    (d) $\cot\frac{2\pi}{3} = \frac{\cos 2\pi/3}{\sin 2\pi/3} = \frac{-1/2}{\sqrt{3}/2} = -\frac{\sqrt{3}}{3}$

5.   If $\sin\theta = \frac{4}{5}$, the corresponding $3-4-5$ right triangle shows that $\tan\theta = \frac{4}{3}$.

6.   If $\cot\theta = \frac{\sqrt{5}}{2}$, the corresponding right triangle shows that $\csc\theta = \frac{3}{2}$.

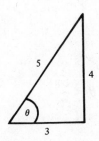

Triangle for Problem 5.

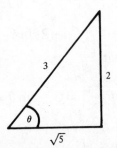

Triangle for Problem 6.

7.   Since $\sin 2\theta = 2\sin\theta\cos\theta$, the equation can be rewritten as

$$0 = 2\cos\theta + \sin 2\theta = 2\cos\theta + 2\sin\theta\cos\theta = 2\cos\theta(1 + \sin\theta)$$

which is zero if

$$\cos \theta = 0 \quad \text{or} \quad \theta = \frac{\pi}{2} \quad \text{and} \quad \theta = \frac{3\pi}{2}$$

or

$$1 + \sin \theta = 0 \qquad \sin \theta = -1 \quad \text{or} \quad \theta = \frac{3\pi}{2}$$

Hence, the solutions are $\theta = \frac{\pi}{2}$ and $\theta = \frac{3\pi}{2}$.

8. Since $\cos^2 \theta = 1 - \sin^2 \theta$, the equation can be rewritten as

$$2 = 3 \sin^2 \theta - \cos^2 \theta = 3 \sin^2 \theta - (1 - \sin^2 \theta) = 4 \sin^2 \theta - 1$$

and so

$$\sin^2 \theta = \frac{3}{4} \qquad \text{or} \qquad \sin \theta = \pm \frac{\sqrt{3}}{2}$$

On the interval $0 \le \theta \le \pi$, the equation $\sin \theta = \frac{\sqrt{3}}{2}$ implies $\theta = \frac{\pi}{3}$ and $\theta = \frac{2\pi}{3}$, while the equation $\sin \theta = -\frac{\sqrt{3}}{2}$ has no solution on this interval. Hence, the only solutions are $\theta = \frac{\pi}{3}$ and $\theta = \frac{2\pi}{3}$.

9. Since $\cos 2\theta = \cos^2 \theta - \sin^2 \theta = 1 - \sin^2 \theta - \sin^2 \theta = 1 - 2 \sin^2 \theta$, the equation can be rewritten as

$$0 = 2 \sin^2 \theta - (1 - 2 \sin^2 \theta) = 4 \sin^2 \theta - 1 \quad \text{or} \quad \sin \theta = \pm \frac{1}{2}$$

On the interval $0 \le \theta \le \pi$, the equation $\sin \theta = \frac{1}{2}$ implies $\theta = \frac{\pi}{6}$ or $\theta = \frac{5\pi}{6}$, while the equation $\sin \theta = -\frac{1}{2}$ has no solution. Hence, the solutions are $\theta = \frac{\pi}{6}$ and $\theta = \frac{5\pi}{6}$.

10. Since $\cos^2 \theta = 1 - \sin^2 \theta$, the equation can be rewritten as

$$5 \sin \theta - 2(1 - \sin^2 \theta) = 1 \qquad 2 \sin^2 \theta + 5 \sin \theta - 3 = 0$$

or

$$(2 \sin \theta - 1)(\sin \theta + 3) = 0$$

which implies that

$$2 \sin \theta - 1 = 0 \quad \sin \theta = \frac{1}{2} \quad \text{or} \quad \theta = \frac{\pi}{6} \quad \text{and} \quad \theta = \frac{5\pi}{6}$$

or

$$\sin \theta + 3 = 0$$

which has no solutions. Hence, the solutions are $\theta = \frac{\pi}{6}$ and $\theta = \frac{5\pi}{6}$.

11. Start with $\sin^2 \theta + \cos^2 \theta = 1$ and divide by $\sin^2 \theta$ to get

$$1 + \frac{\cos^2 \theta}{\sin^2 \theta} = \frac{1}{\sin^2 \theta}, \quad 1 + \left[\frac{\cos \theta}{\sin \theta} \right]^2 = \left[\frac{1}{\sin \theta} \right]^2 \quad \text{or} \quad 1 + \cot^2 \theta = \csc^2 \theta$$

12.   Since $\sin 2\theta = 2 \sin \theta \cos \theta$ and $\cos 2\theta = \cos^2 \theta - \sin^2 \theta$,

$$\tan 2\theta = \frac{\sin 2\theta}{\cos 2\theta} = \frac{2 \sin \theta \cos \theta}{\cos^2 \theta - \sin^2 \theta}$$

$$= \frac{\left[\dfrac{2 \sin \theta \cos \theta}{\cos^2 \theta}\right]}{\left[\dfrac{\cos^2 \theta}{\cos^2 \theta}\right] - \left[\dfrac{\sin^2 \theta}{\cos^2 \theta}\right]}$$

$$= \frac{\left[\dfrac{2 \sin \theta}{\cos \theta}\right]}{1 - \left[\dfrac{\sin \theta}{\cos \theta}\right]^2} = \frac{2 \tan \theta}{1 - \tan^2 \theta}$$

13.   (a)   Apply the identities

$$\cos(a + b) = \cos a \cos b - \sin a \sin b$$

and

$$\sin(a + b) = \sin a \cos b + \sin b \cos a$$

with $a = \frac{\pi}{2}$ and $b = \theta$ to get

$$\cos \left[\frac{\pi}{2} + \theta\right] = \cos \frac{\pi}{2} \cos \theta - \sin \frac{\pi}{2} \sin \theta$$

$$= 0(\cos \theta) - 1(\sin \theta) = -\sin \theta$$

and

$$\sin \left[\frac{\pi}{2} + \theta\right] = \sin \frac{\pi}{2} \cos \theta + \sin \theta \cos \frac{\pi}{2}$$

$$= 1(\cos \theta) + (\sin \theta)(0) = \cos \theta$$

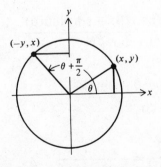

Figure for Problem 13b.

(b)   Notice from the figure that $x = \cos\theta$ and $y = \sin\theta$, where (x, y) are the coordinates associated with θ. The coordinates associated with $\theta + \frac{\pi}{2}$ are $(-y, x)$, and so

$$\cos\left[\frac{\pi}{2} + \theta\right] = -y = -\sin\theta \quad \text{and} \quad \sin\left[\frac{\pi}{2} + \theta\right] = x = \cos\theta$$

14.   If $f(\theta) = \sin(3\theta + 1)^2$, then

$$f'(\theta) = \cos(3\theta + 1)^2 \frac{d}{d\theta}(3\theta + 1)^2 = 6(3\theta + 1)\cos(3\theta + 1)^2$$

15.   If $f(\theta) = \cos^2(3\theta + 1) = [\cos(3\theta + 1)]^2$, then

$$f'(\theta) = 2\cos(3\theta + 1)\frac{d}{d\theta}\cos(3\theta + 1)$$

$$= 2\cos(3\theta + 1)[-3\sin(3\theta + 1)] = -6\cos(3\theta + 1)\sin(3\theta + 1)$$

16.   If $f(\theta) = \tan(3\theta^2 + 1)$, then

$$f'(\theta) = \sec^2(3\theta^2 + 1)\frac{d}{d\theta}(3\theta^2 + 1) = 6\theta\sec^2(3\theta + 1)$$

17.   If $f(\theta) = \tan^2(3\theta^2 + 1) = [\tan(3\theta^2 + 1)]^2$, then

$$f'(\theta) = 2\tan(3\theta^2 + 1)\frac{d}{d\theta}\tan(3\theta^2 + 1)$$

$$= 2\tan(3\theta^2 + 1)\sec^2(3\theta^2 + 1)(6\theta)$$

$$= 12\theta\tan(3\theta^2 + 1)\sec^2(3\theta^2 + 1)$$

18.   If $f(\theta) = \frac{\sin\theta}{1 - \cos\theta}$, then, by the quotient rule,

$$f'(\theta) = \frac{(1 - \cos\theta)\frac{d}{d\theta}(\sin\theta) - \sin\theta\frac{d}{d\theta}(1 - \cos\theta)}{(1 - \cos\theta)^2}$$

$$= \frac{(1 - \cos\theta)(\cos\theta) - \sin\theta(\sin\theta)}{(1 - \cos\theta)^2}$$

$$= \frac{\cos\theta - \cos^2\theta - \sin^2\theta}{(1 - \cos\theta)^2}$$

$$= \frac{\cos\theta - 1}{(1 - \cos\theta)^2} = -\frac{1}{1 - \cos\theta}$$

19.   If $f(\theta) = \ln\cos^2\theta = 2\ln\cos\theta$, then

$$f'(\theta) = 2\left[\frac{1}{\cos\theta}\right]\frac{d}{d\theta}(\cos\theta) = -\frac{2\sin\theta}{\cos\theta} = -2\tan\theta$$

20. By the product rule,

$$\frac{d}{d\theta}(sin\theta\cos\theta) = (\sin\theta)\frac{d}{d\theta}(\cos\theta) + (\cos\theta)\frac{d}{d\theta}(\sin\theta)$$

$$= \sin\theta(-\sin\theta) + \cos\theta(\cos\theta) = \cos^2\theta - \sin^2\theta = \cos 2\theta$$

21. Let x denote the distance (in feet) from the ground to the ball, let t denote time (in minutes), and draw a diagram representing the situation.

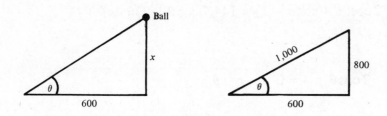

Figure and right triangle for Problem 21.

It is given that $\frac{dx}{dt} = -20$ feet per minute. From the right triangle in the first figure,

$$\tan\theta = \frac{x}{600}$$

Differentiating both sides of this equation with respect to t yields

$$\sec^2\theta\frac{d\theta}{dt} = \frac{1}{600}\frac{dx}{dt}$$

or

$$\frac{d\theta}{dt} = \frac{1}{600}\frac{1}{\sec^2\theta}\frac{dx}{dt} = \frac{\cos^2\theta}{600}\frac{dx}{dt}$$

When $x = 800$, the corresponding $600 - 800 - 1,000$ triangle gives

$$\cos\theta = \frac{600}{1,000} = \frac{3}{5}$$

Hence,

$$\frac{d\theta}{dt} = \frac{1}{600}\left[\frac{3}{5}\right]^2(-20) = -0.012 \text{ radian per minute}$$

22.   Let V denote the volume of the trough. Then,

$$V = \text{(area of trapezoidal cross section)(length of trough)}$$

The length of the trough is 9 meters. As in Problem 7 of Section 12.3 (with 5 replaced by 4), the area A of the cross section is

$$A = (4\cos\theta)(4\sin\theta) + 4(4\sin\theta) = 16\cos\theta\sin\theta + 16\sin\theta$$

Hence,

$$V(\theta) = 9(16)(\cos\theta\sin\theta + \sin\theta)$$

The derivative is

$$V'(\theta) = 144(\cos^2\theta - \sin^2\theta + \cos\theta) = 144[\cos^2\theta - (1 - \cos^2\theta) + \cos\theta]$$

$$= 144(2\cos^2\theta + \cos\theta - 1) = 144(2\cos\theta - 1)(\cos\theta + 1)$$

which is zero if

$$2\cos\theta - 1 = 0 \qquad \cos\theta = \frac{1}{2} \quad \text{or} \quad \theta = \frac{\pi}{3}$$

and

$$\cos\theta = -1 \qquad \text{or} \qquad \theta = \pi$$

The relevant interval is $0 \le \theta \le \pi$. Since,

$$V(0) = 0 \qquad V(\pi) = 0 \quad \text{and} \quad V(\tfrac{\pi}{3}) > 0$$

it follows that the maximum volume is achieved when $\theta = \frac{\pi}{3}$ radians.

23.   Let C denote the horizontal clearance at the corner, as illustrated in the figure. Then,

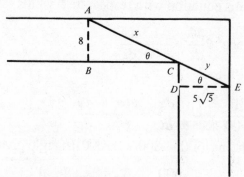

Figure for Problem 23.

From the right triangle ABC,

$$\sin\theta = \frac{8}{x} \qquad \text{or} \qquad x = \frac{8}{\sin\theta}$$

and from the right triangle CDE,

$$\cos \theta = \frac{5\sqrt{5}}{y} \qquad \text{or} \qquad y = \frac{5\sqrt{5}}{\cos \theta}$$

Hence,

$$C(\theta) = \frac{8}{\sin \theta} + \frac{5\sqrt{5}}{\cos \theta} = 8(\sin \theta)^{-1} + 5\sqrt{5}(\cos \theta)^{-1}$$

where the relevant interval is $0 \le \theta \le \frac{\pi}{2}$. Since the function is undefined at the endpoints, the miminum must occur when the derivative is zero. By the chain rule,

$$C'(\theta) = -8(\sin \theta)^{-2}(\cos \theta) - 5\sqrt{5}(\cos \theta)^{-2}(-\sin \theta)$$

$$= \frac{5\sqrt{5}\sin \theta}{\cos^2 \theta} - \frac{8\cos \theta}{\sin^2 \theta}$$

$$= \frac{5\sqrt{5}\sin^3 \theta - 8\cos^2 \theta}{\cos^2 \theta \sin^2 \theta}$$

which is zero if

$$5\sqrt{5}\sin^3 \theta - 8\cos^3 \theta = 0 \qquad \frac{\sin^3 \theta}{\cos^3 \theta} = \frac{8}{5\sqrt{5}}$$

$$\tan^3 \theta = \frac{8}{5^{3/2}} \qquad \text{or} \qquad \tan \theta = \frac{8^{1/3}}{(5^{3/2})^{1/3}} = \frac{2}{\sqrt{5}}$$

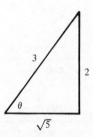

Right triangle for Problem 23.

From the corresponding right triangle,

$$\sin \theta = \frac{2}{3} \qquad \text{and} \qquad \cos \theta = \frac{\sqrt{5}}{3}$$

Hence, the minimal clearance is

$$C = 8\left[\frac{1}{\sin\theta}\right] + 5\sqrt{5}\left[\frac{1}{\cos\theta}\right] = 8\left[\frac{3}{2}\right] + 5\sqrt{5}\left[\frac{3}{\sqrt{5}}\right] = 12 + 15 = 27 \text{ feet}$$

24. Let x and y be the distances indicated in the figure, and let C denote the cost of installing the cable. Then, since it costs \$5 per meter under water and \$4 per meter on land,

$$C = 5y + 4(3,000 - x)$$

From the right triangle ABP,

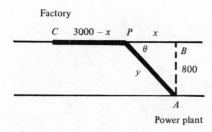

Factory

Figure for Problem 24.

$$\sin\theta = \frac{800}{6} \quad \text{or} \quad y = \frac{800}{\sin\theta} \quad \text{and} \quad \tan\theta = \frac{800}{x} \quad \text{or} \quad x = \frac{800}{\tan\theta}$$

Hence,

$$C(\theta) = 5\left[\frac{800}{\sin\theta}\right] + 4\left[3,000 - \frac{800}{\tan\theta}\right] = \frac{4,000}{\sin\theta} + 12,000 - \frac{3,200}{\tan\theta}$$

The derivative is

$$C'(\theta) = -\frac{4,000\cos\theta}{\sin^2\theta} + \frac{3,200\sec^2\theta}{\tan^2\theta}$$

$$= -\frac{4,000\cos\theta}{\sin^2\theta} + \frac{3,200/\cos^2\theta}{\sin^2\theta/\cos^2\theta}$$

$$= -\frac{4,000\cos\theta}{\sin^2\theta} + \frac{3,200}{\sin^2\theta} = \frac{-4,000\cos\theta + 3,200}{\sin^2\theta}$$

which is zero (and hence C is minimal) if

$$-4,000\cos\theta + 3,200 = 0 \quad \text{or} \quad \cos\theta = \frac{3,200}{4,000} = \frac{4}{5}$$

APPENDIX

ALGEBRA REVIEW

Appendix, Section A

1. $1 < x \leq 5$

3. $x > -5$

5.

7.

9.   Distance between 0 and $-4 = |0 - 4| = 4$

11.   Distance between -2 and $3 = |-2 - 3| = 5$

13.   If $|x| \leq 3$, then $-3 \leq x \leq 3$.

15.   If $|x + 4| \leq 2$, then

$$-2 \leq x + 4 \leq 2 \qquad -2 - 4 \leq x \leq 2 - 4 \quad \text{or} \quad -6 \leq x \leq -2$$

17.   If $|x + 2| \geq 5$, then

$$x + 2 \geq 5 \qquad \text{or} \qquad x + 2 \leq -5$$

or, equivalently,

$$x \geq 3 \qquad \text{or} \qquad x \leq -7$$

19. $5^3 = 5 \cdot 5 \cdot 5 = 125$

21. $16^{1/2} = \sqrt{16} = 4$

23. $8^{2/3} = 2^2 = 4$

25. $(\frac{1}{4})^{1/2} = \sqrt{\frac{1}{4}} = \frac{1}{2}$

27. $\frac{2^5(2^2)}{2^8} = \frac{2^{5+2}}{2^8} = \frac{2^7}{2^8} = \frac{1}{2}$

29. $\frac{2^{4/3}(2^{5/3})}{2^5} = \frac{2^{4/3+5/3}}{2^5} = \frac{2^{9/3}}{2^5} = \frac{2^3}{2^5} = \frac{1}{2^2} = \frac{1}{4}$

31. $\frac{2(16^{3/4})}{2^3} = \frac{2(16^{1/4})^3}{2^3} = \frac{2(2^3)}{2^3} = 2$

33. $\left[\sqrt{8}(2^{5/2})\right]^{-1/2} = [(2^3)^{1/2}(2)^{5/2}]^{-1/2} = [2^{3/2+5/2}]^{-1/2}$

$= [2^{8/2}]^{-1/2} = 2^{-2} = \frac{1}{4}$

35. $a^3 a^7 = a^n$ $a^{3+7} = a^n$ $a^{10} = a^n$ $n = 10$

37. $a^4 a^{-3} = a^n$ $a^{4-3} = a^n$ $a^1 = a^n$ $n = 1$

39. $(a^3)^n = a^{12}$ $a^{3n} = a^{12}$ $3n = 12$ $n = 4$

41. $a^{3/5} a^{-n} = \frac{1}{a^2}$ $a^{3/5-n} = a^{-2}$ $\frac{3}{5} - n = -2$ $n = \frac{3}{5} + 2 = \frac{13}{5}$

43. $x^4 - 4x^4 = x^4(x - 4)$

45. $100 - 25(x - 3) = 25[4 - (x - 3)] = 25(4 - x + 3) = 25(7 - x)$

47. $8(x + 1)^3(x - 2)^2 + 6(x + 1)^2(x - 2)^3$

$= 2(x + 1)^2(x - 2)^2[4(x + 1) + 3(x - 2)]$

$= 2(x + 1)^2(x - 2)^2(4x + 4 + 3x - 6)$

$= 2(x + 1)^2(x - 2)^2(7x - 2)$

49. Factor the numerator and then cancel the common factors from the numerator and denominator to get

$$\frac{(x + 3)^3(x + 1) - (x + 3)^2(x + 1)^2}{(x + 3)(x + 1)} = \frac{(x + 3)^2(x + 1)[(x + 3) - (x + 1)]}{(x + 3)(x + 1)}$$

$$= \frac{(x + 3)^2(x + 1)(x + 3 - x - 1)}{(x + 3)(x + 1)}$$

$$= \frac{2(x + 3)^2(x + 1)}{(x + 3)(x + 1)} = 2(x + 3)$$

51. Factor the numerator and then cancel the common factors from the numerator and denominator to get

$$\frac{4(1-x)^2(x+3)^2 + 2(1-x)(x+3)^4}{(1-x)^4} = \frac{2(1-x)(x+3)^3[2(1-x)+(x+3)]}{(1-x)^4}$$

$$= \frac{2(1-x)(x+3)^3(2-2x+x+3)}{(1-x)^4}$$

$$= \frac{2(1-x)(x+3)^3(5-x)}{(1-x)^4}$$

$$= \frac{2(x+3)^3(5-x)}{(1-x)^3}$$

53. $x^2 + x - 2 = (x+2)(x-1)$

55. $x^2 - 7x + 12 = (x-3)(x-4)$

57. $x^2 - 2x + 1 = (x-1)(x-1) = (x-1)^2$

59. $x^2 - 4 = (x+2)(x-2)$

61. $x^3 - 1 = (x-1)(x^2+x+1)$

63. $x^7 - x^5 = x^5(x^2-1) = x^5(x+1)(x-1)$

65. $2x^3 - 8x^2 - 10x = 2x(x^2-4x-5) = 2x(x-5)(x+1)$

67. Factor the equation to get

$$x^2 - 2x - 8 = (x-4)(x+2) = 0 \qquad \text{or} \qquad x = 4 \quad \text{and} \quad x = -2$$

69. Factor the equation to get

$$x^2 + 10x + 25 = (x+5)^2 = 0 \qquad \text{or} \qquad x = -5$$

71. Factor the equation to get

$$x^2 - 16 = (x+4)(x-4) = 0 \qquad \text{or} \qquad x = -4 \quad \text{and} \quad x = 4$$

73. Factor the equation to get

$$2x^2 + 3x + 1 = (2x+1)(x+1) = 0 \qquad \text{or} \qquad x = -\frac{1}{2} \quad \text{and} \quad x = -1$$

75. Factor the equation to get

$$4x^2 + 12x + 9 = (2x+3)(2x+3) = (2x+3)^2 = 0 \qquad \text{or} \qquad x = -\frac{3}{2}$$

77. Rewrite the equation as

$$1 + \frac{4}{x} - \frac{5}{x^2} = \frac{x^2+4x-5}{x^2} = \frac{(x+5)(x-1)}{x^2} = 0$$

or

$$x = -5 \quad \text{and} \quad x = 1$$

79. Rewrite the equation as

$$2 + \frac{2}{x} - \frac{4}{x^2} = \frac{2x^2 + 2x - 4}{x^2} = \frac{2(x^2 + x - 2)}{x^2}$$

$$= \frac{2(x+2)(x-1)}{x^2} = 0 \quad \text{or} \quad x = -2 \quad \text{and} \quad x = 1$$

81. For the equation $2x^2 + 3x + 1 = 0$, use the quadratic formula with $a = 2$, $b = 3$, and $c = 1$ to get

$$x = \frac{-3 \pm \sqrt{3^2 - 4(2)(1)}}{2(2)} = \frac{-3 \pm \sqrt{1}}{4}$$

or

$$x = \frac{-3 + 1}{4} = -\frac{1}{2} \quad \text{or} \quad x = \frac{-3 - 1}{4} = -1$$

83. For the equation $x^2 - 2x + 3 = 0$, use the quadratic formula with $a = 1$, $b = -2$, and $c = 3$ to get

$$x = \frac{-(-2) \pm \sqrt{(-2)^2 - 4(1)(3)}}{2(1)} = \frac{2 \pm \sqrt{-8}}{2}$$

Since $\sqrt{-8}$ is not a real number, the equation has no real solutions.

85. For the equation $4x^2 + 12x + 9 = 0$, use the quadratic formula with $a = 4$, $b = 12$, and $c = 9$ to get

$$x = \frac{-12 \pm \sqrt{(12)^2 - 4(4)(9)}}{2(4)} = \frac{-12 \pm \sqrt{0}}{8} = -\frac{3}{2}$$

87. Starting with the system

$$x + 5y = 13$$

$$3x - 10y = -11$$

multiply the first equation by 2 to get

$$2x + 10y = 26$$

$$3x - 10y = -11$$

and add to get

$$5x = 15 \quad \text{or} \quad x = 3$$

Substitute $x = 3$ into the first equation to get

$$3 + 5y = 13 \qquad \text{or} \qquad y = 2$$

Hence, the solution of the system is $x = 3$ and $y = 2$.

89. To eliminate x in the system

$$5x - 4y = 12$$
$$2x - 3y = 2$$

multiply the first equation by 2 and the second by -5 so that the system becomes

$$10x - 8y = 24$$
$$-10x + 15y = -10$$

and add to get

$$0 + 7y = 14 \qquad \text{or} \qquad y = 2$$

Substitute $y = 2$ into the second equation and get

$$2x - 3(2) = 2 \qquad 2x = 8 \quad \text{or} \quad x = 4$$

Hence, the solution of the system is $x = 4$ and $y = 2$.

91. To solve the system

$$2y^2 - x^2 = 1$$
$$x - 2y = 3$$

solve the second equation for x to get

$$x = 3 + 2y$$

and substitute this into the first equation to get

$$2y^2 - (3 + 2y)^2 = 1 \qquad\qquad 2y^2 - (9 + 12y + 4y^2) = 1$$

$$2y^2 - 9 - 12y - 4y^2 = 1 \qquad\qquad -2y^2 - 12y - 10 = 0$$

$$y^2 + 6y + 5 = 0 \qquad (y + 5)(y + 1) = 0 \quad \text{or} \quad y = -5 \text{ and } y = -1$$

If $y = -5$, the second equation gives

$$x - 2(-5) = 3 \qquad \text{or} \qquad x = -7$$

and if $y = -1$, the second equation gives

$$x - 2(-1) = 3 \qquad \text{or} \qquad x = 1$$

Hence, the solutions are $x = -7$, $y = -5$ and $x = 1$, $y = -1$.

93. $\displaystyle\sum_{j=1}^{4}(3j+1) = [3(1)+1]+[3(2)+1]+[3(3)+1]+[3(4)+1]$

$= 4+7+10+13 = 34$

95. $\displaystyle\sum_{j=1}^{10}(-1)^j = (-1)^1 + (-1)^2 + (-1)^3 + (-1)^4 + (-1)^5 + (-1)^6$

$+(-1)^7 + (-1)^8 + (-1)^9 + (-1)^{10}$

$= -1+1-1+1-1+1-1+1-1+1 = 0$

97. $1 + \frac{1}{2} + \frac{1}{3} + \frac{1}{4} + \frac{1}{5} + \frac{1}{6} = \displaystyle\sum_{j=1}^{6}\frac{1}{j}$

99. $2x_1 + 2x_2 + 2x_3 + 2x_4 + 2x_5 + 2x_6 = \displaystyle\sum_{j=1}^{6} 2x_j$

101. $1 - 2 + 3 - 4 + 5 - 6 + 7 - 8 = \displaystyle\sum_{j=1}^{8}(-1)^{j+1}j$ since

$$(-1)^{j+1} = \begin{cases} 1 & \text{if } j \text{ is odd} \\ -1 & \text{if } j \text{ is even} \end{cases}$$